T0305604

Third Edition

Electrical Power Cable Engineering

Third Edition

Electrical Power Cable Engineering

Edited by
William Thue

CRC Press is an imprint of the
Taylor & Francis Group, an **informa** business

CRC Press
Taylor & Francis Group
6000 Broken Sound Parkway NW, Suite 300
Boca Raton, FL 33487-2742

First issued in paperback 2017

ISBN 13: 978-1-4398-5643-7 (hbk)
ISBN 13: 978-1-138-07400-2 (pbk)

Visit the Taylor & Francis Web site at
http://www.taylorandfrancis.com

and the CRC Press Web site at
http://www.crcpress.com

Contents

Foreword to the First Edition, 1999

Electrical cable might be considered to be just a conductor, overlying insulation, and, often times, an exterior shield or jacket. Perhaps this naive, simplistic concept is part of the reason that cable engineering, especially for power cable, has been largely neglected by recent electrical engineering education in the United States with its emphasis on computers, electronics, and communication. But power cable does electrically connect the world! The history, so interestingly presented, shows how the subject evolved with both great success and, sometimes, unexpected failure.

As this book emphasizes, cable engineering is technically very complex. Certainly, electrical, mechanical, and, even to some extent, civil engineering are involved in interrelated ways. Many other disciplines—physics, inorganic chemistry, organic (primarily polymer) chemistry, physical chemistry, metallurgy, corrosion with tests and standards in all of these areas—are of concern. Of course, it is impossible in one book to deal with all of these aspects in a completely comprehensive way. However, the various components of power cables are considered with sufficient detail to provide an understanding of the basic considerations in each area. Reference to detailed sources provides a means for those with greater interest to pursue specific subjects.

The importance of factors involved in different types of cable installation is stressed. Long vertical cable runs have special problems. Installation in ducts may lead to problems with joints, terminations, elbows, and pulling stresses. At first, cables with extruded insulation were buried directly in trenches without recognizing the then unknown problem of "water treeing" in polyethylene, which was originally thought to be unaffected by moisture. After massive field failures, well over a thousand papers have been written on water treeing! Field failures can involve many factors, i.e., lightning, switching surges, repeated mechanical stressing, and swelling of voltage grading shields in contact with organic solvents such as oil and gasoline. It is important to recognize how such diverse factors can affect the performance of cables in the field.

Electrical Power Cable Engineering meets a need to consider its complex subject in a readable fashion, especially for those with limited background and experience. Yet, sufficient detail is provided for those with greater needs in evaluating different cables for specific applications. Most of all, the supplier of materials for cables can obtain a better understanding of overall problems. On the other side, an experienced cable engineer may come to recognize some of the parameters in materials with which he or she previously may not have had sufficient knowledge.

Kenneth N. Mathes
Consulting Engineer, Schenectady, New York

[Editor's Note: Even though Ken passed away many years ago, his words are still appropriate today, at the publication time of the Third Edition.]

Foreword to the First Edition 1969

Preface

The authors would like to acknowledge the almost 40 years of dedicated work by Professor Willis F. Long of the Department of Engineering Professional Development at the University of Wisconsin–Madison, which has made this book possible. It was because of his efforts that the first course of a series, Power Cable Engineering Clinic, was presented in the early 1970s, where Dr. Eugene Greenfield gave all his lectures for 8 hours a day during the 5 days of the course. He later added a few lecturers to share the work load.

The course was reorganized in 1999 and called "How to Design, Install, Operate, and Maintain Reliable Power Cable Systems." It was also divided into two sessions: "Understanding Power Cable Characteristics and Applications" and "Assessing and Extending the Life of Shielded Power Cable Systems." The present course is known as "Understanding Power Cable Characteristics and Applications." Over the years, numerous lecturers have produced copious class notes that form the basis for much of the materials, which has been rearranged into a book format. The contributors hope that their team effort will be a useful addition to the library of all dedicated cable engineers.

The dynamics of the cable industry produces many new materials, products, and concepts, which are incorporated into new editions. New chapters have been prepared (Low Voltage Cables and Thermal Resistivity of Soils) while others have been updated and expanded (Conductors now includes metric designations and Acronyms have been added to the Glossary) for this Third Edition.

Emphasis of this book remains on low and medium voltages since they comprise the majority of cables in service throughout the world. Transmission cables have the greater sophistication from an engineering standpoint, but all the basic principles that apply to transmission cables also apply to the lower voltage cables.

The audience that will benefit from the highly knowledgeable writings and diversity of the backgrounds of the contributors to this book include:

- Cable engineers and technicians employed by investor-owned utilities, rural electric cooperatives, industrial users, and power production personnel
- Universities that offer electrical power courses
- Cable manufacturers who would like to provide their employees with an oversight and understanding of power cables
- Professionals seeking information to provide an understanding of the terminology, engineering characteristics, and background information of power cables to assist in making sound decisions for specification, purchase, installation, maintenance, and operation of electrical power cables

William A. Thue
Editor

Preface

Editor

William A. Thue graduated with a BS in electrical engineering in 1946 while in the US Navy. He was employed by Florida Power & Light Co. (FP&L) for 38 years, where he specialized in engineering and operations of underground cable systems for their generation, distribution, and transmission systems. He was active in the development of industry standards for cables and accessories. Thue has authored and coauthored many AIEE and IEEE papers on cable shielding, fireproofing, performance, and life estimation. He also authored a chapter in the *1957 Underground Systems Reference Book* on duct construction.

He is a life fellow in the IEEE, past chair of the Insulated Conductors Committee of IEEE (where he has been a member for over 55 years and received their Outstanding Service Award in 1985), past chair of the Cable Engineering Section of the AEIC, and past chair of two subcommittees of the National Electrical Safety Code (ANSI C-2), and a member of EPRI's Distribution Task Force. He is a member of the Institution of Engineering and Technology (previously, the IEE in the UK) and the Royal Institution. After taking early retirement from FP&L, he began a consulting engineer career in 1984, where he has been involved in research projects for EPRI, taught short courses at the University of California at Los Angles and the University of Wisconsin–Madison during the past 30 years, was editor of *Electrical Power Cable Engineering*, and involved in over 150 legal cases as an expert witness.

Contributors

Bruce S. Bernstein
Consultant
Rockville, Maryland

James D. Medek
Consultant
Mt. Dora, Florida

Lauri J. Hiivala
Consultant
Toronto, Ontario, Canada

Deepak Parmar
President
Geotherm Inc.
Aurora, Ontario, Canada

Carl C. Landinger
Consultant
Longview, Texas

William A. Thue
Consultant
Hendersonville, North Carolina

1 Historical Perspective of Electrical Cables

Bruce S. Bernstein and William A. Thue

CONTENTS

1.1 DEVELOPMENT OF UNDERGROUND CABLES

In order to trace the history of underground cable systems, it is necessary to examine the early days of the telegraph [1,2]. The telegraph was the first device utilizing electrical energy to become of any commercial importance and its development necessarily required the use of underground construction. Accordingly, experimentation with underground cables was carried on contemporaneously with the development of the apparatus for sending and receiving signals. Underground construction was planned for most of the earliest commercial lines. A number of these early installations are of considerable interest as marking steps in the development of the extensive underground power systems in operation around the world.

1.2 EARLY TELEGRAPH LINES

In 1812, Baron Schilling detonated a mine under the Neva River at St. Petersburg, Russia, by using an electrical pulse sent through a cable insulated with strips of India rubber. This is probably the earliest use of a continuously insulated conductor on record. One of the earliest experiments with an underground cable was carried out by Francis Ronalds in 1816. This work was in conjunction with a system of telegraphy consisting of 500 feet of bare copper conductor drawn into glass tubes, joined together with sleeve joints, and sealed with wax. The tubes were placed in a creosoted wooden trough buried in the ground. Ronalds was very enthusiastic over the success of this line, predicting that underground conductors would be widely used for electrical purposes and outlining many of the essential characteristics of a modern distribution system.

The conductor in this case was first insulated with cotton saturated with shellac before being drawn into the tubes. Later, strips of India rubber were used. This installation had many insulation failures and was abandoned. No serious attempt was made to develop the idea commercially.

In 1837, W. R. Cooke and Charles Wheatstone laid an underground line along the railroad right-of-way between London's Euston and Camden stations for their five-wire system of telegraphy. The wires were insulated with cotton saturated in rosin and were installed in separate grooves in a piece of timber coated with pitch. This line operated satisfactorily for a short time, but a number of insulation failures due to the absorption of moisture led to its abandonment. The next year, Cooke and Wheatstone installed a line between Paddington and Drayton stations in London, but iron pipe was substituted for timber to give better protection from moisture. Insulation failures also occurred on this line after a short time, and it was also abandoned.

In 1842, S. F. B. Morse laid a cable insulated with jute, saturated in pitch, and covered with strips of India rubber, between Governor's Island and Castle Garden in New York harbor. The next year, a similar line was laid across a canal in Washington, DC. The success of these experiments induced Morse to write to the Secretary of the Treasury that he believed "telegraphic communications on the electro-magnetic plan can with a certainty be established across the Atlantic Ocean."

In 1844, Morse obtained an appropriation from the U.S. Congress for a telegraph line between Washington and Baltimore. An underground conductor was planned and several miles were actually laid before the insulation was proved to be defective. The underground project was abandoned and an overhead line erected. The conductor was originally planned to be a #16 gage copper insulated with cotton and saturated in shellac. Four insulated wires were drawn into a close-fitting lead pipe, which was then passed between rollers and drawn down into close contact with the conductors. The cable was coiled on drums in 300-foot lengths and laid by means of a specially designed plow.

Thus, the first attempts at underground construction were unsuccessful, and overhead construction was necessary to ensure satisfactory performance of the lines. After the failure of Morse's line, no additional attempts were made to utilize underground construction in the United States until Thomas A. Edison's time.

Gutta-percha—a natural, thermoplastic rubber—was introduced in Europe in 1842 by Dr. W. Montgomery, and in 1846 was adopted upon the recommendation of Dr. Werner Siemens for the telegraph line that the Prussian government was installing. Approximately 3,000 miles of such wire were laid from 1847 to 1852. Unfortunately, the perishable nature of the material was not known at the time and no adequate means of protecting it from oxidation was provided. Insulation troubles soon began to develop and eventually became so serious that the entire installation was abandoned.

However, gutta-percha provided a very satisfactory material for insulating telegraph cables when properly protected from oxidation. It was used extensively for both underground and submarine installations.

In 1860, vulcanized rubber was used for the first time as insulation for wires. Unvulcanized rubber had been used on several of the very early lines in strips applied over fibrous insulation for moisture protection. This system had generally been unsatisfactory because of difficulties in closing the seam. Vulcanized rubber proved to be a much better insulating material, but did not become a serious competitor of gutta-percha until some years later.

1.3 ELECTRIC LIGHTING

While early telegraph systems were being developed, other experimenters were solving the problems related to the commercial development of electric lighting. An electric light required a steady flow of a considerable amount of energy, and was consequently dependent upon the development of the dynamo. The first lamps were designed to utilize the electric arc that had been demonstrated by Sir Humphry Davy as early as 1810. Arc lights were brought to a high state of development by Paul Jablochkoff in 1876 and by C. R. Brush in 1879. Both men developed systems for lighting the streets with arc lamps connected in series supplied from a single generating station.

Lighting by incandescence was principally the result of the work of Thomas A. Edison, who developed a complete system of such lighting in 1879. His lights were designed to operate in parallel instead of in series, as had been the case with the previously developed arc-lighting systems. This radical departure from precedent permitted the use of low voltage and greatly simplified the distribution problems.

1.4 DISTRIBUTION OF ENERGY FOR LIGHTING

Edison planned his first installation in a densely populated area of lower Manhattan in New York City, and decided that an underground system of distribution would be necessary. This took the form of a network supplied by feeders radiating from a centrally located direct current (DC)-generating station to various feed points in the network. Pilot wires were taken back to the generating station from the feed points to give the operator an indication of voltage conditions on the system. Regulation was controlled by cutting feeders in, or out, as needed. At a later date, a battery was connected in parallel with the generator to guard against a station outage (Figures 1.1 and 1.2).

Gutta-percha, which had proved to be a satisfactory material for insulating the telegraph cables, was not suitable for the lighting feeders because of the softening of the material (a natural thermoplastic) at relatively high operating temperature. Experience with other types of insulation had not been sufficient to provide any degree of satisfaction with their use. The development of a cable sufficiently flexible to be drawn into ducts was accordingly considered a rather remote possibility. Therefore, Edison designed a rigid, buried system consisting of copper rods insulated with a wrapping of jute. Two or three insulated rods were drawn into iron pipes and a heavy bituminous compound was forced in and around them. They were then laid in 20-foot sections and joined together with specially designed tube joints from

FIGURE 1.1 Early Edison cable. Photos courtesy of Robert Lobenstein/IEEE Power & Energy.

which taps could be taken if desired. The Edison tube gave a remarkably satisfactory performance for this class of low voltage service.

The low voltage and heavy current characteristics of DC distribution were limited to the area capable of being supplied from one source if the regulation was to be kept within reasonable bounds. The high first cost and heavy losses made such systems uneconomical for general distribution. Accordingly, they were developed in limited areas of high-load density such as the business districts of large cities.

In the outlying districts, alternating current (AC) distribution was universally employed. This type of distribution was developed largely as a result of the work in 1882 of Lucien Gaulard and J. D. Gibbs, who designed a crude AC system using induction coils as transformers. The coils were first connected in series, but satisfactory performance could not be obtained. However, they were able to distribute electrical energy at a voltage considerably higher than that required for lighting and demonstrate the economics of the AC system. This system was introduced in the United States in 1885 by George Westinghouse, and served as the basis for the development of workable systems. An experimental installation went into service at Great Barrington, Massachusetts, early in 1886. The first large-scale commercial installation was built in Buffalo, New York, the same year.

The early installations operated at 1,000 volts. Overhead construction was considered essential for their satisfactory performance and almost universally employed. This was also true of the street-lighting feeders, that operated at about 2,000 volts. In Washington and Chicago, overhead wires were prohibited, so a number of underground lines were installed. Many different types of insulation and methods of installation were tried with little success. Experiments with underground conductors were

FIGURE 1.2 Splice box. Photos courtesy of Robert Lobenstein/IEEE Power & Energy.

also carried out in Philadelphia. The 1884 enactment of a law forcing the removal of all overhead wires from the streets of New York City mandated the development of a type of construction that could withstand such voltages. It was some time, however, before the overhead high-voltage wires disappeared. In 1888, the situation was summarized in a paper before the National Electric Light Association [1] as follows:

> No arc wires had been placed underground in either New York or Brooklyn. The experience in Washington led to the statement that no insulation could be found that would operate two years at 2,000 volts. In Chicago, all installations failed with the exception of lead covered cables which appeared to be operating successfully. In Milwaukee, three different systems had been tried and abandoned. In Detroit, a cable had been installed in Dorsett conduit, but later abandoned. In many of the larger cities, low voltage cables were operating satisfactorily and in Pittsburgh, Denver and Springfield, Mass., some 1,000 volt circuits were in operation. (*Underground Systems Reference Book* 1931, 2).

1.5 PAPER INSULATED CABLES

The first important line insulated with paper was installed by Sebastian de Ferranti in 1890 between Deptford (on the south side of the River Thames) and the City of London, for single-phase operation at 10,000 volts [3]. Some of these mains were still in use at the original voltage after more than 50 years. The cables consisted of two concentric copper conductors insulated with wide strips of paper applied helically around the conductor and saturated with a rosin-based oil. The insulated conductors were forced into an iron pipe filled with bitumen and installed in 20-foot lengths inside train tunnels under the river. This system operated successfully for 43 years and may be the source of the "40 year life" of power cables [4].

In the period between 1885 and 1887, cables insulated with helically applied narrow paper strips saturated with paraffin and later in a rosin compound and covered with a lead sheath (very similar in design to those used at the present time) were manufactured in the United States by the Norwich Wire Company. These were the first flexible paper-insulated cables, and all subsequent progress has been made through improvements in the general design.

Paper-insulated cables were improved through the following years by:

1. The introduction of the shielded design of multiple conductor cables by Martin Hochstadter in 1914. This cable is still known as Type H.
2. Luigi Emanueli's demonstration in 1920 that voids due to expansion and contraction could be controlled by the use of a thin oil impregnating fluid and reservoirs. This permitted the voltages to be raised to 69 kV and higher.
3. The 1927 patent by H. W. Fisher and R. W. Atkinson revealed that the dielectric strength of impregnated paper-insulated cable could be greatly increased by maintaining the insulating system under pressure. This system was not used commercially until the 1932 installation of a 200 psi pressurized cable in London.

Impregnated paper became the most common form of insulation for cables used for bulk transmission and distribution of electrical power, particularly for operating voltages of 12.5 kV and above, where low dielectric loss, low dissipation factor, and high ionization level are important factors in determining the cable life.

Impregnated paper insulation consists of multiple layers of paper tapes, each tape from 2.5 to 7.5 mils in thickness, wrapped helically around the conductor to be insulated. The entire wall of paper tapes is then heated, vacuum dried, and impregnated with an insulating fluid. The quality of the impregnated paper insulation depends not only on the properties and characteristics of the paper and impregnating fluid, but also on the mechanical application of the paper tapes over the conductor, the thoroughness of the vacuum drying, and the control of the saturating and cooling cycles during the manufacturing.

Originally, most of the paper used was made from Manila-rope fiber. This was erratic in its physical properties and not always susceptible to adequate oil penetration. Increased knowledge of the chemical treatment of the wood (in order to obtain pure cellulose by the adjustment of the fiber content and removal of lignin), the control of tear resistance, and the availability of long fiber stock resulted in the almost universal use of wood pulp paper in cables after 1900.

The impregnating compound was changed from a rosin-based compound to a pure mineral oil circa 1925, or oil blended to obtain higher viscosity, until polybutene replaced oil circa 1983.

Paper-insulated, lead-covered cables were the predominant power cables of all the large, metropolitan transmission and distribution systems in the United States, and the rest of the world, throughout the twentieth century. Their reliability was excellent. It was, however, necessary to have a high degree of skill for proper splicing and terminating. A shift toward extruded dielectric cables began about 1975 in those metropolitan areas, but the majority of the distribution cables of the large cities remained paper-insulated, lead-covered cables as the century ended.

Considerable research has been carried out by the utilities, technical organizations, and manufacturers of cables to obtain improved paper and laminated polypropylene-paper-polypropylene (PPP, now used in transmission cables) tapes and insulating fluids that are able to withstand high, continuous operating temperatures.

Impregnated paper insulation has excellent electrical properties, such as high dielectric strength, low dissipation factor, and dielectric loss. Because of these properties, the thickness of impregnated paper insulation was considerably less than for rubber or varnished cambric insulations for the same working voltages. Polyethylene and cross-linked polyethylene cables in the distribution classes are frequently made with the same wall thickness as today's impregnated paper cables.

1.6 UNDERGROUND RESIDENTIAL DISTRIBUTION SYSTEMS

The development of modern underground residential distribution (URD) systems may be viewed as the result of drastically lowering first costs through technology. Post-war URD systems were basically the same as the earlier systems except that there were two directions of feed (the loop system.) System voltages rose from 2,400/4,160 to 7,620/13,200 volts. The pre-1950 systems were very expensive because they utilized items such as paper insulated cables, vaults, switches, and submersible transformers. Those systems had an installation cost of $1,000 to $1,500. Expressed in terms of buying power at that time, you could buy a luxury car for the same price! Underground service was, therefore, limited to the most exclusive housing developments.

But for three developments in the 1960s, the underground distribution systems that exist today might not be in place. First, in 1958–1959, a large Midwestern utility inspired the development of the pad-mounted transformer; the vault was no longer necessary, nor was the submersible transformer. Second, the polyethylene cable with its bare concentric neutral did not require cable splicers, and the cable could be directly buried. While possibly not as revolutionary, the load-break elbow (separable connector) allowed the transformer to be built with a lower, more pleasing appearance.

The booming American economy and the environmental concerns of the nation made underground utility systems for new residential subdivisions the watchword of the Great Society. In a decade, URD had changed from a luxury to a necessity. The goal for the utility engineer was to design a URD system at about the same cost as the equivalent overhead system. There was little or no concern about costs over the system's life because the polyethylene cable was expected to last 100 years!

1.7 EXTRUDED DIELECTRIC POWER CABLES

The use of natural and synthetic polymers for industrial applications (during and after World War II) led to this technology being applied for cable insulation; in contrast to paper-insulated cables, these polymers could be extruded. Natural rubber was used first and synthetic rubber followed; as developments continued, butyl rubber became a material of choice for cable insulation for a while. Further developments led to newer synthetic elastomeric (rubbery) polymers, such as Neoprene (1931), chloroprene, and Hypalon (1951). The newer polymers, as they were developed, facilitated improvements in processing or properties (e.g., longer term reliability on aging, or flame resistance) and all were used concurrently depending on the application. Ethylene-propylene polymers (EPR) were employed in the 1960s as replacement for butyl rubber, but their usage, while steady, did not increase until the 1980s. It is to be noted that improvement in the insulation properties of elastomers was related not only to the polymer itself but also to the nature of the additives used.

High molecular weight polyethylene (HMWPE), which is not an elastomer, was extrudable and its development (starting in 1941) triggered a dramatic change in the insulation of cables for the transmission and distribution of electrical energy. Thermoplastic polyethylene was actually introduced during World War II for high-frequency cable insulation. By 1947, HMWPE was furnished as 15 kV cable insulation. Wide usage began with the advent of URD systems in the early 1960s. In the mid-1960s, conventional HMWPE was the material of choice for the rapidly expanding URD systems in the United States. It was superior to butyl rubber for electrical properties and moisture resistance. These cables were used as loop circuits for #2 and #1/0 AWG cables. The domination of polyethylene lasted until the mid-1970s.

Further discussion of the more modern insulation materials such as HMWPE, cross-linked polyethylene (XLPE), and EPR is in Chapter 5, "Electrical Insulation Materials."

1.8 TROUBLE IN PARADISE

During the mid-1970s, reports of early cable failures in extruded dielectric systems began to be documented in many parts of the world. "Treeing" was reintroduced to the cable engineer's vocabulary. This time it did not have the same meaning as with paper-insulated cables. See Chapter 17 for additional information on treeing.

By 1976, reports from utilities [4] and results of Electric Power Research Institute (EPRI) research [5] confirmed the fact that some thermoplastic polyethylene insulated cables (HMWPE) were failing in service in less than 5 years and failures were occurring at a rapidly increasing rate. By 1980, production of medium voltage HMWPE ceased in North America.

XLPE exhibited a much lower failure rate which was not escalating nearly as rapidly. Data from Europe confirmed the same facts in a report prepared by UNIPEDE-DISCAB. XLPE had been commercially available since 1963, but its slightly higher first cost had confined its use to the heavier loaded feeder cables. XLPE became the medium voltage insulation of choice by 1977.

The realization of the magnitude and significance of the problem led to a series of changes and improvements to the primary voltage cables:

- Research work was initiated to concentrate on solutions to the problem
- Utilities began replacing the poorest performing cables
- Suppliers of component materials improved their products
- Cable manufacturers improved their handling and processing techniques

1.9 MEDIUM VOLTAGE CABLE DEVELOPMENT

In the mid-1960s, conventional polyethylene became the material of choice for the rapidly expanding URD systems in the United States [6]. It was known to be superior to butyl rubber for moisture resistance, and could be readily extruded. It was used with cloth taped conductor and insulation shields, which achieved their semiconducting properties because of carbon black. By 1968, virtually all of the URD installations consisted of polyethylene-insulated medium voltage cables. The polyethylene was referred to as "high molecular weight" (HMWPE); this simply meant that the insulation used had a high "average" molecular weight. The higher the molecular weight, the better the electrical properties. The highest molecular weight polyethylene that could be readily extruded was adopted. Jacketed construction was seldom employed at that time.

Extruded thermoplastic shields were introduced between 1965 and 1975, leading to both easier processing and better reliability of the cable.

XLPE was first patented in 1959 for a carbon filled compound and in 1963 as unfilled by Dr. Frank Percopio. It was not widely used because of the tremendous pressure to keep the cost of URD down near the cost of an overhead system. This higher cost was caused by the need for additives (cross-linking agents) and the cost of manufacturing based on the need for massive, continuous vulcanizing (CV) tubes. EPR was introduced at about the same time. The significantly higher initial cost of these cables slowed their acceptance for utility purposes until the 1980s.

The superior operating and allowable emergency temperatures of XLPE and EPR made them the choice for feeder cables in commercial and industrial applications. These materials did not melt and flow as did the HMWPE material.

In order to facilitate removal for splicing and terminating, those early 1970-era XLPE cables were manufactured with thermoplastic insulation shields as had been used over the HMWPE cables. A reduction in ampacity was required until deformation resistant and then cross-linkable insulation shields became available during the later part of the 1970s.

A two-pass extrusion process was also used where the conductor shield and the insulation were extruded in one pass. The unfinished cable was taken up on a reel and then sent through another extruder to install the insulation shield layer. This resulted in possible contamination in a very critical zone. When cross-linked insulation shield materials became available, cables could be made in one pass utilizing "triple" extrusion of those three layers. "True triple" soon followed, where all layers were extruded in a single head fed by three extruders.

In the mid-1970s, a grade of tree-retardant polyethylene (TR-HMWPE) was introduced. This had limited commercial application and never became a major factor in the market.

Around 1976, another option became available—suppliers provided a grade of "deformation resistant" thermoplastic insulation shield material. This was an attempt to provide a material with "thermoset properties" and thus elevate the allowable temperature rating of the cable. This approach was abandoned when a true thermosetting shield material became available.

By 1976, the market consisted of approximately 45% XLPE, 30% HMWPE, 20% TR-HMWPE, and 5% EPR.

In the late 1970s, a strippable thermosetting insulation shield material was introduced. This allowed the user to install a "high temperature" XLPE that could be stripped for splicing with less effort than the earlier, inconsistent materials.

Jackets became increasingly popular by 1980. Since 1972–1973, there had been increasing recognition of the fact that water presence under voltage stress was causing premature loss of cable life due to "water treeing." Having a jacket reduced the amount of water penetration. This led to the understanding that water treeing could be "finessed" or delayed by utilizing a jacket. By 1980, 40% of the cables sold had a jacket.

EPR cables became more popular in the 1980s. A breakthrough had occurred in the mid-1970s with the introduction of a grade of EPR that could be extruded on the same type of equipment as XLPE insulation. The higher cost of EPR cables, when compared with XLPE, was a deterrent to early acceptance even with this new capability.

In 1981, another significant change took place: the introduction of "dry cure" cables. Until this time, the curing, or cross-linking, process was performed by using high-pressure steam. Because water was a problem for long cable life, the ability to virtually eliminate water became imperative. It was eventually recognized that the "dry cure" process enabled faster processing as well as elimination of the steam process for XLPE production.

Another major turning point occurred in 1982 with the introduction of tree-resistant cross-linked polyethylene (TR-XLPE). This product, which has supplanted conventional XLPE in market volume today, shows superior water tree resistance when compared with conventional XLPE. HMWPE and TR-HMWPE were virtually off the market by 1983.

By 1984, the market was approximately 65% XLPE, 25% TR-XLPE, and 10% EPR. Half the cables, sold had a jacket by that time.

During the second half of the 1980s, a major change in the use of filled strands took place. Although the process had been known for about 10 years, the control of the extruded "jelly-like" material was better understood by a large group of manufacturers. This material prevents water movement between the strands along the cable length and eliminates most of the conductor's air space, which can be a water reservoir.

In the late 1980s, another significant improvement in the materials used in these cables resulted in smoother and cleaner conductor shields. Vast improvements in the materials and processing of extruded, medium voltage power cables in the 1980s have led to cables that can be expected to function for 30, 40, or perhaps even 60 years when all of the proper choices are utilized. In 1995, the market was approximately 45% TR-XLPE, 35% XLPE, and 20% EPR.

REFERENCES

1. *Underground Systems Reference Book*, 1931, National Electric Light Association, Publication # 050, New York, NY.
2. Thue, W. A., 2001, adapted from class notes for "Understanding Power Cable Characteristics and Applications," University of Wisconsin–Madison.

3. *Underground Systems Reference Book,* 1957, Edison Electric Institute, Publication # 55–16, New York, NY.
4. Thue, W. A., Bankoske, J. W. and Burghardt, R. R., 1980, "Operating Experience on Solid Dielectric Cable," *CIGRE Proceedings,* Report 21-11, Paris.
5. *Underground Transmission Systems Reference Book,* 1992, Electric Power Research Institute, P O Box 10412, Palo Alto, CA 94303-0813.
6. Electric Power Research Institute EL-3154, January 1984, Estimation of Life Expectancy of Polyethylene Insulated Cables, Project 1357-1.

2 Basic Dielectric Theory of Cable

Carl C. Landinger

CONTENTS

2.1 INTRODUCTION

Whether being used to convey electric power or signals, it is the purpose of a wire or cable to convey the electric current to the intended device or location [1]. In order to accomplish this, a conductor is provided, which is adequate to convey the imposed electric current. Equally important is the need to keep the current from flowing in unintended paths rather than the conductor provided. Electrical insulation (dielectric) is provided to largely isolate the conductor from other paths or surfaces through which the current might flow. Therefore, it may be said that any conductor conveying electric signals or power is an insulated conductor.

2.2 ELECTRIC FIELDS AND VOLTAGE

Current flow is charge in motion. We might consider the simple case of a conductor carrying current out to a load and then a return conductor as two separate parallel

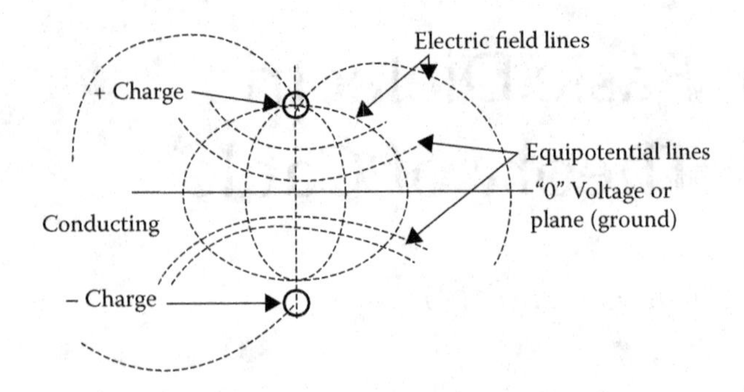

FIGURE 2.1 Electric field lines and equipotential lines for two lines of charge or a line of charge above a conducting plane.

cylinders of charge. If we neglect the conductor diameter (line of charge), there are electric field lines represented by circles of diameters such that the center of the circles are on the "0" line and each circle passes through the center of the cylinders. Everywhere perpendicular to the electric field lines are equipotential (equal voltage) lines due to each charge. The voltage at any location is the sum of the voltages due to each charge. Since the circle centers lie on the straight line equally distant from the charges, the equipotential lines from each charge exactly cancel on this line and the net voltage is "0" (see Figure 2.1).

If the "0" voltage line is now replaced with a conducting plane (such as the earth) and only the conductor above the plane remains, the locations of the electric field lines and equipotential lines are not changed. However, the portion below the "0" voltage line is simply an image of that above the line (method of images). This then, neglecting the conductor diameter, represents the electric field lines and equipotential (equal voltage lines) lines for an energized, current carrying conductor above ground. Of course, in this case the insulation (dielectric) is air.

2.3 AIR INSULATED CONDUCTORS

A metallic conductor suspended from insulating supports, surrounded by air, and carrying electric signals or power may be considered as the simplest case of an insulated conductor. It also presents an opportunity to easily visualize the parameters involved.

In Figure 2.2, the voltage is between the conductor and the ground [2,3], where the ground is at "0" potential as shown in Figure 2.1. The charge separation between the conductor and the ground, results in a capacitor and because there is some (generally very small) conduction from the conductor to the ground, a large resistance also exists between the conductor and the ground. As long as the ground is well away from the conductor, the electric field lines leave the conductor outer surface as reasonably straight lines emanating from the center of the conductor. We know that all the electric field lines bend to ultimately terminate at ground (see Figure 2.1).

Air is not a very good insulating material since it has a lower voltage breakdown strength than many other insulating materials. It is low in cost and if space is not a constraint, then it is a widely used dielectric. As the voltage between the conductor

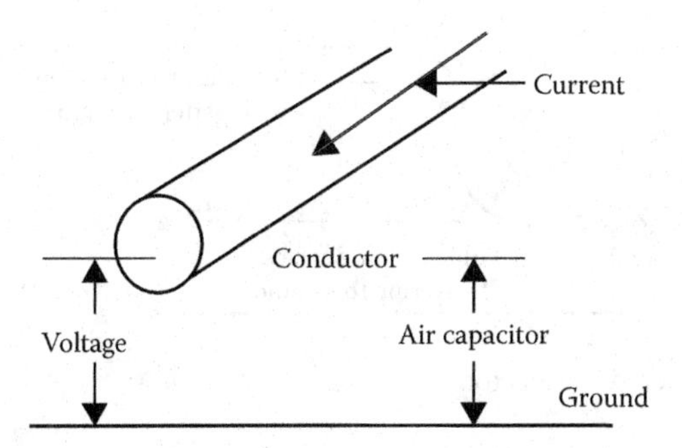

FIGURE 2.2 Location of voltage and current.

and the ground is increased, a point is reached where the electric stress at the conductor exceeds the voltage breakdown strength of air. At this point, the air literally breaks down producing a layer of ionized, conducting air surrounding the conductor. The term for this is corona. It represents power loss and can cause interference to radio, TV, and other signals. It is not uncommon for this condition to appear at isolated spots where a rough burr appears on the conductor or at a connector. This is simply because the electric stress is locally increased by the sharpness of the irregularity or protrusion from the conductor. In air or other gasses, the effect of the ionized gas layer surrounding the conductor is to increase the electrical diameter of the conductor to a point where the air beyond the ionized boundary is no longer stressed to breakdown at the prevailing temperature, pressure, and humidity. This ionized air might be considered as an unintentional conductor shield. The unlimited supply of fresh air and the conditions just mentioned preclude the progression of the ionization of air all the way to the ground. It is possible that the stress level is so high that an ionized channel can breach the entire gap between the conductor and the earth, but this generally requires a very high voltage source such as lightning.

This raises another important fact about dielectrics: that is, their ability to not break down under voltage is thickness dependent. In Chapter 6, the breakdown strength of air is given as 79 V for a 1/1,000 inch (1 mil) thickness (3.1 kV/mm). However, as thickness increases, the breakdown strength does not increase proportionately. If it did, a case could be made that lightning could not occur in the usual case.

2.4 INSULATING TO SAVE SPACE

Space is a common constraint that precludes the use of air as an insulator. Imagine the space requirements to wire a house or apartment using bare conductors on supports with air as the insulation. Let us consider the next step where some of the air surrounding the previous conductor is replaced with a better insulating material (dielectric).

In Figure 2.3, we see that the voltage from the conductor to the ground is the same as before. A voltage divider that is made up of impedance from the conductor to the outside covering surface and also impedance from the covering surface to the ground has been created. The distribution of voltage from the conductor to the covering surface

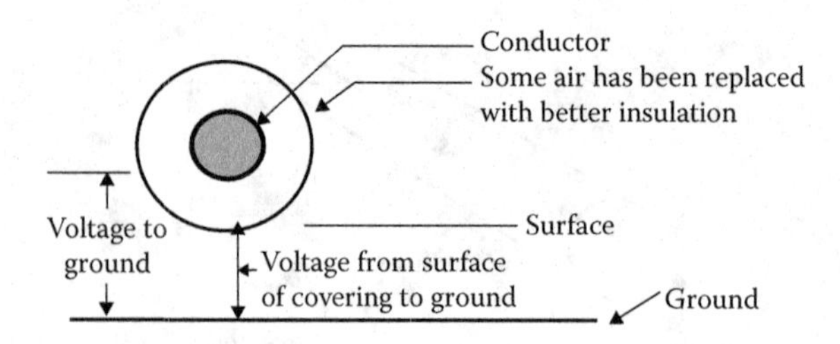

FIGURE 2.3 "Covered" conductor.

and from the covering surface to the ground will be in proportion to these impedances. It is important to note that with ground relatively far away from the covered conductor, the majority of the voltage exists from the covering surface to the ground. Putting this in another way, the outer surface of the covering has a voltage that is within a few percent of the voltage on the conductor (95% to 97% is a common value).

The amount of current that can flow from an intact covering to the ground in the event of contact by a grounded object is limited by the thickness, dielectric constant, and surface impedance of the covering as well as the area of contact. If the covering is made of an excellent insulation, the majority of the current will be due to capacitive charging current, which can be released from the covering surface by the contacting object.

So little current is available at the covering surface of a low voltage cable (600 V or less), which is imperceptible. When this condition exists with some level of confidence, the "covering" is then considered to be "insulation" and suitable for continuous contact by a grounded surface as long as such contact does not result in chemical or thermal degradation. The question arises as to what is considered to be low voltage. The voltage rating of insulated cables is based on the phase-to-phase voltage. Low voltage is generally considered to be less than 600 V phase-to-phase (1 kV is also common internationally). See Chapters 4 and 9 for additional information.

Because of the proximity and contact with other objects, the thickness of insulating materials used for low voltage cables is generally based on mechanical rather than electrical requirements. The surrounding environment, the need for special properties such as sunlight, or flame resistance, and rigors of installation often make it difficult for a single material to satisfy all related requirements. Designs involving two or more layers are commonly used in low voltage cable designs. The outer layer, though commonly insulating, may sacrifice some of the insulating quality to achieve toughness, sunlight resistance, flame resistance, chemical resistance, and more. In this case, the outer layer may serve as both insulation and jacket.

2.5 RISING VOLTAGE

Return to the metallic conductor that is covered with an insulating material and suspended in air. When the ground plane is brought close or touches the covering, the electric field lines, as depicted in Figure 2.1, must bend more sharply to terminate at right angles to the ground plane.

In Figure 2.4, we see considerable bending of the electric field lines. Recognizing that equipotential lines are perpendicular to the field lines, the bending results in

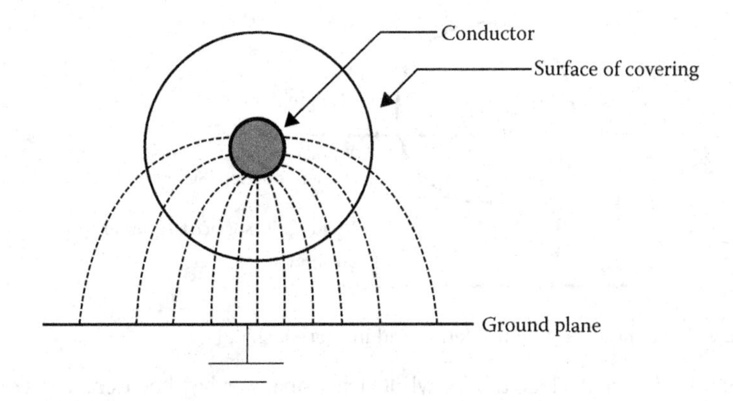

FIGURE 2.4 Electric field lines bend to terminate at ground plane.

potential differences on the covering surface. At low voltages, the effect is negligible. As the voltage increases, the point is reached where the potential gradients are sufficient to cause the current to flow across the surface of the covering. This is commonly known as "tracking." Even though the currents are small, the high surface resistance causes heating to take place, which ultimately damages the covering. If this condition is allowed to continue, the erosion may progress to significant covering damage, and, if in contact with the ground, results in failure.

It is important to note that the utilization of spacer cable systems and heavy walled tree wires depends on this ability of the covering to reduce current flow to a minimum. When sustained contact with branches, limbs, or other objects occurs, damage may result with time—hence such contacts may not be left permanently, but must be removed from the cable periodically as a maintenance practice.

At first, it might be thought that the solution is to continuously increase the insulating covering thickness as the operating voltage increases. Cost and complications involved in overcoming this difficulty would make this a desirable first choice. Unfortunately, breakdown strength, surface erosion, and personnel hazards are not linear functions of voltage versus thickness and this approach becomes impractical.

2.6 INSULATION SHIELD

In Figure 2.1, imagine that the ground plane was "wrapped" around the conductor with the same thickness of air separating the two. Barring surface irregularities at the conductor or ground, the electric field lines would be straight lines taking the shortest path from the conductor to the ground and the equipotential lines would be concentric cylinders around the conductor. This would form a cylindrical capacitor and would make the most effective use of the dielectric.

In order to make this ground contact possible, a semiconducting or resistive layer may be placed over the insulation surface. This material forces the bending of the field lines in the semiconducting layer. This layer creates some complications, however.

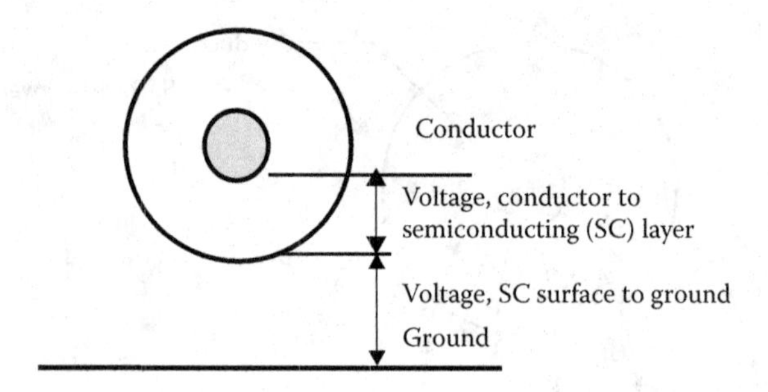

FIGURE 2.5 Conductor with insulation and insulation shield.

In Figure 2.5, it is clear that a cylindrical capacitor has been created from the conductor to the surface of the semiconducting layer, and a noncylindrical capacitor from the semiconducting layer to ground. A great deal of charge can be contained in the capacitor involving the ground because the outer plate is semiconducting, allowing for greater charge mobility in the layer. This charging current must be controlled so that a path to the ground is not established along the surface of the semiconducting layer. This path can lead to burning and ultimate failure of that layer. Accidental human contact would be a very serious event. It is clearly necessary to provide a continuous contact with the ground, which provides an adequate conducting path to drain the capacitive charging current to the ground without damage to the cable. This is done by adding a metallic path in contact with the semiconducting shield and making a relatively low resistance connection to the ground.

Once a metallic member has been added to the shield system, there is simply no way to avoid its presence underground fault conditions. This must be considered by providing either adequate conductive capacity in the shield to handle the fault currents or supplemental means to accomplish this. This is a critical factor in cable design.

Electric utility cables have fault current requirements that are sufficiently large that it is common to provide for a neutral in the design of the metallic shield. These cables have been known as underground residential distribution (URD) and underground distribution (UD) style cables. It is important that the functions of the metallic shield system are understood, since many serious errors and accidents have occurred because the functions were misunderstood.

2.7 CONDUCTOR SHIELD IS NEEDED

The presence of an insulation shield creates another complication. The grounded insulation shield results in the entire voltage stress being placed across the insulation.

Just as in the case of the air-insulated conductor, there is concern about exceeding the maximum stress that the insulating layer can withstand. The problem is magnified by stranded conductors or burrs and scratches that may be present in both stranded and solid conductors.

In Figure 2.6, a semiconducting layer has been added over the conductor to smooth out any irregularities. This reduces the probability of protrusions into the insulating layer. Protrusions into the insulation or into the semiconducting layer increase

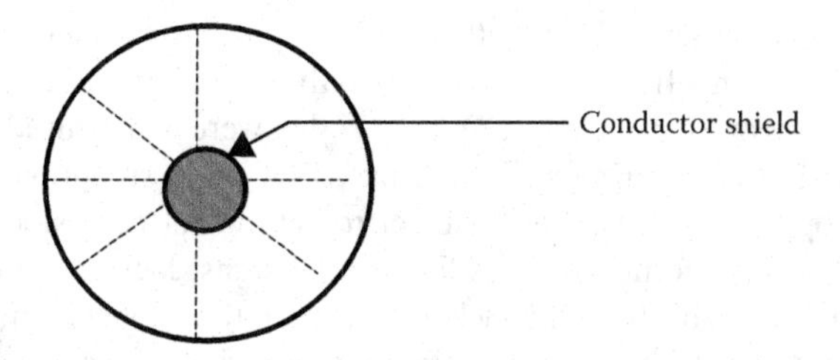

FIGURE 2.6 A conductor shield is added to provide a smooth inner electrode.

the localized stress (stress enhancement) that may exceed the long-term breakdown strength of the insulation. This is especially critical in the case of extruded dielectric insulations. Unlike air, there can be no fresh supply of insulation. Any damage will be progressive and lead to total breakdown of the insulating layer. There will be more discussion about "treeing" in Chapters 6 and 19.

2.8 SHIELDING LAYER REQUIREMENTS

There are certain requirements inherent in shielding layers to reduce stress enhancement. First, protrusions, whether by material smoothness or manufacturing, must be minimized. Such protrusions defeat the very purpose of a shield by enhancing electrical stress. The insulation's shield layer has a further complication in that it is desirable to have it easily removable to facilitate splicing and terminating. This certainly is the case at medium voltage (5 to 35 kV). At higher voltages, the inconveniences of a bonded insulation shield can be tolerated to gain the additional probability of a smooth, void-free insulation–insulation shield interface for cables with a bonded shield.

2.9 INSULATION LAYER REQUIREMENTS

At medium and higher voltages, it is critical that both the insulation and insulation-shield interfaces be contamination free. Contamination at the interface results in stress enhancement just as a protrusion that can increase the probability of breakdown. Voids can do the same with the additional possibility of capacitive–resistive (CR) discharges in the gas-filled void as voltage gradients appear across the void. Such discharges can be destructive of the surrounding insulating material and lead to progressive deterioration and breakdown.

2.10 JACKETS

In low voltage applications, jackets are commonly used to protect the underlying layers from physical abuse, sunlight, flame, or chemical attack. In medium voltage shielded cables, chemical attack includes corrosion of underlying metallic layers for shielding and armoring. In multiconductor designs, overall jackets are common for the same purposes. For medium and high voltage cables, jackets have been almost universally used throughout the history of cable designs. They are used for the same purposes as for low voltage cables, with special emphasis on protecting the underlying metallic components from corrosion. The only exceptions were paper-insulated, lead-covered cables and early URD/UD designs that were widely used by the electric utility industry. Both "experiments" were based on the assumption that lead, and subsequently copper wires, was not subject to significant corrosion. Both experiments resulted in elevated failure rates for these designs. Jackets are presently used for these designs. The ability of the jackets to reduce the ingress of moisture, which has been shown to have a deleterious effect on most dielectrics, causes them to be mentioned here.

2.11 TERMINOLOGY

To better understand the terminology that is used throughout this book, a brief introduction of the terms follows.

2.11.1 Nonshielded Power Cable

A nonshielded cable may consist of one or several conductors and one or several insulating layers. The cable may contain a jacket. The cable may also include a conductor shield. A cable is not considered fully shielded until both conductor and insulation shields are present. In the USA, nonshielded cables are common in the 0 to 2 kV voltage range although nonshielded power cables through 5 kV were common in the past and as high as 8 kV have been available.

2.11.2 Medium Voltage Shielded Cables

Medium voltage cables generally are fully shielded (having both conductor and insulation shields) cables in the 5 kV (6 kV internationally) through 35 kV (30 kV internationally) voltage range.

2.11.3 Conductors

Conductors may be solid or stranded. Metals used are commonly copper or aluminum. An attempt to use sodium was short-lived. The strand can be concentric, compressed, compacted, segmental, or annular to achieve desired properties of flexibility, diameter, and current density. The introduction of steel strands, higher strength aluminum alloys, hard drawn copper, Alumoweld, and Copperweld is common in overhead applications requiring greater tensile strength or other applications with the same requirements.

Assuming the same cross-sectional area of the conductor, there is a difference in the diameters between solid and various stranded conductors. This diameter differential is an important consideration in selecting the methods to effect joints, terminations, and fill of conduits.

Conductors are covered in Chapter 3.

2.11.4 Electrical Insulation (Dielectric)

The insulation (dielectric) provides sufficient separation between the conductor and the nearest electrical ground to adjacent phase to preclude dielectric failure. For low voltage cables, the required thickness of insulation to physically protect the conductor is more than adequate for the required dielectric strength.

2.11.5 Electric Field

Emphasis will be on 60 Hz alternating current fields (50 Hz is more common internationally). In all cables, regardless of their kilovolt ratings, there exists an electric field whenever the conductor is energized. This electric field can be visualized as electric field lines and lines of equipotential (see Figure 2.1).

2.11.6 Equipotential Lines

Equipotential lines represent points of equal potential difference between electrodes having different electrical potentials.

REFERENCES

1. Landinger, C. C., 2001, adapted from class notes for "Understanding Power Cable Characteristics and Applications," University of Wisconsin–Madison.
2. Clapp, A., Landinger, C. C., and Thue, W. A., September 15–20, 1996, "Design and Application of Aerial Systems Using Insulating and Covered Wire and Cable," *Proceedings of the 1996 IEEE/PES Transmission and Distribution Conference,* 96CH35968, Los Angeles, CA.
3. Clapp, A., Landinger, C. C., and Thue, W. A., September 15–20, 1996, "Safety Considerations of Aerial Systems Using Insulating and Covered Wire and Cable," *Proceedings of the 1996 IEEE/PES Transmission and Distribution Conference,* 96CH35968, Los Angeles, CA.

3 Conductors

Lauri J. Hiivala and Carl C. Landinger

CONTENTS

3.1 INTRODUCTION

The fundamental concern of power cable engineering is to transmit electrical current (power) economically and efficiently. The choice of the conductor material, size, and design must take into consideration such items as:

- Ampacity (current carrying capacity)
- Voltage stress at the conductor
- Voltage regulation
- Conductor losses
- Bending radius and flexibility
- Overall economics
- Material considerations
- Mechanical properties

3.2 MATERIAL CONSIDERATIONS

There are several low resistivity (or high conductivity) metals that may be used as conductors for power cables. Examples of these as ranked in order of increased resistivity at 20°C are shown in Table 3.1 [1].

Considering these resistivity figures and the cost of each of these materials, copper and aluminum become the logical choices. As such, they are the dominant metals used in the power cable industry today.

When choosing between copper and aluminum conductors, one should carefully compare the properties of the two metals, as each has advantages that may outweigh

TABLE 3.1
Resistivity of Metals at 20°C

Metal	Ohm-mm²/m × 10^{-8}	Ohm-cmil/ft × 10^{-6}
Silver	1.629	9.80
Copper, annealed	1.724	10.371
Copper, hard drawn	1.777	10.69
Copper, tinned	1.741–1.814	10.47–10.91
Aluminum, soft, 61.2% cond.	2.803	16.82
Aluminum, 1/2 hard to full hard	2.828	16.946
Sodium	4.3	25.87
Nickel	7.8	46.9

the other under certain conditions. The properties most important to the cable designer are shown in the following sections.

3.2.1 Direct Current Resistance

The conductivity of aluminum is about 61.2% to 62.0% of that of copper. Therefore, an aluminum conductor must have a cross-sectional area about 1.6 times that of a copper conductor to have the equivalent DC resistance. This difference in area is approximately equal to two American wire gauge (AWG) sizes.

3.2.2 Weight

One of the most important advantages of aluminum, other than economics, is its low density. A unit length of bare aluminum wire weighs only 48% as much as the same length of copper wire having an equivalent DC resistance. However, some of this weight advantage is lost when the conductor is insulated, because more insulation volume is required over the equivalent aluminum wire to cover the greater circumference.

3.2.3 Ampacity

The current carrying capacity (ampacity) of aluminum versus copper conductors can be compared by referring to many documents. See Chapter 14 for details and references, but obviously a larger aluminum cross-sectional area is required to carry the same current as a copper conductor as can be seen from Table 3.1.

3.2.4 Voltage Regulation

In alternating current (AC) circuits having small conductors (up to #2/0 AWG), and in all DC circuits, the effect of reactance is negligible. Equivalent voltage drop results with an aluminum conductor that has about 1.6 times the cross-sectional area of a copper conductor.

In AC circuits having larger conductors, however, skin and proximity effects influence the resistance value (AC to DC ratio, later written as AC/DC ratio), and the effect of reactance becomes important. Under these conditions, the conversion factor drops slightly, reaching a value of approximately 1.4.

3.2.5 Short Circuits

Consideration should also be given to possible short circuit conditions, since copper conductors have higher capabilities in short circuit operation. However, when making this comparison, the thermal limits of the materials in contact with the conductor (e.g., shields, insulation, coverings, jackets, etc.) must be considered.

3.2.6 Other Important Factors

Additional care must be taken when making connections with aluminum conductors. Not only does the metal tend to creep, but it also oxidizes rapidly. When aluminum is exposed to air, a thin, corrosion-resistant, high dielectric strength film quickly forms.

When copper and aluminum conductors are connected together, special techniques are required in order to make a satisfactory connection. See the discussion in Chapter 13.

Aluminum is not used extensively in generating station, substation, or portable cables because the lower bending life of small strands of aluminum does not always meet the mechanical requirements of those cables. Space is frequently a consideration at such locations also. However, aluminum is the overwhelming choice for aerial conductors because of its high conductivity-to-weight ratio and for underground distribution for economy where space is not a consideration.

The 8000 series aluminum alloys have found good acceptance in large commercial, institutional, and some industrial applications. These alloys offer reduced cold flow and improved creep resistance. This offers greater retention of torque at "screw down" terminals commonly used in "indoor plant." In the US, the National Electrical Code (NEC) calls for the use of these alloys if aluminum is to be used in a number of wire types recognized by the NEC.

Economics of the cost of the two metals must, of course, be considered, but always weighed after the cost of the overlying materials is added.

3.3 CONDUCTOR SIZES

3.3.1 American Wire Gauge (AWG)

Just as in any industry, a standard unit must be established for measuring the conductor sizes. In the US and Canada, electrical conductors are sized using the AWG system. This system is based on the following definitions:

- The diameter of size #0000 AWG (usually written #4/0 AWG and said as "four ought") is 0.4600 inches for a solid conductor.
- The diameter of size #36 AWG is 0.0050 inches.
- There are 38 intermediate sizes governed by a geometric progression.

The ratio of any diameter to that of the next smaller size is:

$$\sqrt[39]{\frac{0.4600}{0.0050}} = 1.122932 \qquad (3.1)$$

3.3.1.1 Shortcuts for Estimations

The square of the above ratio (the ratio of diameters of successive sizes) is 1.2610. Thus, an increase of one AWG size yields a 12.3% increase in diameter and an increase of 26.1% in area. An increase of two AWG sizes results in a change of 1.261 (or 26.1%) in diameter and 59% increase in area.

The sixth power of 1.122932 is 2.0050 or very nearly 2. Therefore, changing six AWG sizes will approximately double (or halve) the diameter. Another useful shortcut is that a #10 AWG wire has a diameter of approximately 0.1 inch, for copper a resistance of 1 ohm per 1,000 feet and a weight of about 10π or 31.4 pounds per 1,000 feet.

Another convenient rule is based on the fact that the tenth power of 1.2610 is 10.164 or approximately 10. Thus, for every increase or decrease of 10 gauge numbers (starting anywhere in the table), the cross-sectional area, resistance, and weight are divided or multiplied by about 10.

From a manufacturing standpoint, the AWG sizes have the convenient property that successive sizes represent approximately one reduction in die size in the wire drawing operation.

The AWG sizes were originally known as the Brown and Sharpe gage (B&S). The Birmingham wire gage (BWG) is used for steel armor wires. In Britain, wire sizes were specified by the standard wire gage (SWG), and were also known as the new British standard (NBS).

3.4 CIRCULAR MIL SIZES

Sizes larger than #4/0 AWG are specified in terms of the total cross-sectional area of the conductor and are expressed in circular mils. This method uses an arbitrary area of a conductor that is achieved by *squaring the diameter* of a solid conductor. This drops the $\pi/4$ multiplier required for the actual area of a round conductor. A circular mil is a unit of area equal to the area of a circle having a diameter of 1 mil (1 mil = 0.001 inch). Such a circle has an area of 0.7854 (or $\pi/4$) square mils. Thus, a wire 10 mils in diameter has a cross-sectional area of 100 circular mils. Likewise, 1 square inch = $4/\pi$ times 1,000,000 = 1,273,000 circular mils. For convenience, this is usually expressed in thousands of circular mils and abbreviated kcmil. Thus, an area of 1 square inch = 1,273 kcmil.

$$A = \pi r^2 \tag{3.2}$$

where A = area in circular mils; $\pi = 3.1416$; r = radius in 1/1,000 of an inch.

The abbreviation used in the past for thousand circular mils was MCM. The SI abbreviations for million, M, and for coulombs, C, are easily confused with this older term. Thus, the preferred abbreviation is kcmil for "thousand circular mils."

The AWG/kcmil system is prevalent in North America (population over 425 million) and is also used to some extent in over 65 countries in the world. The market share of cable products sized with this system is estimated at over 30% ($17.3 billion) of the world market for power and control wires and cables. The AWG/kcmil system is also the reference sizing system for all electrical products and installations in North America. As such it represents a basic element of the infrastructure. When considered in conjunction with wiring devices, connectors, and other related products, the market affected by the AWG/kcmil system is substantially larger [7].

Tables 3.2 and 3.3 provide nominal DC resistance and nominal diameter values for solid and concentric-lay-stranded copper and aluminum conductors.

TABLE 3.2A
Nominal DC Resistance in Ohms per 1,000 Feet at 25°C of Solid and Concentric-Lay-Stranded Conductor

Conductor Size	Solid			Concentric-Lay-Stranded*				
	Aluminum	Copper		Aluminum	Copper			
					Uncoated	Coated		
AWG or kcmil		Uncoated	Coated	Class B, C, D	Class B, C, D	Class B	Class C	Class D
8	1.05	0.640	0.659	1.07	0.652	0.678	0.678	0.680
7	0.833	0.508	0.522	0.851	0.519	0.538	0.538	0.538
6	0.661	0.403	0.414	0.675	0.411	0.427	0.427	0.427
5	0.524	0.319	0.329	0.534	0.325	0.338	0.339	0.339
4	0.415	0.253	0.261	0.424	0.258	0.269	0.269	0.269
3	0.329	0.201	0.207	0.334	0.205	0.213	0.213	0.213
2	0.261	0.159	0.164	0.266	0.162	0.169	0.169	0.169
1	0.207	0.126	0.130	0.211	0.129	0.134	0.134	0.134
1/0	0.164	0.100	0.102	0.168	0.102	0.106	0.106	0.106
2/0	0.130	0.0794	0.0813	0.133	0.0810	0.0842	0.0842	0.0842
3/0	0.103	0.0630	0.0645	0.105	0.0642	0.0667	0.0669	0.0669
4/0	0.0819	0.0500	0.0511	0.0836	0.0510	0.0524	0.0530	0.0530
250	0.0694	—	—	0.0707	0.0431	0.0448	0.0448	0.0448
300	0.0578	—	—	0.0590	0.0360	0.0374	0.0374	0.0374
350	0.0495	—	—	0.0505	0.0308	0.0320	0.0320	0.0320
400	0.0433	—	—	0.0442	0.0269	0.0277	0.0280	0.0280
450	0.0385	—	—	0.0393	0.0240	0.0246	0.0249	0.0249
500	0.0347	—	—	0.0354	0.0216	0.0222	0.0224	0.0224
550	—	—	—	0.0321	0.0196	0.0204	0.0204	0.0204
600	—	—	—	0.0295	0.0180	0.0187	0.0187	0.0187

650	—	—	—	0.0272	0.0166	0.0171	0.0172	0.0173
700	—	—	—	0.0253	0.0154	0.0159	0.0160	0.0160
750	—	—	—	0.0236	0.0144	0.0148	0.0149	0.0150
800	—	—	—	0.0221	0.0135	0.0139	0.0140	0.0140
900	—	—	—	0.0196	0.0120	0.0123	0.0126	0.0126
1,000	—	—	—	0.0177	0.0108	0.0111	0.0111	0.0112
1,100	—	—	—	0.0161	0.00981	0.0101	0.0102	0.0102
1,200	—	—	—	0.0147	0.00899	0.00925	0.00934	0.00934
1,250	—	—	—	0.0141	0.00863	0.00888	0.00897	0.00897
1,300	—	—	—	0.0136	0.00830	0.00854	0.00861	0.00862
1,400	—	—	—	0.0126	0.00771	0.00793	0.00793	0.00801
1,500	—	—	—	0.0118	0.00719	0.00740	0.00740	0.00747
1,600	—	—	—	0.0111	0.00674	0.00694	0.00700	0.00700
1,700	—	—	—	0.0104	0.00634	0.00653	0.00659	0.00659
1,750	—	—	—	0.0101	0.00616	0.00634	0.00640	0.00640
1,800	—	—	—	0.00982	0.00599	0.00616	0.00616	0.00622
1,900	—	—	—	0.00931	0.00568	0.00584	0.00584	0.00589
2,000	—	—	—	0.00885	0.00539	0.00555	0.00555	0.00560
2,500	—	—	—	0.00715	0.00436	0.00448	—	—
3,000	—	—	—	0.00596	0.00363	0.00374	—	—

Source: ANSI/ICEA S-94-649, "Standard for Concentric Neutral Cables Rated 5,000–46,000 Volts," 2004.

* Concentric-lay-stranded includes compressed and compact conductors.

TABLE 3.2B (METRIC)
Nominal DC Resistance in Milliohms per Meter at 25°C of Solid and Concentric-Lay-Stranded Conductor

Conductor Size		Solid			Concentric-Lay-Stranded*				
		Aluminum	Copper		Aluminum	Copper			
						Uncoated	Coated		
AWG or kcmil	mm²		Uncoated	Coated	Class B, C, D	Class B, C, D	Class B	Class C	Class D
8	8.37	3.44	2.10	2.16	3.51	2.14	2.22	2.22	2.23
7	10.6	2.73	1.67	1.71	2.79	1.70	1.76	1.76	1.76
6	13.3	2.17	1.32	1.36	2.21	1.35	1.40	1.40	1.40
5	16.8	1.72	1.05	1.08	1.75	1.07	1.11	1.11	1.11
4	21.1	1.36	0.830	0.856	1.39	0.846	0.882	0.882	0.882
3	26.7	1.08	0.659	0.679	1.10	0.672	0.699	0.699	0.699
2	33.6	0.856	0.522	0.538	0.872	0.531	0.554	0.554	0.554
1	42.4	0.679	0.413	0.426	0.692	0.423	0.440	0.440	0.440
1/0	53.5	0.538	0.328	0.335	0.551	0.335	0.348	0.348	0.348
2/0	67.4	0.426	0.260	0.267	0.436	0.266	0.276	0.276	0.276
3/0	85.0	0.338	0.207	0.212	0.344	0.211	0.219	0.219	0.219
4/0	107	0.269	0.164	0.168	0.274	0.167	0.172	0.174	0.174
250	127	0.228	—	—	0.232	0.141	0.147	0.147	0.147
300	152	0.190	—	—	0.194	0.118	0.123	0.123	0.123
350	177	0.162	—	—	0.166	0.101	0.105	0.105	0.105
400	203	0.142	—	—	0.145	0.0882	0.0909	0.0918	0.0918
450	228	0.126	—	—	0.129	0.0787	0.0807	0.0817	0.0817
500	253	0.114	—	—	0.116	0.0708	0.0728	0.0735	0.0735
550	279	—	—	—	0.105	0.0643	0.0669	0.0669	0.0669
600	304	—	—	—	0.0968	0.0590	0.0613	0.0613	0.0613

650	329	—	—	—	0.0892	0.0544	0.0561	0.0564	0.0567
700	355	—	—	—	0.0830	0.0505	0.0522	0.0525	0.0525
750	380	—	—	—	0.0774	0.0472	0.0485	0.0489	0.0492
800	405	—	—	—	0.0725	0.0443	0.0456	0.0459	0.0459
900	456	—	—	—	0.0643	0.0394	0.0403	0.0413	0.0413
1,000	507	—	—	—	0.0581	0.0354	0.0364	0.0364	0.0367
1,100	557	—	—	—	0.0528	0.0322	0.0331	0.0335	0.0335
1,200	608	—	—	—	0.0482	0.0295	0.0303	0.0306	0.0306
1,250	633	—	—	—	0.0462	0.0283	0.0291	0.0294	0.0294
1,300	659	—	—	—	0.0446	0.0272	0.0280	0.0282	0.0283
1,400	709	—	—	—	0.0413	0.0253	0.0260	0.0260	0.0263
1,500	760	—	—	—	0.0387	0.0236	0.0243	0.0243	0.0245
1,600	811	—	—	—	0.0364	0.0221	0.0228	0.0230	0.0230
1,700	861	—	—	—	0.0341	0.0208	0.0214	0.0216	0.0216
1,750	887	—	—	—	0.0331	0.0202	0.0208	0.0210	0.0210
1,800	912	—	—	—	0.0322	0.0196	0.0202	0.0202	0.0204
1,900	963	—	—	—	0.0305	0.0186	0.0192	0.0192	0.0193
2,000	1,013	—	—	—	0.0290	0.0177	0.0182	0.0182	0.0184
2,500	1,266	—	—	—	0.0235	0.0143	0.0147	—	—
3,000	1,520	—	—	—	0.0195	0.0119	0.0123	—	—

Source: ANSI/ICEA S-94-649, "Standard for Concentric Neutral Cables Rated 5,000–46,000 Volts," 2004.

* Concentric-lay-stranded includes compressed and compact conductors.

TABLE 3.3A
Nominal Diameters for Copper and Aluminum Conductors

Conductor Size		Nominal Diameters (Inches)							
			Concentric-Lay-Stranded					Combination	Unilay
AWG	kcmil	Solid	Compact*	Compressed	Class B**	Class C	Class D	Unilay	Compressed
8	16.51	0.1285	0.134	0.141	0.146	0.148	0.148	0.143	—
7	20.82	0.1443	—	0.158	0.164	0.166	0.166	0.160	—
6	26.24	0.1620	0.169	0.178	0.184	0.186	0.186	0.179	—
5	33.09	0.1819	—	0.200	0.206	0.208	0.208	0.202	—
4	41.74	0.2043	0.213	0.225	0.232	0.234	0.235	0.226	—
3	52.62	0.2294	0.238	0.252	0.260	0.263	0.264	0.254	—
2	66.36	0.2576	0.268	0.283	0.292	0.296	0.297	0.286	—
1	83.69	0.2893	0.299	0.322	0.332	0.333	0.333	0.321	0.313
1/0	105.6	0.3249	0.336	0.362	0.373	0.374	0.374	0.360	0.352
2/0	133.1	0.3648	0.376	0.406	0.419	0.420	0.420	0.404	0.395
3/0	167.8	0.4096	0.423	0.456	0.470	0.471	0.472	0.454	0.443
4/0	211.6	0.4600	0.475	0.512	0.528	0.529	0.530	0.510	0.498
	250	0.5000	0.520	0.558	0.575	0.576	0.576	0.554	0.542
	300	0.5477	0.570	0.611	0.630	0.631	0.631	0.607	0.594
	350	0.5916	0.616	0.661	0.681	0.681	0.682	0.656	0.641
	400	0.6325	0.659	0.706	0.728	0.729	0.729	0.701	0.685
	450	0.6708	0.700	0.749	0.772	0.773	0.773	0.744	0.727
	500	0.7071	0.736	0.789	0.813	0.814	0.815	0.784	0.766
	550	—	0.775	0.829	0.855	0.855	0.855	—	0.804
	600	—	0.813	0.866	0.893	0.893	0.893	—	0.840

650	—	0.845	0.901	0.929	0.930	0.930	—	0.874
700	—	0.877	0.935	0.964	0.965	0.965	—	0.907
750	—	0.908	0.968	0.998	0.999	0.998	—	0.939
800	—	0.938	1.000	1.031	1.032	1.032	—	0.969
900	—	0.999	1.061	1.094	1.093	1.095	—	1.028
1,000	—	1.060	1.117	1.152	1.153	1.153	—	1.084
1,100	—	—	1.173	1.209	1.210	1.211	—	1.137
1,200	—	—	1.225	1.263	1.264	1.264	—	1.187
1,250	—	—	1.251	1.289	1.290	1.290	—	1.212
1,300	—	—	1.276	1.315	1.316	1.316	—	1.236
1,400	—	—	1.323	1.364	1.365	1.365	—	1.282
1,500	—	—	1.370	1.412	1.413	1.413	—	1.327
1,600	—	—	1.415	1.459	1.460	1.460	—	1.371
1,700	—	—	1.459	1.504	1.504	1.504	—	1.413
1,750	—	—	1.480	1.526	1.527	1.527	—	1.434
1,800	—	—	1.502	1.548	1.548	1.549	—	1.454
1,900	—	—	1.542	1.590	1.590	1.591	—	1.494
2,000	—	—	1.583	1.632	1.632	1.632	—	1.533
2,500	—	—	1.769	1.824	1.824	1.824	—	—
3,000	—	—	1.938	1.998	1.999	1.999	—	—

Source: ANSI/ICEA S-94-649, "Standard for Concentric Neutral Cables Rated 5,000–46,000 Volts," 2004.

* Diameters shown are for compact round, compact modified concentric, and compact single input wire.

** Diameters shown are for concentric round and modified concentric.

TABLE 3.3B (METRIC)
Nominal Diameters for Copper and Aluminum Conductors

Conductor Size		Nominal Diameters (mm)							
			Concentric-Lay-Stranded					Combination	Unilay
AWG or kcmil	mm^2	Solid	Compact*	Compressed	Class B**	Class C	Class D	Unilay	Compressed
8	8.37	3.26	3.40	3.58	3.71	3.76	3.76	3.63	—
7	10.6	3.67	—	4.01	4.17	4.22	4.22	4.06	—
6	13.3	4.11	4.29	4.52	4.67	4.72	4.72	4.55	—
5	16.8	4.62	—	5.08	5.23	5.28	5.31	5.13	—
4	21.1	5.19	5.41	5.72	5.89	5.94	5.97	5.74	—
3	26.7	5.83	6.05	6.40	6.60	6.68	6.71	6.45	—
2	33.6	6.54	6.81	7.19	7.42	7.52	7.54	7.26	—
1	42.4	7.35	7.59	8.18	8.43	8.46	8.46	8.15	7.95
1/0	53.5	8.25	8.53	9.19	9.47	9.50	9.50	9.14	8.94
2/0	67.4	9.27	9.55	10.3	10.6	10.7	10.7	10.3	10.0
3/0	85.0	10.4	10.7	11.6	11.9	12.0	12.0	—	11.3
4/0	107	11.7	12.1	13.0	13.4	13.4	13.5	—	12.6
250	127	12.7	13.2	14.2	14.6	14.6	14.6	—	13.8
300	152	13.9	14.5	15.5	16.0	16.0	16.0	—	15.1
350	177	15.0	15.6	16.8	17.3	17.3	17.3	—	16.3
400	203	16.1	16.7	17.9	18.5	18.5	18.5	—	17.4
450	228	17.0	17.8	19.0	19.6	19.6	19.6	—	18.5
500	253	18.0	18.7	20.0	20.7	20.7	20.7	—	19.5
550	279	—	19.7	21.1	21.7	21.7	21.7	—	20.4
600	304	—	20.7	22.0	22.7	22.7	22.7	—	21.3

650	329	—	21.5	22.9	23.6	23.6	23.6	—	22.2
700	355	—	22.3	23.7	24.5	24.5	24.5	—	23.0
750	380	—	23.1	24.6	25.3	25.4	25.3	—	23.9
800	405	—	23.8	25.4	26.2	26.2	26.2	—	24.6
900	456	—	25.4	26.9	27.8	27.8	27.8	—	26.1
1,000	507	—	26.9	28.4	29.3	29.3	29.3	—	27.5
1,100	557	—	—	29.8	30.7	30.7	30.8	—	28.9
1,200	608	—	—	31.1	32.1	32.1	32.1	—	30.1
1,250	633	—	—	31.8	32.7	32.8	32.8	—	30.8
1,300	659	—	—	32.4	33.4	33.4	33.4	—	31.4
1,400	709	—	—	33.6	34.6	34.7	34.7	—	32.6
1,500	760	—	—	34.8	35.9	35.9	35.9	—	33.7
1,600	811	—	—	35.9	37.1	37.1	37.1	—	34.8
1,700	861	—	—	37.1	38.2	38.2	38.2	—	35.9
1,750	887	—	—	37.6	38.8	38.8	38.8	—	36.4
1,800	912	—	—	38.2	39.3	39.3	39.3	—	36.9
1,900	963	—	—	39.2	40.4	40.4	40.4	—	37.9
2,000	1,013	—	—	40.2	41.5	41.5	41.5	—	38.9
2,500	1,266	—	—	44.9	46.3	46.3	46.3	—	—
3,000	1,520	—	—	49.2	50.7	50.8	50.8	—	—

Source: ANSI/ICEA S-94-649, "Standard for Concentric Neutral Cables Rated 5,000–46,000 Volts," 2004.

* Diameters shown are for compact round, compact modified concentric, and compact single input wire.

** Diameters shown are for concentric round and modified concentric.

3.5 METRIC DESIGNATIONS

Except as noted above, most of the world uses the SI unit of square millimeters (mm^2) to designate conductor size. The International Electrotechnical Commission has adopted IEC 60228 [8] to define these sizes. An important consideration is that these are not *precise* sizes. For instance, their 50 mm^2 conductor is actually 47 mm^2. To accommodate everyone, the IEC standard allows as much as a 20% variation in conductor area from the size designated.

In Canada, metric designations are used for all cable dimensions except for the conductor size. The variations in the two systems are too great to use any of the SI sizes as a direct substitute for standard sizes.

Conductors described in IEC 60228 are specified in metric sizes. North America and certain other regions at present use conductor sizes and characteristics according to the AWG system, and thousands of circular mils for larger sizes. The use of these sizes is currently prescribed across North America and elsewhere for installations by subnational regulations. IEC TC 20 cable product standards do not prescribe cables with AWG/kcmil conductors.

IEC TC 20 recognizes the need to produce a single, harmonized standard for conductors that is truly international. Harmonization, in this respect, is understood as the merging of AWG-based and metric-based sizes to produce one rationalized range of conductor sizes for power cables. TC 20 also recognizes that the development of such a harmonized standard is a long-term project.

A three-stage approach, which will culminate in a single International Standard for conductors, has been agreed.

Stage one of the approach is to produce a technical report that defines the range of AWG/kcmil sizes that are to be considered in the harmonization process.

Stage two of the process is to develop this technical report by starting the rationalization process. The test methods and requirements in this technical report are to be aligned with those in IEC 60228.

The third and final stage will be to produce a harmonized standard, based on IEC 60228 and the work of the first two stages, with a single, rationalized range of conductor sizes. The present expectation is that the third stage will not be achieved before 2020.

IEC Technical Report 62602 provides resistance and dimensional details for AWG and thousands of circular mils sizes as well as approximate equivalent metric nominal cross-sectional areas [9].

Table 3.4 provides maximum resistance values for solid Class 1 circular, annealed copper and circular or shaped aluminum and aluminum alloy conductors for use in single-core and multicore cables. Such conductors are available in nominal cross-sectional areas up to 1,200 mm^2.

Likewise, Table 3.5 provides maximum resistance values for stranded Class 2 circular, circular compacted and shaped annealed copper and aluminum and aluminum alloy conductors for use in single-core and multicore cables.

It should be noted that maximum or minimum diameter requirements are not specified in IEC 60228. Instead, it gives guidance on dimensional limits for the following types of conductors, which are included in this standard:

TABLE 3.4
Class 1 Solid Conductors for Single-Core and Multicore Cables

1	2	3	4
	Maximum Resistance of Conductor at 20°C		
	Circular, Annealed Copper Conductors		
Nominal Cross-Sectional Area mm²	Plain Ω/km	Metal-Coated Ω/km	Aluminum and Aluminum Alloy Conductors, Circular or Shaped[c] Ω/km
0.5	36.0	36.7	—
0.75	24.5	24.8	—
1.0	18.1	18.2	—
1.5	12.1	12.2	—
2.5	7.41	7.56	—
4	4.61	4.70	—
6	3.08	3.11	—
10	1.83	1.84	3.08[a]
16	1.15	1.16	1.91[a]
25	0.727[b]	—	1.20[a]
35	0.524[b]	—	0.868[a]
50	0.387[b]	—	0.641
70	0.268[b]	—	0.443
95	0.193[b]	—	0.320[d]
120	0.153[b]	—	0.253[d]
150	0.124[b]	—	0.206[d]
185	0.101[b]	—	0.164[d]
240	0.0775[b]	—	0.125[d]
300	0.0620[b]	—	0.100[d]
400	0.0465[b]	—	0.0778
500	—	—	0.0605
630	—	—	0.0469
800	—	—	0.0367
1,000	—	—	0.0291
1,200	—	—	0.0247

Source: IEC 60228 (Edition 3.0 2004-11), "Conductors of insulated cables," Copyright © 2004 IEC Geneva, Switzerland. www.iec.ch

[a] Aluminum conductors 10 mm² to 35 mm² circular only; see 5.1.1c.

[b] See note to 5.1.1b.

[c] See note to 5.1.2.

[d] For single-core cables, four sectoral shaped conductors may be assembled into a single circular shaped conductor. The maximum resistance of the assembled conductor shall be 25% of that of the individual component conductors.

TABLE 3.5
Class 2 Stranded Conductors for Single-Core and Multicore Cables

1	2	3	4	5	6	7	8	9	10
	Minimum Number of Wires in the Conductor						Maximum Resistance of Conductor at 20°C		
	Circular		Circular Compacted		Shaped		Annealed Copper Conductor		
Nominal Cross-Sectional Area mm²	Cu	Al	Cu	Al	Cu	Al	Plain Wires Ω/km	Metal-Coated Wires Ω/km	Aluminum or Aluminum Alloy Conductor[c] Ω/km
0.5	7	—	—	—	—	—	36.0	36.7	
0.75	7	—	—	—	—	—	24.5	24.8	
1.0	7	—	—	—	—	—	18.1	18.2	
1.5	7	—	6	—	—	—	12.1	12.2	
2.5	7	—	6	—	—	—	7.41	7.56	
4	7	—	6	—	—	—	4.61	4.70	
6	7	—	6	—	—	—	3.08	3.11	
10	7	7	6	6	—	—	1.83	1.84	3.08
16	7	7	6	6	—	—	1.15	1.16	1.91
25	7	7	6	6	6	6	0.727	0.734	1.20
35	7	7	6	6	6	6	0.524	0.529	0.868
50	19	19	6	6	6	6	0.387	0.391	0.641
70	19	19	12	12	12	12	0.268	0.270	0.443

95	19	19	15	15	15	15	0.193	0.195	0.320
120	37	37	18	15	18	15	0.153	0.154	0.253
150	37	37	18	15	18	15	0.124	0.126	0.206
185	37	37	30	30	30	30	0.0991	0.100	0.164
240	37	37	34	30	34	30	0.0754	0.0762	0.125
300	61	61	34	30	34	30	0.0601	0.0607	0.100
400	61	61	53	53	53	53	0.0470	0.0475	0.0778
500	61	61	53	53	53	53	0.0366	0.0369	0.0605
630	91	91	53	53	53	53	0.0283	0.0286	0.0469
800	91	91	53	53	—	—	0.0221	0.0224	0.0367
1,000	91	91	53	53	—	—	0.0176	0.0177	0.0291
1,200			b				0.0151	0.0151	0.0247
1,400			b				0.0129	0.0129	0.0212
1,600			b				0.0113	0.0113	0.0186
1,800			b				0.0101	0.0101	0.0165
2,000			b				0.0090	0.0090	0.0149
2,500			b				0.0072	0.0072	0.0127

Source: IEC 60228 (Edition 3.0 2004-11), "Conductors of insulated cables," Copyright © 2004 IEC Geneva, Switzerland. www.iec.ch

[a] These sizes are nonpreferred. Other nonpreferred sizes are recognized for some specialized applications but are not within the scope of this standard.

[b] The minimum number of wires for these sizes is not specified. These sizes may be constructed from 4, 5, or 6 equal segments (Milliken).

[c] For stranded aluminum alloy conductors having the same nominal cross-sectional area as an aluminum conductor, the resistance value should be agreed between the manufacturer and the purchaser.

1. circular solid conductors (Class 1) of copper, aluminum, and aluminum alloy;
2. circular and compacted circular stranded conductors (Class 2) of copper, aluminum, and aluminum alloy.

This is intended as a guide to the manufacturers of cables and cable connectors to assist in ensuring that the conductors and connectors are dimensionally compatible.

3.6 STRANDING

Larger sizes of solid conductors become too rigid to install, form, and terminate. Stranding becomes the solution to these difficulties. The point at which stranding should be used is dependent on the type of metal as well as the temper of that metal. Copper conductors are frequently stranded at #6 AWG and greater. Aluminum, in the half-hard temper, can be readily used as a solid conductor up to a #2/0 AWG size.

3.6.1 Concentric Stranding

This is the typical choice for power cable conductors. This consists of a central wire or core surrounded by one or more layers of helically applied wires. Each additional layer has six more wires than the preceding layer. Except in unilay-stranded conductors, each layer is applied in a direction opposite to that of the layer underneath. In the case of power cable conductors, the core is a single wire and all of the strands have the same diameter. As shown in Figure 3.1, the first layer over the core contains 6 wires; the second, 12; the third, 18; etc. The distance that it takes for one strand of the conductor to make one complete revolution of the layer is called the length of

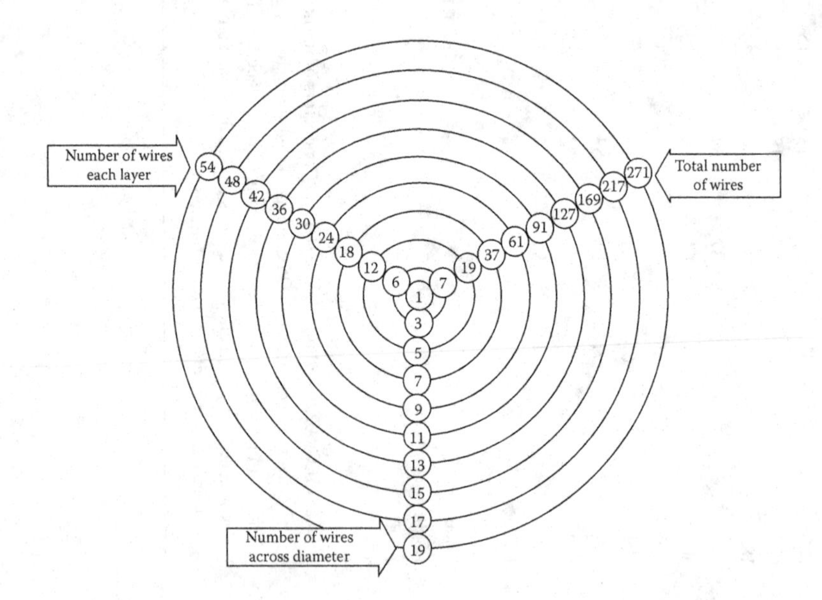

FIGURE 3.1 Concentric standing relationships.

TABLE 3.6
Examples of Class B, C, and D Stranding

Size	Class B	Class C	Class D
#2 AWG	7 × 0.0974	19 × 0.0591	37 × 0.0424
#4/0 AWG	19 × 0.1055	37 × 0.0756	61 × 0.0589
500 kcmil	37 × 0.1162	61 × 0.0905	91 × 0.0741
750 kcmil	61 × 0.1109	91 × 0.0908	127 × 0.0768

lay. The requirement for the length of lay is set forth in ASTM standards [6] to be neither less than 8 nor more than 16 times the overall diameter (OD) of that layer.

In power cables, the standard stranding is Class B. Standards require that the outermost layer be of a left hand lay. This means that as you look along the axis of the conductor, the outermost layer of strands roll toward the left as they recede from the observer. More flexibility is achieved by increasing the number of wires in the conductor. Class C has one more layer than Class B; Class D has one more layer than C. The class designation goes up to M (normally used for welding cables, etc.). These are covered by ASTM standards.

Class C and D conductors have approximately the same weight as a Class B and an OD within 3 mils of Class B. Examples of Class B (standard), Class C (flexible), and Class D (extra flexible) are shown in Table 3.6 with the number of strands and diameter of each strand.

The following formula may be used to calculate the number of wires in a concentric stranded conductor:

$$n = 1 + 3N\left(N + 1\right) \tag{3.3}$$

where n = total number of wires in stranded conductor and N = number of layers around the center wire.

3.6.2 Compressed Stranding

This is the term that is used to describe a slight deformation of the layers to allow the layer being applied to close tightly. There is no reduction in conductor area. The diameter of the finished conductor can be reduced no more than 3% of the equivalent

TABLE 3.7
Gaps in Outer Layer of a Stranded Conductor

Total Number of Strands	Angle of Gap at 16 × OD
19	8.3^0
37	10^0
61	10^0

concentric strand. A typical reduction is about 2.5%. Examples of gaps in the outer layer for concentric stranded conductors are shown in Table 3.7.

Shortening the length of lay on the outer layer could solve the problem but would result in higher resistance and would require more conductor material.

Compressed stranding is often the preferred construction, because concentric stranding, with its designated lay length, creates a slight gap between the outer strands of such a conductor. Lower viscosity materials that are extruded over such a conductor tend to "fall in" to any gap that forms. This results in surface irregularities that create increased voltage stresses and make it more difficult to strip off that layer.

3.6.3 Compact Stranding

This is similar to compressed stranding except that additional forming is given to the conductor so that the reduction in diameter is typically 9% less than the concentric stranded conductor. This results in a diameter nearing that of a solid conductor. Some air spaces that can serve as channels for moisture migration are still present. The main advantage of compact conductors is the reduced conductor diameter.

3.6.4 Bunch Stranding

This term is applied to a collection of strands twisted together in the same direction without regard to the geometric arrangement. This construction is used when extreme flexibility is required for small AWG sizes, such as portable cables. Examples of bunch-stranded conductors are cords for vacuum cleaners, extension cords for lawn mowers, etc. Examples of bunch stranding are shown in Table 3.8.

Note that in Class K and M conductors, the individual wire diameters are constant and the cross-sectional area is developed by adding a sufficient number of wires to provide the total conductor area required.

3.6.5 Rope Stranding

This term is applied to a concentric-stranded conductor, each of whose component strands is stranded. This is a combination of the concentric conductor and a bunch-stranded conductor. The finished conductor is made up of a number of groups of

TABLE 3.8
Examples of Class K and M Stranding

Conductor Size	Class K	Class M
#16 AWG	26 × 0.0100	65 × 0.0063
#14 AWG	41 × 0.0100	104 × 0.0063
#12 AWG	65 × 0.0100	168 × 0.0063

Note in Class K and M that the individual wire diameters are constant and the area is developed by adding a sufficient number of wires to provide the total conductor area required.

bunch- or concentric-stranded conductors assembled concentrically together. The individual groups are made up of a number of wires rather than a single, individual strand. A rope-stranded conductor is described by giving the number of groups laid together to form the rope and the number of wires in each group.

Classes G and H are generally used on portable cables for mining applications. Classes I, L, and M utilize bunch-stranded members assembled into a concentric arrangement. The individual wire size is the same with more wires added as necessary to provide the area. Class I uses #24 AWG (0.020 inch) individual wires, Class L uses #30 AWG (0.010 inch) individual wires, and Class M uses #34 AWG (0.0063 inch) individual wires. Class I stranding is generally used for railroad applications and Classes L and M are used for extreme portability such as welding cable and portable cords.

3.6.6 Sector Conductors

These have a cross section approximating the shape of a sector of a circle. A typical three-conductor cable has three 120° segments that combine to form the basic circle of the finished cable. Such cables have a smaller OD than the corresponding cable with concentric round conductors, and exhibit lower AC resistance due to a reduction of the proximity effect.

For paper-insulated cables, the sector conductor was almost always stranded and then compacted in order to achieve the highest possible ratio of conductor area to cable area. The precise shape and dimensions varied somewhat between the manufacturers.

Figure 3.2 and Table 3.9 show the nominal dimensions of typical compact sector conductors.

For the calculation of cable capacitance, for instance, an "equivalent round conductor" is required. Over the 2/0 AWG to 750 kcmil size range, the following formula holds:

$$D = 1.337\sqrt{A} \tag{3.4}$$

where D = equivalent round diameter in mils and A = area of sector conductor in circular mils.

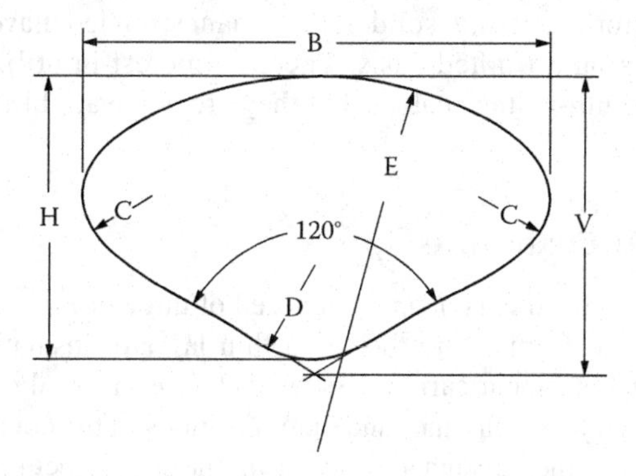

FIGURE 3.2 Outline of typical compact sector.

TABLE 3.9
Nominal Dimensions of 3/c Compact Sector Conductor

Cond. AWG/kcmil	V-Gage Inch	V-Gage* Inch	B Inch	C Inch	D Inch	E Inch
1/0	0.288		0.462	0.080	0.080	0.504
2/0	0.323		0.520	0.085	0.085	0.540
3/0	0.364		0.592	0.100	0.100	0.584
4/0	0.417		0.660	0.111	0.090	0.595
4/0		0.410	0.660	0.117	0.090	0.770
250	0.455		0.720	0.118	0.220	0.635
250		0.447	0.720	0.125	0.220	0.812
300	0.497		0.784	0.130	0.179	0.678
300		0.490	0.784	0.138	0.179	0.852
350	0.539		0.834	0.151	0.259	0.718
350		0.532	0.834	0.151	0.259	0.890
400	0.572		0.902	0.147	0.244	0.754
400		0.566	0.902	0.158	0.244	0.928
500	0.642		1.018	0.155	0.207	0.820
500		0.635	1.018	0.167	0.207	1.000
600	0.700		1.120	0.165	0.210	0.882
600		0.690	1.120	0.178	0.210	1.050
750	0.780		1.280	0.163	0.284	0.970
750		0.767	1.280	0.185	0.284	1.140
800	0.806		1.324	0.164	0.224	0.890
800		0.795	1.324	0.176	0.224	1.083
900	0.854		1.405	0.170	0.236	1.040
900		0.842	1.405	0.180	0.236	1.110
1,000	0.900		1.500	0.137	0.300	1.008
1,000		0.899	1.500	0.192	0.300	1.266

* Denotes the column that applies for insulation thickness over 200 mils.

Sector conductors that are solid rather than stranded have been used for low-voltage cables on a limited basis. There is interest in utilizing this type of conductor for medium-voltage cables, but they are not available on a commercial basis at this time.

3.6.7 Segmental Conductors

These are round, stranded conductors composed of three or more segments that are electrically separated from each other by a thin layer of insulation around every other segment. Each segment carries less current than the total conductor, and the current is transposed between inner and outer positions in the completed conductor. This construction has the advantage of lowering the skin effect ratio and hence the AC resistance by having less skin effect than a conventionally stranded conductor.

This type of conductor should be considered for large sizes such as 1,000 kcmil and above that are to carry large amounts of current.

The diameters of four-segment conductors are approximately the same as that of Class B concentric-stranded conductors (Table 3.10).

3.6.8 Annular Conductors

These are round, stranded conductors whose strands are laid around a core of rope, fibrous material, helical metal tube, or a twisted I-beam. This construction has the advantage of lowering the total AC resistance for a given cross-sectional area of conductor by eliminating the greater skin effect at the center of the completed conductor. Where space is available, annular conductors may be economical to use for

TABLE 3.10
Nominal Diameters for Segmental Copper and Aluminum Conductors

Conductor Size		Segmental Conductor Diameter* (Four Segments)	
kcmil	mm²	Inches	mm
1,000	507	1.140–1.152	29.0–29.3
1,100	557	1.195–1.209	30.4–30.7
1,200	608	1.235–1.263	31.4–32.1
1,250	633	1.260–1.289	32.0–32.7
1,300	659	1.285–1.315	32.6–33.4
1,400	709	1.325–1.364	33.7–34.6
1,500	760	1.375–1.412	34.9–35.9
1,600	811	1.420–1.459	36.1–37.1
1,700	861	1.460–1.504	37.1–38.2
1,750	887	1.480–1.526	37.6–38.8
1,800	912	1.500–1.548	38.1–39.3
1,900	963	1.530–1.590	38.9–40.4
2,000	1,013	1.570–1.632	39.9–41.5
2,250	1,140	1.665–1.730	42.3–43.9
2,500	1,266	1.740–1.824	44.2–46.3
2,750	1,393	1.830–1.913	46.5–48.6
3,000	1,520	1.910–1.998	48.5–50.7
3,250	1,647	1.985–2.080	50.4–52.8
3,500	1,773	2.085–2.159	53.0–54.8
3,750	1,900	2.150–2.234	54.6–56.7
4,000	2,027	2.225–2.309	56.5–58.6
4,250	2,154	2.245–2.378	57.0–60.4
4,500	2,280	2.315–2.448	58.8–62.2
4,750	2,407	2.375–2.516	60.3–63.9
5,000	2,534	2.435–2.581	61.8–65.6

Source: ANSI/ICEA S-108-720, "Standard for Extruded Insulation Power Cables Rated Above 46 Through 345 kV," 2004.

* Diameter over binder tape.

1,000 kcmil cables and above at 60 hertz and for 1,500 kcmil cables and above for lower frequencies such as 25 hertz.

3.6.9 Unilay Conductors

Unilay has, as the name implies, all of its strands applied in the same direction of lay. A design frequently used for low-voltage power cables is the combination unilay where the outer layer of strands are partially comprised of strands having a smaller diameter than the other strands. This makes it possible to attain the same diameter as a compact stranded conductor. The most common unilay conductor is a compact, 8,000 series aluminum alloy.

3.7 PHYSICAL AND MECHANICAL PROPERTIES

3.7.1 Conductor Properties

Although high conductivity is one of the important features of a good conductor material, other factors must be taken into account. Silver is an interesting possibility for a cable conductor. Its high cost is certainly one of the reasons to look for other candidates. Silver has another disadvantage, which is its lack of physical strength that is necessary for pulling the cables into conduits.

3.7.1.1 Copper

Impurities have a very deleterious effect on the conductivity of copper. The specified purity of copper for conductors is 100%. Small amounts of impurities, such as phosphorous or arsenic, can reduce the conductivity to as low as 80%.

3.7.1.2 Aluminum

Electrical conductor (EC) grade aluminum is also low in impurities, 99.5% purity or better. ASTM B 233 specifies the permissible impurity levels for aluminum [6].

TABLE 3.11
Comparative Properties, Copper versus Aluminum

Property	Unit	Copper, Annealed	Alum, Hard Drawn
Density at 20°C	Pounds/in^3	0.32117	0.0975
	Grams/cm^3	8.890	2.705
Linear Temp. Coef.	per °F	9.4×10^{-6}	12.8×10^{-6}
of Expansion	per °C	17.0×10^{-6}	23.0×10^{-6}
Melting Point	°F	1981	1205–1215
Melting Point	°C	1083	652–657

3.7.1.3 Comparative Properties, Copper versus Aluminum

Table 3.11 compares the properties of annealed copper and hard-drawn aluminum, which are typically used for power cable conductors.

3.7.2 Temper

Drawing of the copper and aluminum rod into wire results in work hardening of both. This results in a slightly lower conductivity as well as a higher temper. Stranding and compacting also increase the temper of the conductor. If a more flexible conductor is required, annealing the metal may be desirable. This can be done either while the strand is being drawn or the finished conductor may be annealed by placing a reel of the finished conductor in an oven usually having a nitrogen atmosphere and at an elevated temperature for a specified period of time.

3.7.2.1 Copper

ASTM Standards B1, B2, and B3 cover three tempers for copper conductors: hard-drawn, medium-hard-drawn, and soft or annealed, respectively. Soft-drawn is usually specified for insulated conductors because of its flexibility and ease of handling in the field. Medium-hard-drawn and hard-drawn are usually specified for overhead conductors.

3.7.2.2 Aluminum

ASTM Standards B231 and B400 cover concentric-lay and compact-round stranded aluminum conductors, respectively. ASTM has five designations for aluminum tempers as shown in Table 3.12. Note that some of the values overlap. Half-hard aluminum is usually specified for solid and for 8,000 series alloy conductors because of the need for greater flexibility. Three-quarter and full-hard are usually specified for stranded cables.

It is important to consider two factors before deciding which temper should be specified:

- The increased cost of the energy and equipment required to anneal the conductor.
- Even with a more flexible conductor, the overall stiffness of the insulated cable may only be marginally improved.

TABLE 3.12
Aluminum Temper

1350 Aluminum Tempers	PSI × 10^3
Full Soft (H–0)	8.5–14.0
1/4 Hard (H–12 or –22)	12.0–17.0
1/2 Hard (H–14 or –24)	15.0–20.0
3/4 Hard (H–16 or –26)	17.0–22.0
Full Hard (H–19)	22.5–29.0

Overhead conductors and cables that will be pulled in to long lengths frequently utilize higher tempers in order to increase the tensile strength of the conductor. Examples of cables that might require high tensile strength conductors are bore hole cables, mineshaft cables, or extremely long pulls of large conductors.

3.8 STRAND BLOCKING

Moisture in an insulated conductor has been shown to cause several problems. Aluminum, in the presence of water and in the absence of oxygen, will hydrolyze. Thus, if water enters an insulated cable having an aluminum conductor, the aluminum and water combine chemically to form aluminum hydroxide and hydrogen gas. This condition is aggravated by a deficiency in oxygen in the insulated conductor. The chemical reaction is:

$$2\text{Al} + 6\text{H}_2\text{O} \rightarrow 2\text{Al}(\text{OH})_3 + 3\overset{\uparrow}{\text{H}_2}$$

Aluminum hydroxide is a white, powdery material which is a good insulator. Many users of stranded aluminum conductors now require blocked conductors for this reason. Water blocking components, such as water-swellable tapes and yarns or sealants, incorporated into the interstices of the stranded conductor act as an impediment to longitudinal water penetration and thus help retard this form of deterioration. Copper conductors may, of course, also be water-blocked in the same manner.

Regardless of the conductor material and degree of compaction, there is still some air space remaining in the interstices of the stranded conductor. This space can act as a reservoir for moisture to collect and hence provide a source of water for water treeing. Water-blocked stranded conductors are frequently specified for underground cables to reduce the possibility of this happening. Solid conductors, of course, are typically specified for the same reason for #2/0 AWG and smaller aluminum conductors.

3.9 ELECTRICAL CALCULATIONS

3.9.1 CONDUCTOR DC RESISTANCE

$$R_{\text{DC}} \text{ at } 25°C = 1{,}000\rho/A \tag{3.5}$$

where R_{DC} = DC resistance of conductor in ohms per 1,000 feet at 25°C; ρ = resistivity of metal in ohm circular mils per foot; ρ for copper = 10.575 Ω·cmil/ft (100% conductivity) at 25°C; ρ for aluminum = 17.345 Ω·cmil/ft (61.0% conductivity) at 25°C; A = conductor area in circular mils.

The resistance of a stranded conductor is more difficult to calculate. It is generally assumed that the current is evenly divided among the strands and does not transfer from one strand to the next. For this reason, the DC resistance is based on:

- Multiply the number of strands by the cross-sectional area of each taken perpendicular to the axis of that strand. The product is then the cross-sectional area of the conductor.

- Compare the length of each strand to the axial length of the conductor. This increased length is arithmetically averaged.
- The DC resistance of a solid conductor having the same effective cross-sectional area is multiplied by the average increase in length of the strand. The resultant is the calculated resistance of the stranded conductor.

Since resistance is based on temperature, the following formulae correct for other temperatures in the range most commonly encountered:

Copper:

$$R_2 = R_1 \frac{234.5 + T_2}{234.5 + T_1} \tag{3.6}$$

Aluminum:

$$R_2 = R_1 \frac{228.1 + T_2}{228.1 + T_1} \tag{3.7}$$

where R_2 = conductor resistance at temperature T_2 in °C; R_1 = conductor resistance at temperature T_1 in °C.

These formulas are based on the resistance coefficient of copper having 100% conductivity and of aluminum having 61.2% conductivity (International Annealed Copper Standard).

3.9.2 Conductor AC Resistance

A conductor offers a greater resistance to the flow of AC than it does to DC. This increased resistance is generally expressed as the AC/DC resistance ratio. The two major factors for this increase are the skin effect and the proximity effect of closely spaced current carrying conductors. Other magnetic effects can also cause an additional increase in AC/DC resistance ratios.

$$R_{AC} = \text{AC/DC ratio} \times R_{DC} \tag{3.8}$$

The AC/DC resistance ratio is increased by larger conductor sizes and higher AC frequencies.

3.9.3 Skin Effect

In AC circuits, the current tends to distribute itself within a conductor so that the current density near the surface of the conductor is greater than that at its core. This phenomenon is known as skin effect. A longitudinal element of the conductor near the center of the axis is surrounded by more lines of magnetic force than near the rim. This results in an increase in inductance toward the center. The decreased area of conductance causes an apparent increase in resistance. At 60 hertz, the phenomenon is negligible in copper conductor sizes of #2 AWG and smaller and aluminum sizes of #1/0 AWG and smaller. As the conductor size increases, this effect becomes more significant.

The following formula can be used to give an approximation of skin effect for round conductors at 60 hertz; another approximation will be given in Chapter 14.

$$Y_{CS} = \frac{11.18}{R_{DC}^2 + 8.8} \tag{3.9}$$

where Y_{CS} = skin effect expressed as a number to be added to the DC resistance; R_{DC} = DC resistance of the conductor in micro-ohms per foot at operating temperature.

3.9.4 Proximity Effect

In closely spaced AC conductors, there is a tendency for the current to shift to the portion of the conductor that is away from the other conductors of that cable. This phenomenon is known as proximity effect. The alternating magnetic field linking the current in one isolated conductor is distorted by the current in an adjacent conductor. This in turn causes an uneven distribution of the current across the conductor cross section.

Since skin and proximity effects are cumbersome to calculate, tables have been established to give these values for common modes of operation [5].

3.9.5 Cables in Magnetic Metallic Conduit

Due to excessive hysteresis and eddy current losses, individual phases of an AC circuit should not be installed in separate magnetic metal conduits under any circumstances. This is because of the high inductance of such an installation. In fact, separate phases should not pass through magnetic structures since overheating can occur in such a situation. All phases should pass through any magnetic enclosure simultaneously, so that maximum cancellation of the resultant magnetic field occurs. This greatly reduces the magnetic effect. However, even under these conditions, an increase in skin and proximity effects will occur because of the proximity of the magnetic material. There can be significant losses when large conductors are simply placed near the magnetic materials.

Cables in 50 or 60 hertz AC circuits should not be installed with each phase in a separate nonmagnetic metal conduit when their conductor size is #4/0 AWG or larger due to high circulating currents in the conduit. This causes a significant decrease in the cable ampacity.

3.9.6 Resistance at Higher Frequencies

Cables operating at frequencies higher than 60 hertz may need to be evaluated for ampacity and AC/DC ratios because they can cause higher voltage drops than might be anticipated. Also at higher frequencies, an increase in the inductive reactance may affect voltage drops. Insulated conductors should not be installed in metallic conduits, nor should they be run close to magnetic materials.

For frequencies other than 60 hertz, a correction factor is provided by:

$$x = 0.027678\sqrt{f/R_{DC}} \tag{3.10}$$

where f = frequency in hertz, R_{DC} = conductor DC resistance at operating temperature, in ohms per 1,000 feet.

For additional information on the effects of higher frequency, see the ICEA report in Reference [3] and the cable manufacturer's manuals [4,5].

REFERENCES

1. Kelly, L. J., 1995, adapted from class notes for "Power Cable Engineering Clinic," University of Wisconsin–Madison.
2. Landinger, C. C., 2001, adapted from class notes for "Understanding Power Cable Characteristics and Applications," University of Wisconsin–Madison.
3. ICEA P-34-359, 1973, "AC/DC Resistance Ratios at 60 Hz," Global Engineering Documents, 15 Inverness Way East, Englewood, CO 80112.
4. "Engineering Data for Copper and Aluminum Conductor Electrical Cables," 1990, The Okonite Company, Bulletin EHB-90.
5. *Southwire Company Power Cable Manual*, Second Edition, 1997, Carrollton, GA.
6. *Annual Book of ASTM Standards*, Vol. 02.03: Electrical Conductors. Section 2: Nonferrous Metal Products, 2010, ASTM International, 100 Barr Harbor Drive, PO Box C700, West Conshohocken, PA, 19428-2959 USA.
7. IEC 20/680/RVC Result of voting on 20/633/CDV: IEC 60228 Ed. 3: "Conductors of Insulated Cables," 2004, IEC Central Office, 3 rue de Varembé, P.O. Box 131, CH-1211 Geneva 20, Switzerland.
8. IEC 60228 (Edition 3.0 2004-11), "Conductors of Insulated Cables," 2004, IEC Central Office, 3 rue de Varembé, P.O. Box 131, CH-1211 Geneva 20, Switzerland.
9. IEC/TR 62602 (Edition 1.0 2009-09), "Conductors of Insulated Cables – Data for AWG and kcmil Sizes," 2009, IEC Central Office, 3 rue de Varembé, P.O. Box 131, CH-1211 Geneva 20, Switzerland.
10. ANSI/ICEA S-94-649, "Standard for Concentric Neutral Cables Rated 5,000-46,000 Volts," 2004, Global Engineering Documents, 15 Inverness Way East, Englewood, CO 80112.
11. ANSI/ICEA S-108-720, "Standard for Extruded Insulation Power Cables Rated Above 46 Through 345 kV," 2004, Global Engineering Documents, 15 Inverness Way East, Englewood, CO 80112.

4 Cable Characteristics: Electrical

William A. Thue

CONTENTS

4.1 VOLTAGE RATING OF CABLES

The rating, or voltage class, of a cable is based on the phase-to-phase voltage of the system even though it is in a single- or three-phase circuit. For example, a 15 kV rated cable (or a higher value) must be specified on a system that operates at 7,200 or 7,620 V to ground on a grounded wye 12,500 or 13,200 V system. This is based on the fact that the phase-to-phase voltage on a wye system is 1.732 (the square root of 3) times the phase-to-ground voltage. Another example is that a cable for operation at 14.4 kV to ground must be rated at 25 kV or higher since 14.4 times 1.732 is 24.94 kV (Figure 4.1).

The wye systems described above are usually protected by fuses or fast-acting relays. This is generally known as the 100% voltage level and was previously known as a "grounded" circuit. Additional insulation thickness is required for systems that are not grounded, such as found in some delta systems, impedance or resistance

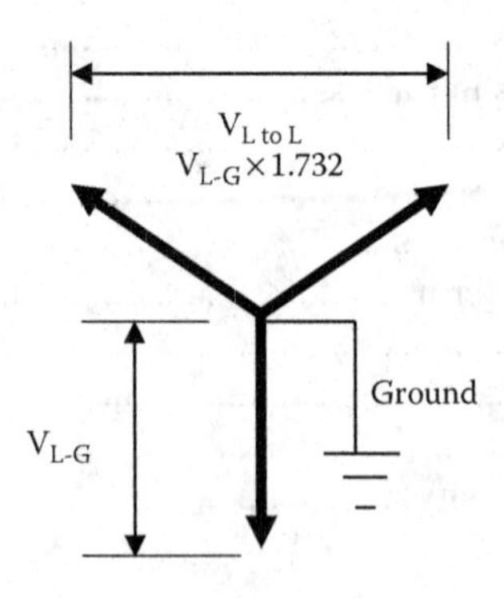

FIGURE 4.1 Voltage rating.

grounded systems, or systems that have slow-acting isolation schemes. The following voltage levels are found in AEIC specifications [1] and many other technical documents.

4.2 100 PERCENT LEVEL

Cables in this category may be applied where the system is provided with relay protection such that ground faults will be cleared as rapidly as possible, but in any case within 1 minute. While these cables are applicable to the great majority of cable installations that are on grounded systems, they may be used also on other systems for which the application of cables is acceptable, provided the above clearing requirements are met in completely de-energizing the faulted section.

[*Editor's note: There have been incidents where devices designed to quickly de-energize a circuit have been installed but not utilized. Such reluctance to de-energize quickly can result in serious problems for those cable systems.*]

4.3 133 PERCENT LEVEL

This insulation level corresponds to that formerly designated for "ungrounded" systems. Cables in this category may be applied in those situations where the clearing time requirements of the 100 percent category cannot be met, and yet there is adequate assurance that the faulted section will be de-energized in a time not exceeding 1 hour. Also, they may be used when additional insulation strength over the 100 percent level category is desirable.

4.4 173 PERCENT LEVEL

Cables in this category should be applied on systems where the time required to de-energize a section is indefinite. Their use is recommended also for resonant grounded systems. Consult the (cable) manufacturer for insulation thickness.

4.5 CABLES NOT RECOMMENDED

Cables are not recommended for use on systems where the ratio of the zero to positive phase reactance of the system at the point of cable application lies between −1 and −40 since excessively high voltages may be encountered in the case of ground faults.

4.6 LOW-VOLTAGE CABLE RATINGS

Low-voltage cable ratings follow the same general rules as for the medium-voltage cables previously discussed in that they are also based on phase-to-phase operation. The practical point here is that a cable that operates at say 480 V from phase-to-ground on a grounded wye system requires an insulation thickness applicable to 480×1.732 or 831 V phase-to-phase. This, of course, means that a 1,000 V level of insulation thickness should be selected.

There are no categories for low-voltage cables that address the 100, 133, and 173 percent levels. One of the main reasons for the thickness of insulation walls for these low-voltage cables in the applicable standards is that mechanical requirements of these cables dictate the insulation thickness. As a practical matter, all these cables are overinsulated for the actual voltages involved.

4.7 CABLE CALCULATION CONSTANTS

There are four main calculation constants that affect the functioning of a cable on an electric system: resistance, capacitance, inductance, and conductance. Conductor resistance has been addressed in Chapter 3.

4.7.1 Cable Insulation Resistance

The resistance to flow of a direct current (DC) through an insulating material (dielectric) is known as insulation resistance. There are two possible paths for current to flow when measuring insulation resistance:

1. Through the body of the insulation (volume insulation resistance)
2. Over the surface of the insulation system (surface resistivity)

4.7.2 Volume Insulation Resistance

The volume insulation resistance of a cable is the DC resistance offered by the insulation to an impressed DC voltage tending to produce a radial flow of leakage through that insulation material. This is expressed as a resistance value in megohms for 1,000 feet of cable for a given conductor diameter and insulation thickness. *Note that this is for 1,000 feet,* not per *1,000 feet!* This means that the longer the cable, the lower the resistance value that is read on a meter since there are more parallel paths for current to flow to ground. The basic formula for a single conductor cable of cylindrical geometry is:

$$IR = \mathrm{K} \log_{10} D/d \tag{4.1}$$

where

IR = megohms for 1,000 feet of cable
K = insulation resistance constant
D = diameter over the insulation (under the insulation shield)
d = diameter under the insulation (over the conductor shield). *Note:* Both D and d must be expressed in the same units.

In order to measure the insulation resistance of a cable, the insulation must either be enclosed in a grounded metallic shield or immersed in water. Resistance measurements are greatly influenced by temperature—the higher the temperature, the lower the insulation resistance. The cable manufacturer should be contacted for the temperature correction factor for the specific insulation. Equation 4.1 is based on values at 60°F.

TABLE 4.1
Insulation Resistance

Insulation	ICEA Minimum	Typical
HMWPE	50,000	1,000,000
XLPE and EPR, 600 V	10,000	100,000
XLPE and EPR, Med. Voltage	20,000	200,000
PVC at 60°C	2,000	20,000
PVC at 75°C	500	5,000

The values shown in Table 4.1 are also based on this temperature. The ICEA minimum requirements of IR (sometimes referred to as "guaranteed values") are shown as well to represent values that may be measured in the field. The actual values of IR that would be read in a laboratory environment are many times higher than these "minimum" values and approach the "typical" values shown in Table 4.1.

4.7.3 Surface Resistivity

One of our contributors often stated that "all cables have two ends." These terminations or ends, when voltage is applied to the conductor, can have current flow over the surface of that material. This current adds to the current that flows through the volume of insulation, which lowers the *apparent* volume insulation resistance unless measures are taken to eliminate that current flow while the measurements described above are being made. This same situation can occur when samples of insulation are measured in the laboratory. A "guard" circuit is used to eliminate the surface leakage currents from the volume resistivity measurement.

4.7.4 DC Charging Current

The current generated when a cable is energized from a DC source is somewhat complicated because there are several currents that combine to form the total leakage current. These currents are: I_L = leakage current, I_G = charging current, I_A = absorption current.

The DC charging current behaves differently than the alternating current (AC) in that the DC value rises dramatically during the initial inrush. It decreases rather quickly with time, however. The magnitude of the charging and absorption currents is not usually very important except that it may distort the true leakage current reading. The longer the length and the larger the cable size, the greater the inrush current and the longer it will take for the current to recede. This initial current decays exponentially to zero in accordance with the following equation:

$$I_G = (E/R)\varepsilon^{-t/RC} \tag{4.2}$$

where

I_G = charging current in microamperes per 1,000 feet
E = voltage of conductor to ground in volts
R = DC resistance of cable in megohms for 1,000 feet
ε = base of natural logarithm (2.718281...)
t = time in seconds
C = capacitance of circuit in microfarads per 1,000 feet

The absorption current is caused by the polarization and accumulation of electric charges that accumulate in a dielectric under applied voltage stress. The absorption current normally is relatively small and decreases with time. Absorption current represents the stored energy in the dielectric. Short-term grounding of the conductor may not give a sufficient amount of time for that energy to flow to ground. Removing the ground too quickly can result in the charge reappearing as a voltage on the conductor. The general rule is that the ground should be left on for one to four times the time period during which the DC source was applied to the cable. The absorption current is:

$$I_A = \mathrm{A}VCt^{-\mathrm{B}} \tag{4.3}$$

where

I_A = absorption current in microamperes per 1,000 feet
V = incremental voltage change in volts
C = capacitance in microfarads per 1,000 feet
t = time in seconds
A and B are constants depending on the insulation.

A and B are constants that differ with the specific cable since they are dependent on the type and condition of the insulation. They generally vary in a range that limits the absorption current to a small value compared to the other DC currents. This current decays rather rapidly when a steady-state voltage level is reached.

The current that is of most importance is the leakage or conduction current. The leakage current is dependent on the applied voltage, the insulation resistance of the cable insulation, and any other series resistance in the circuit. This value becomes very difficult to read accurately at high voltages because of the possibility of end leakage currents as well as the transient currents. The formula for leakage current is:

$$I_L = E/R_I \tag{4.4}$$

where

I_L = leakage current in microamperes per 1,000 feet
E = voltage between conductor and ground in volts
R_I = insulation resistance in megohms for 1,000 feet.

The total current is:

$$I_T = I_G + I_A + I_L \tag{4.5}$$

The voltage must be raised slowly and gradually because of the rapid rise of I_G and I_A with time. Also, since both of these values are a function of cable length, the longer the cable length, the slower the rise of voltage allowable. Equation 4.5 demonstrates the reason for taking a reading of leakage current after a specified period of time so that the actual leakage current can be determined.

4.8 DIELECTRIC CONSTANT

Dielectric constant, relative permittivity, and specific inductive capacitance all mean the same. They are the ratio of the absolute permittivity of a given dielectric material to the absolute permittivity of free space (vacuum). The symbol for permittivity is ε (epsilon). To put this in another way, these terms refer to the ratio of the capacitance of a given thickness of insulation to the capacitance of the same capacitor insulated with vacuum. (This is occasionally referred to as air rather than vacuum, but the actual dielectric constant of air is 1.0006). Since the calculations are usually not taken out to more than two decimal points, it is practical to use air for comparison. The value of permittivity, dielectric constant, and specific inductive capacitance (SIC) are expressed simply as a number since the dielectric constant of a vacuum is taken as 1.0000.

4.9 DIELECTRIC LOSS IN CABLE INSULATION

The losses in the insulation of a cable may be calculated from the following equation:

$$W = 2\pi f \varepsilon C n e^2 F_p \times 10^{-6} \tag{4.6}$$

where

W = watts loss per foot of cable
ε = dielectric constant of the insulation
f = frequency in hertz
C = capacitance of the insulation per foot
n = number of conductors in the cable
e = voltage conductor to neutral in kilovolts
F_P = power factor of the insulation as a decimal.

4.10 CABLE CAPACITANCE

The property of a cable system that permits the conductor to maintain a potential across the insulation is known as capacitance. Its value is dependent on the permittivity (dielectric constant) of the insulation and the diameters of the conductor and the insulation. A cable is a distributed capacitor. Capacitance is important in cable applications since charging current is proportional to the capacitance as well as to the system voltage and frequency. Since the charging current is also proportional to length, the required current will increase with cable length.

The capacitance of a single conductor cable having an overall grounded shield or immersed in water to provide a ground plane may be calculated from the following formula:

$$C = \frac{0.00736\varepsilon}{\log_{10}\frac{D}{d}} \quad (4.7)$$

where

C = capacitance in microfarads per 1,000 feet
ε = permittivity of the insulating material. Permittivity (ε, epsilon), dielectric constant (K), and SIC terms are used interchangeably. The term permittivity is preferred. See Table 4.2.
D = diameter over the insulation (under the insulation shield)
d = diameter under the insulation (over the conductor shield). *Note:* Both D and d must be expressed in the same units.

In single conductor, low-voltage cables where there is no semiconducting layer over the conductor, a correction factor must be used to compensate for the irregularities of the stranded conductor surface as shown in Table 4.3. This measurement

TABLE 4.2
Permittivity, Dielectric Constant, and SIC (*All Mean the Same Thing*)

Material	Range	Typical
Butyl Rubber	3.0–4.5	3.5
EPR	2.5–3.5	3.0 or 3.5
Halar (ETFE)	2.5	2.5
HMWPE2	1–2.6	2.2
Hypalon	7–10	8
Kynar (PVDF)	6–12	10
Mica	6.9	6.9
Neoprene	9–10	9.5
Paper, impregnated	3.3–3.7	3.5
Polyester (Mylar)	3.3–3.8	3.5
Polyvinyl chloride (PVC)	3.1–10	6.0
Rubber-GRS or Natural	2.7–7	3.5
Silicone Rubber	2.9–6	4.0
Teflon (FEP, TFE)	2.1	2.1
Tefzel (EFTE)	2.6	2.6
TR-XLPE2.	1–2.6	2.3
XLPE2.	1–2.6	2.3
XLPE, filled	3.5–6.0	4.5
Varnished Cambric	4.0–6.0	5.0

TABLE 4.3
Correction Factors for Irregularities

Number of Strands	Factor k
1 (solid)	1.0
7	0.94
19	0.97
37	0.98
61 and 91	0.985

is based on having an insulation shield/sheath or a conducting surface over the insulation.

$$C = \frac{0.00736\varepsilon}{\log_{10} \frac{d}{kd}} \tag{4.8}$$

4.11 CAPACITIVE REACTANCE

The capacitive reactance of a cable is inversely dependent on the capacitance of the cable and the frequency at which it operates.

$$X_c = \frac{1}{2\pi fC} \tag{4.9}$$

where

X_c = ohms per foot
f = frequency in hertz
C = capacitance in picofarads per foot

4.12 CHARGING CURRENT FOR ALTERNATING CURRENT OPERATION

For a single conductor cable, the current may be calculated from the formula:

$$I_C = 2\pi fCE \times 10^{-3} \tag{4.10}$$

where

I_C = charging current in milliamperes per 1,000 feet
f = frequency in hertz
C = capacitance in picofarads per foot
E = voltage from conductor to neutral in kilovolts.

Other leakage currents are also present, but the capacitive current has the largest magnitude. In addition to this, the capacitive charging current flows as long as the system is energized. The resistive component of the charging current is also dependent on the same factors as the capacitive current and is given by the formula:

$$I_R = 2\pi fCE \tan\delta \quad (4.11)$$

where

I_R = resistive component of the charging current
$\tan\delta$ = dissipation factor of the insulation.

The $\tan\delta$ of medium-voltage insulation, such as cross-linked polyethylene and ethylene propylene, has values that are generally below 0.02 so the resistive component of the charging current is only a small fraction of the total charging current. The $\tan\delta$ is sometimes referred to as the insulation power factor since at small angles, these values are approximately equal. Since the capacitive charging current is 90° out of phase with the resistive charging current, the total charging current is generally given as the capacitive component and leads any resistive current flowing in the circuit by 90°. The result of these AC currents generated puts demands on the power required for test equipment.

4.13 CABLE INDUCTIVE REACTANCE

The inductive reactance of an electrical circuit is based on Faraday's law. That law states that the induced voltage appearing in a circuit is proportional to the rate of change of the magnetic flux that links it. The inductance of an electrical circuit consisting of parallel conductors, such as a single-phase concentric neutral cable, may be calculated from the following equation:

$$X_L = 2\pi f\left(0.1404\log_{10} S/r + 0.153\right)\times 10^{-3} \quad (4.12)$$

where

X_L = ohms per 1,000 feet
S = distance from the center of the cable conductor to the center of the neutral
r = radius of the center conductor. S and r must be expressed in the same unit, such as inches.

The inductance of a multiconductor cable mainly depends on the thickness of the insulation over the conductor.

4.13.1 Cable Inductive Reactance at Higher Frequencies

Since the inductive reactance of an insulated conductor is directly proportional to frequency, the inductive reactance is substantially increased in higher frequency applications. Conductors must be kept as close together as possible. Due to the severe increase in inductive reactance at high frequency, many applications will require using two conductors per phase to reduce the inductive reactance to approximately one-half of that of using one conductor per phase. A six-conductor installation should have the same phase conductors 180° apart.

4.14 MUTUAL INDUCTANCE IN CABLES

In single-conductor shielded or metallic-sheathed cables, current in the conductor will cause a voltage to be produced in the shield or sheath. If the shield or sheath forms part of a closed circuit, a current will flow. (Shield and sheath losses are described under Ampacity in Chapter 13).

The approximate mutual inductance between shields or sheaths is given by the following relation:

$$L_m = 0.1404 \log_{10} S/r_m \times 10^{-3} \tag{4.13}$$

where

L_m = henries to neutral per 1,000 feet
S = geometric mean spacing between cable centerlines in inches
r_m = mean shield or sheath radius in inches. See Figure 4.2.

4.15 CABLE CONDUCTOR IMPEDANCE

Conductor impedance of a cable may be calculated from the following equation:

$$Z = R_{AC} + jX_L \tag{4.14}$$

where

Z = conductor impedance in ohms per 1,000 feet
R_{AC} = AC resistance in ohms per 1,000 feet
X_L = conductor reactance in ohms per 1,000 feet.

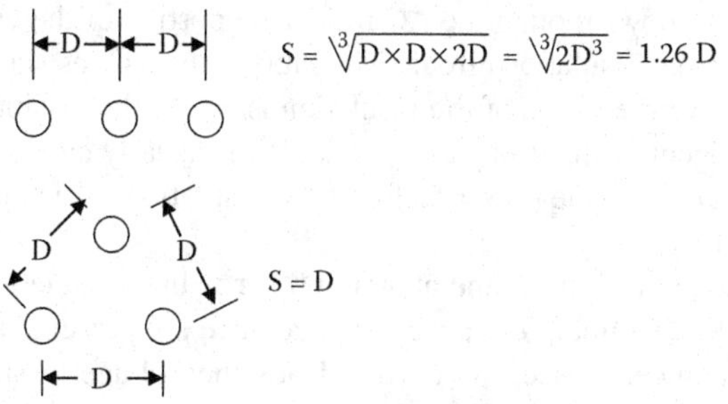

FIGURE 4.2 Geometric spacing.

Conductor impedance becomes an important factor when calculating voltage drop. Since the power factor angle of the load and impedance angle are usually different, the voltage drop calculation can be cumbersome. The following voltage drop equation can be used for a close approximation:

$$V_D = R_{AC} I \cos\theta + X_L I \sin\theta \tag{4.15}$$

where

V_D = voltage drop from phase to neutral in volts
R_{AC} = AC resistance of the length of cable in ohms
$\cos\theta$ = power factor of the load
X_L = inductive reactance of the length of cable in ohms.

4.16 TOTAL CABLE REACTANCE

The total cable reactance (X) is the vector sum of the capacitive reactance and the inductive reactance of the cable in ohms per foot.

$$X = X_C + X_L \tag{4.16}$$

4.17 CABLE DISSIPATION FACTOR

In cable engineering, the small amount of power consumed in the insulation (dielectric absorption) is due to losses. These losses are quite small in medium-voltage cables, but can become more significant in systems operating at 15 kV and above.

A small amount of power is consumed in an insulation due to its presence in an electric field and to the fact that no insulation is a perfect dielectric. In practical cables, there is a large amount of capacitive reactance and virtually no inductive reactance. Hence, the current leads the voltage by almost 90°. Since the cosine of 90° is zero, the cosine of an angle approaching 90° is very small. The actual value of such a small angle is almost equal to tan δ, hence cable engineers use this term to represent this "defect angle."

In addition to the charging current flowing through the capacitive portion of the circuit, current also flows though the AC resistance portion of the circuit. This is the AC loss portion of the insulation circuit. The ratio of the AC resistance of the insulation to the capacitive reactance of the insulation is called the dissipation factor. This is equal to the tangent of the dissipation angle that is usually called tan δ. This tan δ is approximately equal to the power factor of the insulation, which is the cos θ of the complimentary angle.

In practical cable insulations and at 50 to 60 hertz, high insulation resistance and a comparatively large amount of capacitive reactance are present. There is virtually no inductive reactance. Hence, the current leads the voltage by almost 90°. Since the cosine of 90° is zero, the cosine of an angle approaching 90° is small and the dissipation factor (often referred to as power factor of the insulation) also is small.

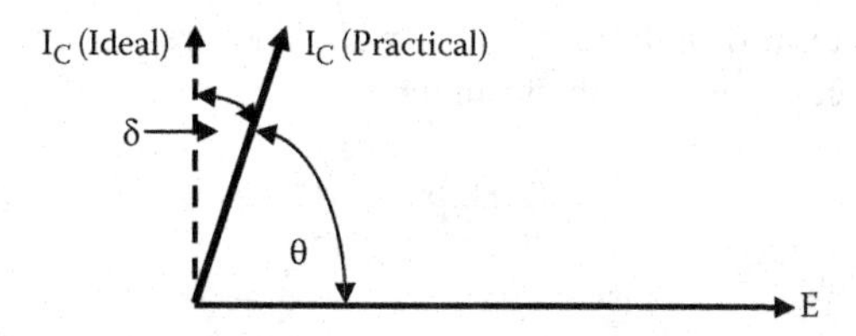

FIGURE 4.3 Cable insulation power factor or dissipation factor. *Note:* Tan δ is approximately equal to cos θ for the small angles involved.

Typical values for insulation power factor are 0.005 to 0.02 or slightly higher for other materials.

Dissipation factor is used in cable engineering to determine the dielectric loss in the insulation, expressed as watts per foot of cable that is dissipated as heat. It is also used to some extent to describe the efficiency or perfection of the insulation as a dielectric. Hence, the term tan δ (delta) was chosen to represent the defect angle of the material (Figure 4.3).

The power dissipation per phase (dielectric loss, W_d) is a function of the voltage and the in-phase component of the current.

$$W_d = E_O I_C \tan \delta W/\text{foot} \tag{4.17}$$

The insulation parameters that determine the dielectric loss of a cable are the dielectric constant (permittivity, ε) and the dissipation factor (tan δ). The product of the permittivity and the dissipation factor is the dielectric loss factor (DLF). The lower the DLF, the better the insulation.

4.18 INSULATION PARAMETERS

4.18.1 Voltage Stress in Cables

Voltage stress in shielded cable insulations with smooth, round conductors is defined as the electrical stress or voltage to which a unit thickness of insulation is subjected. The average stress in volts per mil is determined by dividing the voltage across the insulation by the insulation thickness in mils.

$$S_{avg} = 2V/(D-d) \tag{4.18}$$

where

S_{avg} = average stress in volts per mil
V = voltage across the insulation
D = outside diameter of the insulation in mils
d = inside diameter of the insulation in mils.

The stress is not uniform throughout the wall. The stress at any point in the insulation wall can be calculated by the formula:

$$S = V/2.303r \log_{10}(D/d) \tag{4.19}$$

where

S = stress in volts per mil at a point in the insulation r mils from the cylindrical axis.

The maximum stress occurs at the surface of the conductor shield.

$$S_{max} = 0.868V/d \log_{10}(D/d) \tag{4.20}$$

where the terms and units are the same as in Equations 4.16 through 4.18.

4.18.2 Dielectric Strength

Although maximum and average stresses are important, dielectric strength is usually specified as the average stress at electrical breakdown. The dielectric strength of a material depends on the dimensions and the testing conditions, particularly the time duration of the test. A thin wall of material generally withstands a higher average stress before breaking down than a thicker wall.

4.18.3 AC Dielectric Strength

These measurements are made in two ways: quick-rise or step-rise.

In the quick-rise method, the voltage applied to the insulation is raised at a uniform rate until the insulation breaks down. As an example, a rate of rise of 500 V per second is known as "quick-rise."

In the "step-rise" method, the voltage is raised to a predetermined level and held at that level for an amount of time, such as 5 or 10 minutes at each level, until breakdown occurs. A relatively short time, say the 5 or 10 minutes described above, has the advantage of reaching breakdown in a shorter amount of total test time. In the real world, the time at a voltage level is much longer, so some cable engineers prefer a longer step time such as 30 minutes or 1 hour at each step. With the longer step times, the breakdown voltage is lower than with the quick-rise or short step time methods.

4.18.4 Impulse Strength

Because cable insulation is frequently subjected to lightning or switching surges, it is often desirable to know the impulse strength of the cable. Surges of "standard" wave shape, such as 8 seconds to reach 90% of crest value, and 40 microseconds to drop to one-half of crest value, are frequently used in the laboratory. The increasing voltages are applied to the insulation with several surges at a negative potential and then, at the same voltage level, the same number of surges are applied with positive pulses. The average stress in volts per mil is calculated from the crest voltage of the surge on which breakdown occurs.

4.19 REVIEW OF ELECTRICAL TERMS

These terms apply to all electrical engineering circuits. The actual application of these terms to cables is covered in Section 4.2 of this chapter. The more important equations are:

$$E = IR \tag{4.21}$$

$$O = It \tag{4.22}$$

$$E = LdI/dt \tag{4.23}$$

$$Q = CE \tag{4.24}$$

$$P = EI = I^2R \tag{4.25}$$

$$W = EIt \tag{4.26}$$

where

E = electromotive force in volts
I = current in amperes
R = resistance in ohms
Q = quantity in ohms
L = inductance in henries
C = capacitance in farads
P = power in watts
W = energy in joules
t = time in seconds.

4.19.1 Resistance

(R) is the scalar property of an electric circuit that determines, for a given current, the rate at which electric energy is converted into heat or radiant energy. Its value is such that the product of the resistance and the square of the current gives the rate of conversion of energy.

In a direct-current circuit,

$$R = \frac{E}{I} \tag{4.27}$$

and

$$P = I^2R \tag{4.28}$$

where

R = resistance in ohms
E = electromotive force in volts
I = current in amperes
P = power in watts.

4.19.2 Conductance

(G) is the property of an electric circuit that determines, for a given electromotive force in the circuit or for a given potential difference between the terminals of a part of a circuit, the rate at which energy is converted into heat or radiant energy. This value is such that the product of the conductance and the square of the electromotive force, or potential difference, gives the rate of conversion of energy.

$$P = E^2G \tag{4.29}$$

where

P = power in watts
E = voltage, phase to ground, in kilovolts
G = conductance in mhos.

The unit of conductance is mho. Conductance is the reciprocal of resistance.

$$G = \frac{1}{R} \tag{4.30}$$

where

G = conductance in mhos
R = resistance in ohms.

4.19.3 Conductivity

Conductivity (γ) of a material is the direct-current conductance between the opposite parallel faces of a portion of the material having unit length and unit cross section. It is the reciprocal of resistivity.

$$G = \frac{\gamma\alpha}{l} \tag{4.31}$$

where

G = conductance
a = area
l = length.

4.19.4 Volume Resistivity

Volume resistivity of a material is the reciprocal of conductivity. The unit for volume resistivity is ρ (rho). It is the resistance of a section of material of unit length and unit cross section.

$$R = \frac{\rho l}{a} \tag{4.32}$$

where

R = resistance in ohms
ρ = rho, volume resistivity
a = area
l = length.

4.19.5 Inductance

Unit L represents the scalar property of an electric circuit, or two neighboring circuits, which determines the electromotive force induced in one of the circuits by a change of current in either of them.

4.19.5.1 Self-Inductance

This is the property of an electric circuit that determines, for a given rate of change of current in the circuit, the electromotive force induced in the same circuit.

The unit of inductance is 1 henry. One henry is the self-inductance of a closed circuit in which an electromotive force of 1 V is produced when the electric current traversing the circuit varies uniformly at the rate of 1 ampere per second.

$$e_1 = -L\frac{di_1}{dl} \tag{4.33}$$

where e_1 and i_1 are in the same circuit and L is the coefficient of self-inductance.

4.19.5.2 Mutual Inductance

(L_m) is the common property of two associated electric circuits that determines, for a given rate of change of current in one of the circuits, the electromotive force induced in the other.

4.19.6 Inductance in Multiconductor Cables

This is the same as any other arrangement of conductors and follows the following equation:

$$L = 0.1404 \log_{10} \frac{GMD}{GMR} \times 10^{-3} \quad (4.34)$$

where

GMD = geometric mean distance between conductors in inches
GMR = geometric radius of conductors in inches.

4.19.7 Mutual Inductance in Cables

In single conductor, metallic covered cables, current flowing in the conductor will produce an electromotive force in the sheath. If by any means the sheath forms a closed circuit, current will flow in the sheath based on the following equation:

$$L_m = 0.1404 \log_{10} \frac{GMD}{r_m} \times 10^{-3} \quad (4.35)$$

where

L_m = henries to neutral per 1,000 feet
r_m = mean sheath radius in inches.

4.19.8 Inductance in Coaxial Cables

In coaxial cables, three kinds of inductance must be taken into account: space inductance, inductance within the inner conductor, and inductance within the outer conductor. Above 50 kilohertz, only space inductance needs to be considered for results with less than 0.5% error. The equation for a coaxial cable with a tubular outer conductor becomes:

$$L_f = 4.6 \log_{10} \frac{r_2}{r_1} \times 10^{-9} \quad (4.36)$$

where

L_f = inductance in henries per centimeter
r_2 = inner radius of outer conductor in inches
r_1 = radius of inner conductor in inches.

If the outer conductor is stranded or braided, the inductance is slightly higher.

4.20 CAPACITANCE

This is the property of an electric system comprising insulated conductors and associated dielectric materials that determines, for a given time rate of change of potential difference between the conductors, the displacement currents in the system.

The unit of capacitance is the farad and it is that capacitance of a circuit whose potential difference will be raised 1 V by the addition of a charge of 1 coulomb.

4.20.1 Capacitance of a Cable

The electrostatic capacitance of an insulated conductor 1 cm in length, in absolute units, is:

$$C = \frac{\varepsilon}{2\log_{\varepsilon}\frac{D}{d}} \tag{4.37}$$

where

C = capacitance
ε = dielectric constant of the insulating material
D = outer diameter of the insulation
d = inside diameter of the insulation.

In more common terms, the equation is:

$$C = \frac{7.354\varepsilon}{\log_{10}1 + 2t/d} \tag{4.38}$$

4.21 Reactance

Reactance is the product of the sine of the angular phase difference between the current and potential difference times the ratio of the effective potential difference to the effective current because there is no source of power in the portion of the circuit under consideration. The total reactance of a circuit is the sum of the inductive and capacitive reactances.

4.21.1 Inductive Reactance Is Calculated From

$$X_L = 2\pi fL \tag{4.39}$$

where

X_L = inductive reactance in ohms/foot to neutral
f = frequency in hertz
L = inductance in henries/foot

and

$$X_L = 0.05292\log_{10}\frac{GMD}{GMR} \tag{4.40}$$

4.21.2 Capacitive Reactance Is Calculated From

$$X_C = -\frac{1}{2\pi fC} \tag{4.41}$$

where

X_C = capacitive reactance in ohms/foot
C = capacitance in picofarads/foot.

4.21.3 Total Reactance

The total reactance of a circuit is the sum of the inductive and capacitive reactances:

$$X = 2\pi fL + \frac{1}{2\pi fC} = X_L + X_C \tag{4.42}$$

4.22 IMPEDANCE

Impedance is the ratio of the effective value of the potential difference between the terminals to the effective value of the current, there being no source of power in the portion of the circuit under consideration.

$$Z = \frac{E}{I} = \sqrt{R^2 + X^2} \tag{4.43}$$

4.23 ADMITTANCE

(Y) is the reciprocal of impedance.

4.24 POWER FACTOR IN POWER ENGINEERING

Power factor, as used in power engineering, is not the same as "power factor" as used in cable engineering. In power engineering, power factor is the ratio of active power to apparent power. Apparent power (S) consists of two components; active (in-phase) power P_a, which does useful work, and reactive (out-of-phase) power P_r. Their geometric sum is the apparent power. Power factor is given by the equation:

$$F_P = P_a/S \tag{4.44}$$

In power engineering, power factor is used, among other things, to determine the amount of useful work. In a motor, for example, resistance and a comparatively large amount of inductive reactance are present so that the current lags behind the voltage. If the power factor was 1.00 (unity) and no reactance was present, then with every

10 kVA all the current delivered to the motor (neglecting losses) would be in phase with the voltage and 10 kW would be applied as useful work. With a reactance present and a power factor of, say 0.80, only part of the current (8/10) is in phase with the voltage and only 8 kW would be delivered to the motor. The out-of-phase component of current increases the total current and results in increased heat loss.

Power factor may also be described as a measure of the relationship in time phase between the current and the voltage in any AC circuit. Practically, all AC circuits contain resistance, inductive reactance, and capacitive reactance. These characteristics determine how much the current leads or lags behind the voltage during each cycle. This is usually expressed in degrees by the use of a vector diagram. The angle between them indicates the amount of lead or lag. The cosine of that angle is called the power factor of that circuit. With only capacitive reactance, the current leads the voltage by 90°; with only inductive reactance, the current lags behind the voltage by 90°.

4.25 HOOP STRESS OF CYLINDERS

Cable systems with sealed metal sheaths, such as paper insulated lead covered (PILC) cables, can develop internal pressure or vacuum that may expand or collapse this sheath. These systems can develop this condition when the hydraulic pressure of the impregnate increases above the allowable "hoop stress" level of the sheath. PILC cables with viscous impregnates are generally limited to a height of about 32 feet in order to avoid damage to the sheath at the low point or excess vacuum at the high point of the system—usually at the terminations.

Most such systems are not sealed along their route. This results in the pressure being the same in the cable as in the splice when they are at the same elevation. The result of the same pressure can be dramatic since the diameter of the cable is considerably smaller than the diameter of the splice. Let us consider a cable having an internal sheath diameter of 3 inches and a splice with a 6 inch inner diameter of the sleeve. To keep this simple, the thickness of the sheath and sleeve will both be 0.1 inches. When a pressure of 10 psi is reached, the internal force trying to pull a 1 inch ring or "hoop" of the sheath or sleeve apart is three times the 10 psi or 30 pounds for two pieces of 0.1 square inch sections of lead—hence 150 psi. In the splice, the force is six times the 10 psi or 60 pounds. That translates to 300 psi, trying to swell the lead into a greater diameter. With the same thickness in both places, the force trying to pull the lead apart in the cable is 150 psi, but is 300 psi in the splice. This accounts for swollen splice sleeves. The unfortunate outcome is that if the sleeve increases in diameter, so does the force increase.

The allowable hoop stress for lead pipe (including cable sheath or splice sleeve) may be obtained by using the following formula:

$$S = PD/2t \quad (4.45)$$

where

S = working stress (hoop stress) in pounds per square inch
D = inside diameter of sheath or sleeve in inches

P = internal fluid pressure in pounds per square inch
t = thickness of sheath or sleeve in inches.

The working value of S is obtained by assuming a safety factor based on ultimate strength. Values of S commonly used for lead sheaths and sleeve are: copper bearing lead: 125 pounds per square inch, arsenical lead: 175 pounds per square inch.

The practical solution to minimize lead sheath swelling in the splice is to use a stronger material such as arsenical lead and/or to increase the thickness of the splice sleeve.

REFERENCE

1. Association of Edison Illuminating Companies, Specification CS8-07, 1st Ed., 2007, AEIC, P O Box 2641, Birmingham, AL, 35291-0992, USA.

5 Fundamentals of Electrical Insulation Materials

Bruce S. Bernstein

CONTENTS

5.1 INTRODUCTION

Electrical insulation materials are utilized to provide protection over the metallic conductors of underground cables. The insulating materials physically enclose the conductor and provide a margin of safety. These materials are composed of either synthetic or natural polymers. The polymeric insulation material selected for use may vary with the voltage class of the cable. For medium voltage cable constructions, compatible polymeric shields are employed between the insulation and the conductor, and over the insulation to grade the voltage stress; these are composed of flexible polymers blended with conducting carbon black that imparts the semiconducting characteristics. Metallic neutrals or tapes are applied over this cable core, and polymeric jackets are applied on the outside of the cable core.

Sections 5.2 to 5.10 of this chapter review polymeric insulation materials employed primarily in distribution and transmission cables, with reference to low voltage cable materials as needed. Fundamental principles will be reviewed. Section 5.11 discusses polymeric materials for low voltage cables. Section 5.12 deals with the fundamentals of aged-cable rejuvenation by impregnation. Chapter 6 will focus separately on electrical properties of insulating polymers discussed in this chapter.

Until the early 1980s, transmission class cables (defined as cables operating above 46 kV) had traditionally employed oil-impregnated paper as the insulation. This paper insulation is applied as thin layers wound over the cable core, and is later impregnated with a dielectric fluid. Paper-insulated lead covered (PILC) cable was also used at distribution voltages. Paper-based insulated cables are still being installed today; however, as the application of synthetic polymers to cable technology matured, extruded polyethylene that has been subjected to cross-linking (XLPE) gradually displaced paper and can be considered as the insulation material of choice for transmission voltages. XLPE has traditionally been considered as preferred due to its ease of processing and handling (as well as cost), although paper/oil systems have a much longer history of usage and much more information on service reliability exists. Care in handling of materials and assurance of extreme cleanliness prior to and during manufacturing are required for all extruded cables.

For distribution voltage class cables (mostly 15 to 35 kV), the prime extruded material developed for use in the 1960s was conventional high molecular weight polyethylene (referred to as polyethylene, or HMWPE) [1]. However, this insulation material (also referred to as being a thermoplastic polymer, meaning it can be recycled) was replaced by XLPE (referred to as being a thermoset polymer, meaning it cannot be recycled) as the material of choice during the late 1970s–early 1980s, as a result of unanticipated early failures in service due to the water-treeing problem (see Chapter 19). Installed polyethylene-insulated cables are gradually being replaced (or rejuvenated in-situ for stranded constructions [see Section 5.12]). Elastomeric ethylene–propylene copolymers have also been used (EPR or EPDM) for low and medium voltage cable insulation and are employed for accessories. The term EP has been used to generically describe both EPR and EPDM-insulated cables and "EPR" is the terminology that will be applied here. EPR cables have been available since the 1960s, but their use had been consistently less common as compared with HMWPE or XLPE due to higher costs and operating losses. EPR usage started to increase from the 1970s to 1980s partly due to easier processing as a result of modification of the EPR compound to facilitate easier extrusion (hence reducing the cost). In contrast to XLPE, which is a semicrystalline polymer, EPR is an elastomer (rubber) and therefore requires the incorporation of inorganic mineral fillers (and other additives) in order to allow EPR to serve as a satisfactory functional insulation; the requirement to blend in additional additives leads to additional handling and processing requirements by materials suppliers [2].

Starting in the mid-1980s, XLPE had been gradually replaced by "tree retardant" XLPE (TR-XLPE) as the material of choice for new distribution class cables. From the early 1980s and well into the late 1990s, a single grade of TR-XLPE was employed commercially in North America. Several grades of TR-XLPE have been available over the years as improvements have been incorporated [3].

Although many of the older insulated cables manufactured with HMWPE, EPR, and XLPE insulations are still in service on many utility systems, the insulation choices for new medium voltage cables today are considered to be TR-XLPE and EPR. It should be noted here, however, that all EPRs are not alike. All these subjects will be discussed in greater detail later in this chapter.

The use of present-day XLPE, TR-XLPE, and EPR grew out of prior experience going as far back as World War II. As described in Chapter 1, development and application of the various early insulation materials employed for wire and cable were, progressively, natural rubber, synthetic rubber, and butyl rubber, each representing improvements. Butyl rubber was in common use when it was replaced by EPR for many applications. HMWPE was employed due to its superior electrical properties and ease of handling. Virtually all of the older butyl-based installed cables for distribution applications have now been replaced.

Polymers such as polyethylene, XLPE, polypropylene (PP) (used as jacket material), and ethylene–propylene copolymers and terpolymers are hydrocarbon polymers, and are known as polyolefins. Polyethylene and PP are known as homopolymers; EPR is known as a copolymer, meaning that it is composed of two different polymers in the chemical structure. It is manufactured by copolymerizing ethylene and propylene gases. These polymers are composed exclusively of carbon and hydrogen.

It is also possible to replace propylene with other monomers; examples are vinyl acetate (EVA) and ethyl acrylate (EEA). These copolymers of polyethylene are employed as shield materials (see Section 5.7). Copolymerization with other monomers referred to as alkenes has led to commercialization of materials known as EAM (ethylene alkene copolymers).

Preferred insulation characteristics of this class of polyolefin-based polymers include:

- Excellent electrical properties
  - Low dielectric constant
  - Low power factor
  - High dielectric strength
- Excellent moisture resistance
- Extremely low moisture vapor transmission
- High resistance to chemicals and solvents
- Ease of processing and extrusion

Paper-insulated cables were historically one of the first types of polymer used since paper was, and is, readily available from natural sources. Paper is derived from wood pulp and is a natural polymer based on cellulose. In use, the paper is impregnated with a dielectric fluid (a low molecular weight hydrocarbon) so the practical insulation is actually a two-phase composition. PILC cables have been employed at distribution voltages since the early twentieth century; many of these cables are also still in service, even after 60 to 70 (or more) years. They are highly reliable (partly due to the presence of an outer lead sheath that provides protection from the local environment), which makes them a construction of choice for many urban locations. They may be preferred in specific locations where existing duct size and space are limited. Paper insulation is discussed in more detail in Section 5.10.

The dielectric losses of polyolefins are superior to those of paper/oil insulation systems, and the polymers are considerably more moisture-resistant than paper. A moisture-resistant sheath has always been incorporated into paper cable designs.

At lower operating voltages, the possible choice of polymeric materials widens. Here it is possible to use polyvinyl chloride (PVC), silicone rubber (SIR), or other polymers that are readily available and easily processed. PVC was used for a time in Europe for medium voltage cables in the 10 to 20 kV class, but that practice has been largely discontinued. PVC is actually a tough, rigid polymer, and requires a softening agent (plasticizer) to increase flexibility and render it useful for wire and cable applications.

At the lower voltages, such as those employed for secondary cable, reliability is related primarily to operating temperatures rather than operating voltage stress or losses, and thermal resistance becomes a priority. At low voltage levels, other properties can be addressed, such as ability to impart for example, flame resistance, and polymer insulations such as Neoprene and Hypalon are factors. This is discussed in Section 5.11.

Each insulation type has certain advantages and disadvantages. An overview of present-day medium voltage insulations is summarized as follows:

Insulation Type	Key Property Information
Polyethylene	Low dielectric losses Moisture sensitive under voltage stress
XLPE	Slightly higher dielectric losses than polyethylene Less moisture sensitive than polyethylene; ages more slowly Excellent properties if kept dry
EP (EPR/EPDM)	Higher dielectric losses vs. XLPE or TR-XLPE More flexible, less moisture sensitive than XLPE or polyethylene Requires inorganic filler additives Many different compositions available; some proprietary
TR-XLPE	Losses slightly greater than XLPE Losses less than EPRs. Less moisture sensitive than XLPE; ages more slowly
PILC	High reliability Manufactured with a lead sheath

In addition to use as primary insulation, polymers are employed as components of conductor and insulation shields. These materials are ethylene copolymers that possess controlled quantities of conducting carbon black (and often other ingredients) to provide the semiconducting properties required for shields. It is the use of the conducting material dispersed throughout the polymer matrix that makes the mixture semiconducting in nature; hence, the term "semiconducting" is applied to shield materials. The copolymer itself can be viewed as a "carrier," but this carrier must possess the property of controlled adhesion to the insulation with which it is in contact.

Almost all present extruded cable constructions are covered by an outer extruded jacket. The purpose of the jacket is to reduce moisture ingress, protect the cable mechanically (e.g., during handling and installation and from abrasion), and also to provide resistance to sunlight and ultraviolet light. Jackets are commonly composed of one of several polyethylene types (low, medium, or high density) or PP, and also contain small quantities of carbon black, which provides the resistance to light. Jackets are of two types, insulating and semiconducting; insulating jackets contain a different type of carbon black compared with that used in semiconducting shields (it is nonconducting), and it is present in much smaller quantities. For semiconducting jackets, as might be expected, the carbon black used is similar to that used in shields. Polymeric jackets are discussed in Section 5.8. As noted earlier, the covering used for paper/oil cables is a lead sheath.

This chapter will review the following topics:

1. Properties of semicrystalline polymers such as polyethylene from a fundamental perspective. Basic topics such as molecular weight, molecular weight distribution, branching, crystallinity, and cross-linking are reviewed
2. Methodologies for inducing cross-linking of the polyolefins; function of peroxides and antioxidants

3. Introduction to manufacturing processes for polyolefins
4. XLPE and TR-XLPE, including thermal effects
5. Properties of EPR and how it differs from polyethylene, XLPE, and TR-XLPE
6. Review of extruded shield and jacket technology
7. Fundamentals of paper/fluid cables that employ cellulosic insulation along with low molecular weight hydrocarbon-based fluids, and how the cellulose differs from polyolefins
8. Fundamentals of low voltage cable insulation materials
9. Treatment of extruded cables in the field to extend life (rejuvenation)

The fundamental properties to be focused on here are important to understand since the properties that the electrical insulations possess and display are related to their physicochemical structure; the latter primarily controls the physical as well as the electrical properties. By understanding the fundamentals of their chemical nature as described herein, we will be defining the properties. In essence, we will be providing an overview of "structure-property relationships."

5.2 PHYSICOCHEMICAL PROPERTIES OF MATERIALS USED AS ELECTRICAL INSULATION

5.2.1 Overview

Although use of polyethylene itself as insulation for new medium voltage cables has been essentially discontinued, the properties of this polymer will be reviewed first as a "model" as it is a homopolymer from which the other insulating materials are derived and continue to be used for low voltage insulation. When XLPE is considered, the starting material is polyethylene; when EPR, an ethylene copolymer, is considered, the starting material for discussion is the ethylene homopolymer, i.e., polyethylene. Hence, once the basics of polyethylene are described and understood, it will be easier to understand the properties of EPR, XLPE, and also TR-XLPE.

The chemical structure as it relates to molecular weight and chain branching will be reviewed, followed by the subjects of crystallinity and cross-linking of polyethylene. This in turn will lead to a discussion of the properties of XLPE. Some aspects of polyethylene manufacture will also be covered. With this as background, the discussion will be followed by the properties of copolymers of ethylene such as EPR, why fillers are required, and the differences between different EPRs. Finally, semiconducting shields and then jackets will be discussed.

5.2.2 Polyethylene Chain Length and Molecular Weight

Polyethylene is a hydrocarbon polymer composed exclusively of carbon and hydrogen. It is manufactured from the monomer ethylene (in turn, derived from the cracking of petroleum), as shown in Figure 5.1. Note that the chemical structure is a series of repeating $-CH_2-$ units.

Hence, the individual molecules of ethylene gas combine to produce a polyethylene "chain." During this process, the gas is converted to a solid. The number of ethylene molecules (often referred to as "mers") in the chain is significant.

Ethylene (gas) Polyethylene (solid)

$\sim CH_2{=}CH_2\sim \longrightarrow \sim CH_2{-}CH_2{-}CH_2{-}CH_2{-}CH_2{-}CH_2\sim$

FIGURE 5.1 Ethylene polymerization to polyethylene.

Polyethylene falls into the class of polymers known as polyolefins (PP is another example). Key properties of interest relate to molecular weight, molecular weight distribution, branching, cross-linking, as well as crystallinity [4].

Polyethylene is produced by one of several processes that are summarized in Section 5.3. While details are beyond the scope of this book, it is necessary to note that the method of manufacture of the polyethylene controls whether it is "high density," "medium density," "low density," or "linear low density," terms commonly employed in the cable industry. Density is a measure of crystallinity (discussed later in this chapter), and is a factor that determines what makes the specific polyethylene type applicable as an insulation, semiconducting material, or jacket material. Hence, the method of polyethylene manufacture controls the exact chemical structure, which in turn controls the properties.

The carbon–hydrogen polymeric structure noted in Figure 5.1 is simplified; the chemical structure of polyethylene is actually more complex than is shown there (as might be deduced from the number of key subjects noted above). Figure 5.2 shows the carbon–hydrogen structure and for simplicity, we can depict that structure as a wavy line.

The wavy line is referred to as a "chain" and the length of the chain is significant. The length of the line is related to the molecular weight, and the "wave" indicates that the chain has a tendency to coil. The greater the number of ethylene molecules incorporated into the polyethylene chain, the higher the molecular weight (and the greater the degree of coiling). Hence, a longer chain of polyethylene has a higher molecular weight than a shorter chain, and the molecular weight increases as the number of ethylene groups in the molecule increases. Of interest is the fact that the polyethylene employed as insulation for medium voltage cables in the past was described and commonly referred to as "high molecular weight polyethylene" (or HMWPE). It was not referred to as low molecular weight polyethylene for a reason; the properties of HMWPE are far superior.

However, the actual molecular weight of polyethylene used as cable insulation (or for other applications) is more complex than as described so far. It cannot be

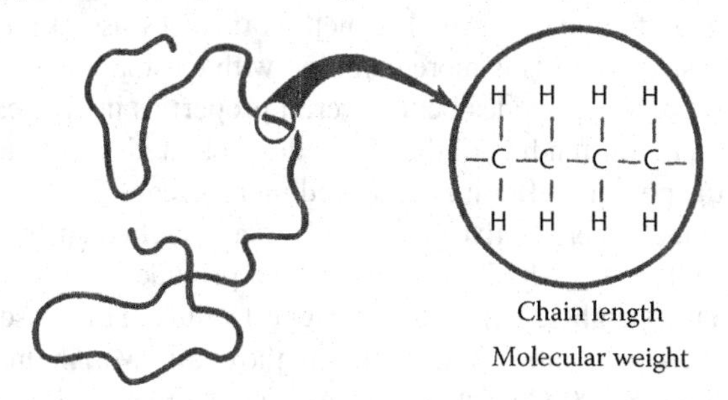

FIGURE 5.2 Depiction of polyethylene chemical structure.

accurately described as a single coiled chain. Conventional polyethylene is actually composed of numerous chains (not a single one as discussed so far), and the chain lengths of individual molecules can vary considerably. Hence, in reality, polyethylene is composed of polymer chains that have a *distribution* of molecular weights (chain lengths). Indeed, the molecular weight distribution is a means of characterizing the polyethylene. This merely means that the "average" chain length is what is referred to and is considered to be "high." As can be inferred, the higher the molecular weight, the better the overall properties.

Since typical polyethylenes that have been employed for electrical insulation contain a variety of individual chains of different lengths (i.e., weights), it is easy to see that there can be a large number of commercially available grades of polyethylene, all varying in "average" molecular weight.

The average molecular weight can be described in several ways. The terms employed most often are "weight average" (Mw) and "number average" (Mn) molecular weight. These values arise from different mathematical methods of averaging the molecular weights in polymer samples possessing molecules of different sizes. The mathematical definitions of the number and weight averages are related to the smaller and larger sized molecules, respectively. Hence, the average molecular weight is always greater than the number average. [When the polymer insulation is cross-linked, the molecular weight determination becomes more complex since the cross-linked fraction can be considered to have an "infinite" molecular weight.] From the perspective of the cable engineer, what is relevant to understand is that there is no single way of characterizing the polymer molecular weight.

The average molecular weight (and distribution) can be determined by a technique referred to as gel permeation chromatography. The molecules of different weights are separated, but the equipment required to perform this is complex, expensive, and special training is required. A simpler alternate method of characterization (to meet most purposes) is to measure the viscosity (resistance to flow) of the polymer; the higher the average molecular weight, the higher the viscosity. The equipment required to measure viscosity is significantly less complex than that required to measure the average molecular weight directly, so focusing on this function of the average molecular weight is common. One method involves determining flow of the polymer through an orifice at a temperature above the crystalline melting point range. Since flow is slower as the average molecular weight increases, due to the higher viscosity, this function of molecular weight is readily characterized. This is referred to as melt flow index. Another method involves use of a rotating disc to measure the viscosity, and this is more common with elastomers. Again, the higher average molecular weight provides better overall properties in application. It is also possible to gain an understanding of the molecular weight distribution by measuring the viscosity of the polymer after it is dissolved in a solvent [5].

Figure 5.3 provides a perspective on the molecular weight distribution of polyethylene [1] and demonstrates why it is so difficult to provide a "single" number. The average based on the weight (Mw) offers the highest value; in this case about 80,000; the average based on the number of molecules in the chain (Mn) in this case provides an estimate of about 8,000 (Mw is always greater than Mn). However, the figure shows that there are a small percentage of molecules having much lower (1,000) and

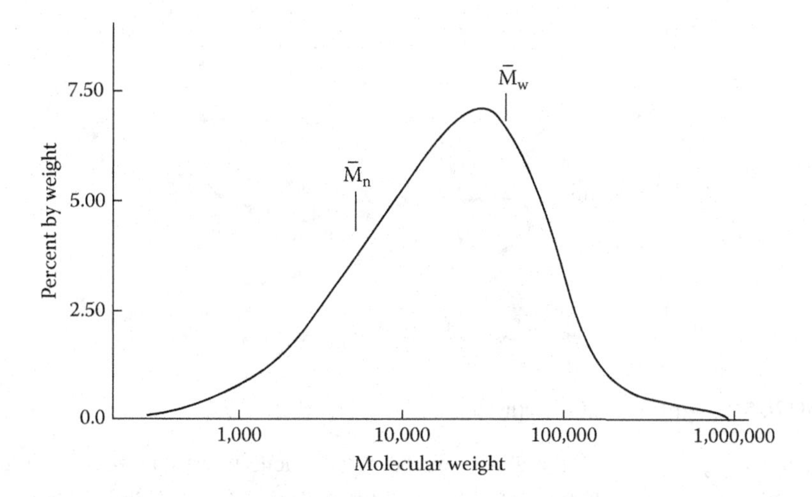

FIGURE 5.3 Typical molecular weight distribution of polyethylene.

much higher (almost 1 million) molecular weights. The Mw/Mn ratio is considered a satisfactory method of characterization.

EPR (which is discussed in detail in Section 5.2.4) is a copolymer of ethylene and propylene (which means that the two gases are blended together prior to polymerization). EPR, also referred to as a polyolefin, is an elastomer, and it is noted here that all the principles reviewed relating to molecular weight, molecular weight distribution, viscosity, and branching (see Section 5.2.3) also apply to EPR.

5.2.3 Branching

When ethylene monomer is converted to ethylene polymer (polyethylene), the polymer chains that form are not always linear as shown in Figure 5.2. There is a tendency to form side chains or "branches." These branches are "hanging" off the main chains as appendages (like "T"s). This is a natural event; when polyethylene is manufactured, the process employed always leads to side chains "hanging" off the long main chain. The chain branching phenomenon contributes to increase in the molecular weight, but does not lead to an increase in the chain length. Branches for various grades of polyethylene are shown in Figure 5.4; note that the chain length of the branches can also vary, and that there are both long and short chain-length branches. It is possible to control the length and distribution of the branches (via the polymerization process).

The significance of branching is that their length and distribution affect the physical properties and also influence the ability to satisfactorily extrude the polyethylene.

It is possible to now visualize that two single molecules may have the same exact molecular weight, but one may have a longer main chain with few branches, and the other a shorter main chain with a longer branches than the first. Therefore, two different polyethylene material batches having many molecules like the two described here (if it were possible to manufacture these) could have different properties, despite having approximately the same molecular weight distributions.

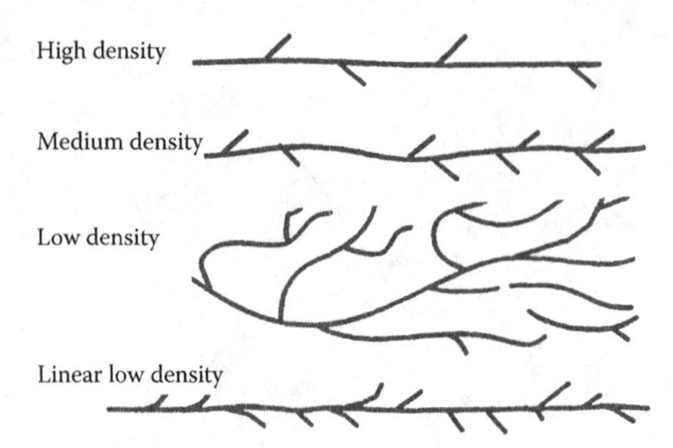

FIGURE 5.4 Structures of polyethylene depicting branches.

Branching also affects the ability of the polyethylene to crystallize (see Figure 5.4). However, branching does not have any meaningful influence on the electrical properties, such as dielectric strength or losses.

In Figure 5.2, we have depicted the polyethylene chain as a wavy (rather than straight) line, and that is because the chains have a tendency to coil. In other words, they have a tendency to achieve a random configuration (like a bowl of spaghetti). This coiled configuration is better shown in Figure 5.5. When depicting the branching in Figure 5.4, we did not focus on the aspect of coiling as the objective was to emphasize the branching. However, it is easy to visualize branches "hanging" from the coiled structure of Figure 5.5. This tendency is independent of the molecular weight, but the configuration that results is influenced by the branching.

The tendency to coil means that the polymer chains also have a tendency to entangle with each other. These entanglements mean that when the chains are pulled apart (as occurs in performing a tensile strength or elongation measurement), there will be some "resistance" to movement. Such chain entanglements influence the mechanical properties of the polymer. These entanglements contribute to the good properties of polyethylene, but not to the qualities that make polyethylene resistant to the penetration of water vapor. Entanglements do not have a major influence on electrical properties.

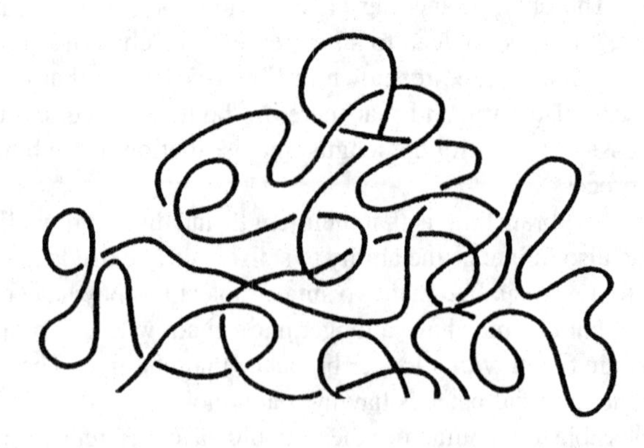

FIGURE 5.5 Simplified depiction of random coiled configuration.

In conclusion, molecular weight, molecular weight distribution, and branching represent several important characteristics of polyethylene that influence properties and also represent methods of describing the characteristics of polyethylene insulation.

5.2.4 Crystallinity

Another very important characteristic of polyethylene is the subject of crystallinity. Polyethylene and some other polyolefins (PP being an example) are known as semi-crystalline polymers. This characteristic results from the fact that the polymer chains not only have a tendency to coil (as previously described), but also have a tendency to align themselves relative to each other (Figure 5.6). Alignment means that there are short- and long-term orders to the chain structure. While the nature of these alignments is quite complex, and the detailed structure is beyond the scope of this chapter, it is important to understand that the alignment contributes to the crystalline nature of the polyethylene, and therefore to the density and ultimately to properties, such as stiffness and resistance to migration of impurities.

For polyethylene, different chain segments also have a tendency to align next to each other. The aligned portions cannot coil. The portions that are not aligned can coil. The chain portions that are aligned are said to be "crystalline." The chain portions that are not aligned are described as "amorphous."

The lower portion of Figure 5.6 shows chain alignment where the polymer chain lengths differ. Some portions of the same chains align with adjacent chains, and some portions of the very same chains are not aligned. Those chain portions where alignment occurs are in the regions called "crystalline." Figure 5.6 shows that such alignment is not necessarily related to molecular weight. It is theoretically possible to have low or HMWPE of the same, or different, degrees of alignment.

The nature of the crystallinity in polyethylene has been the subject of numerous studies over the years. These studies reveal that the crystalline structure is more

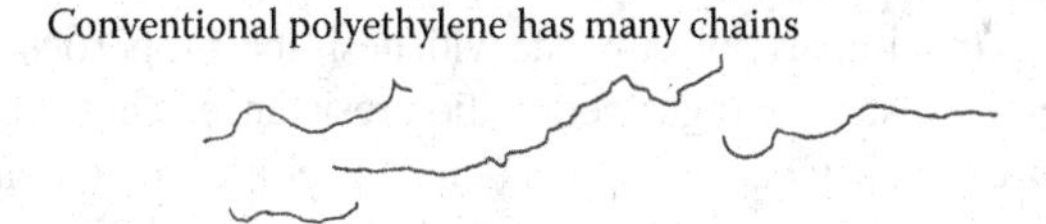

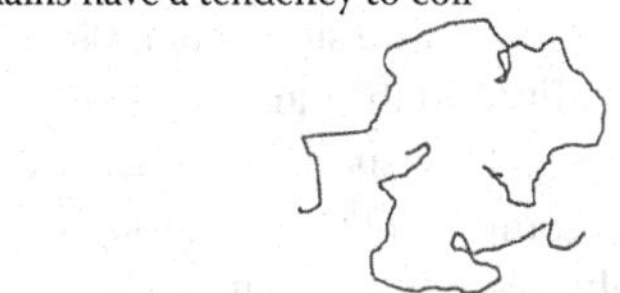

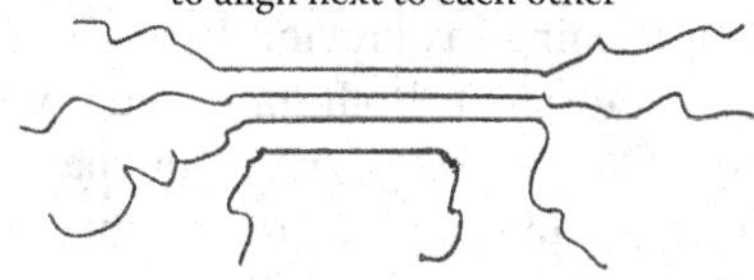

FIGURE 5.6 Polyethylene chain configurations.

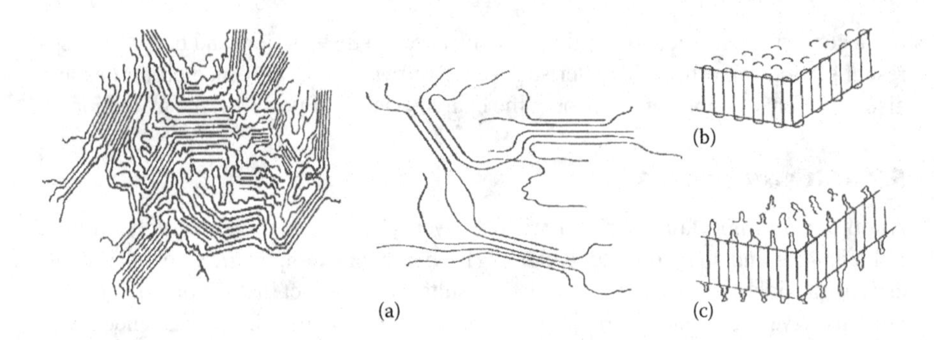

FIGURE 5.7 (a–c) Various depictions of semicrystalline polyethylene.

complex than that described so far. Crystallites, or crystalline regions of the polyethylene, can themselves fold (like a series of connected upright and upside-down "U"s)," Various possible descriptions are shown in Figure 5.7. Semicrystalline polymers are 'tough' at ambient but soften at elevated temperatures (crystalline regions melt and become amorphous)

These regions can, in turn, align into larger structures called spherulites (which can be seen under polarizing light). The size of these spherulites may vary, as can the fold dimensions of the crystallites. The folded regions are referred to as lamellae. While all these structures "disappear" on melting (and re-form in a qualitatively similar manner upon subsequent cooling), it is known that upon annealing below the melting temperature range, changes in the lamellae thickness occurs. It is also known that there may be numerous crystallites in a spherulite; it is possible for one polymer chain within a crystallite to cross through the amorphous region into another crystallite, becoming what is called a "tie molecule." The latter are considered to influence mechanical properties.

Regardless of the "fine" structure previously described (further details are beyond the scope of this chapter) from a practical perspective regarding cable insulation behavior, it is the crystalline regions that impart polyethylene with desirable properties such as toughness, high modulus, and moisture and gas permeation resistance. Those regions that are aligned possess increased density due to "tighter" chain packing, and the increased crystallinity resulting from chain alignment leads to higher density. The alignment process logically means less amorphous regions in the polymer and more polymer per unit volume. Nevertheless, the amorphous regions play a significant role in controlling properties such as increased ductility and flexibility and they also facilitate processing.

For simplification, polyethylene can be visualized as being a "blend" of two materials having different geometrical components, even though the chemical structure of the polymer is comprised solely of $-CH_2-$ groups. The "two materials" are the crystalline and amorphous regions.

As might be surmised, branching (as previously described) influences the ability of the polyethylene chains to align. Both long chain branches and short chain branches hinder the ability of the polyethylene main chain backbone components to crystallize (but not equally). Branching, therefore, due to the "bulky" nature of the chemical structure of the polymer chains, influences the crystallization process. For crystallinity to occur, nonbranched chain segments must be able to approach each other. When branching is present, the ability of the main chain to come in close proximity to another main chain is hindered.

Different polyethylenes have historically been classified into the following general categories due to this phenomenon (see Figure 5.4); the density increases from the 0.91 (g/cc) range for very low density polyethylene (VLDPE) to the 0.94 range for HDPE:

- Very low density
- Low density
- Medium density
- High density
- Linear low density

As the density increases, the degree of chain alignment increases and the "volume" of aligned chains increases. As noted above, the degree of branching is related to the polymerization process. Linear low density polyethylene (LLDPE) approaches the branching structure of high density polyethylene (HDPE), but is referred to differently due to the fact that it is manufactured by a different polymerization process. Branching clearly influences crystallinity, but the latter is minimally affected, if at all, by the conversion of polymer pellets into cable insulation during the extrusion process.

As noted previously, as the degree of crystallinity varies, the properties vary. Increased crystallinity leads to increased density. Hence, in principle, it is theoretically possible to have the following different types of polyethylenes:

- High density, high molecular weight
- High density, low molecular weight
- Low density, high molecular weight
- Low density, low molecular weight

Not all these types are of practical interest for cable (or other) applications. Low density, high molecular weight is, as we have seen, the type commercially provided and employed as cable insulation in the past. (This was known simply as HMWPE.) HDPE (of varying molecular weights) has been and continues to be employed as jackets. (As noted earlier, when referring to molecular weight, we are referring to averages.)

One further characteristic of polyethylene that influences crystallinity is worthy of mention at this point. This is the effect of temperature on the polymer chain alignment and motion at the molecular level as a function of temperature. As the temperature increases, the chains will move farther apart as they absorb heat. This motion disrupts the alignment and crystalline melting takes place (see Section 5.4.5).

5.2.5 Polyethylene Copolymers

Copolymers are insulation materials that are manufactured by incorporating more than one monomer during the polymerization process (see Section 5.3). Ethylene monomer is a gas; when ethylene is polymerized alone, solid polyethylene is produced. If gaseous propylene monomer is mixed with the ethylene prior to polymerization, one obtains ethylene–propylene copolymer(s)—hence EPR. What should be apparent is that the ratio of ethylene to propylene (E/P) employed in the polymerization process should influence the E/P ratio in the ultimate EPR insulation material. It should be possible to manufacture a wide variety of EPR copolymers each with different E/P ratios, and indeed this is so. However, not all E/P ratios in polymers make them suitable as insulation materials; an E/P ratio of 50%–70% may be typical for different insulations. This ratio also influences the method used to extrude the polymer as cable insulation.

Copolymerization, as described here, is different from mixing polyethylene and polypropylene after manufacture of the homopolymers. In the latter case, one does not have a copolymer, but a blend with entirely different properties. Indeed, polymer blends are often incompatible, and phase separation of the different polymers can occur; that does not happen with true copolymers.

It is also possible to manufacture copolymers of ethylene with monomers other than propylene. Common monomers for wire and cable applications include EVA or EEA. These latter copolymers (E-VA or E-EA) are employed in shield compounds. As with polyethylene or EPR, the chain lengths may vary, branching is common, and their chain lengths influence properties. The relative amounts of the second (copolymerized) monomer must also be taken into consideration when evaluating the properties. It is also possible to copolymerize ethylene with various other monomers, possible examples being butene or higher unsaturated hydrocarbon monomers (different monomers would provide different chain lengths on the branches).

When ethylene is copolymerized with other monomers, the result is polymer structure(s) that can disrupt the ability to impart crystallinity. We are now producing polymers that are more rubbery (elastomeric) rather than crystalline in nature. The properties are now drastically changed and this will be discussed in detail in Section 5.6. A question arises as to how to classify such rubbery copolymer materials and the numerous possible compositions based on them. This applies to all elastomers, even those that are not based on polyethylene copolymers. The approach is to apply designations described in ASTM D-1418, where rubbers of the polyethylene type (the ones of interest to us here) are designated "M" and the term EAM has been applied to some copolymers of polyethylene.

It is not uncommon to polymerize ethylene with more than one additional monomer, hence producing a terpolymer. This will be discussed with reference to EPR.

It can be noted that some older polymer systems used for wire and cable are copolymers. For example, butyl rubber (commonly employed prior to the advent of EPR) is composed of a copolymer of two monomers known as isobutylene and isoprene, the latter being present in the 1%–3% range. Even earlier, synthetic rubber (developed to replace natural rubber) is a copolymer of butadiene and styrene, the ratio being 75:25.

5.3 MANUFACTURE OF POLYETHYLENE

5.3.1 Conventional Manufacturing Methods

Historically, low and medium density polyethylenes (MDPEs) have been manufactured by a high-pressure polymerization process [4]. This process induces polymerization of ethylene gas in a reactor vessel under extreme conditions of very high pressure and temperature and leads to the branched polyethylene structures discussed above. It also employs a peroxide initiator to induce the polymerization. The polymer produced in the reactor is extruded through a die, pelletized and cooled after manufacture.

HDPE is manufactured through a low pressure process using a different catalyst concept. The low pressure process, developed later in time, uses nonperoxide catalysts, one of which is called "Ziegler-Natta" (named after the inventors) and allows manufacture of polyethylenes with fewer and shorter branches. This process produces a stiffer, tougher type of polyethylene, and is termed "high density." LLDPE, developed even more recently, is manufactured by a low pressure process; as can be seen from Figure 5.4, it has many short chain branches, rendering it more like HDPE in structure. (That is why it is called "linear low density polyethylene," rather than "high density polyethylene").

The different types of polyethylenes are therefore all manufactured by different processes. Recall that all these processes will provide a polymer with a variety of (different) degrees of crystallinity (hence, density) and also a variety of molecular weight distributions.

The manufacturing technology is continuously improving, as discussed in the next section.

5.3.2 Controlled Molecular Weight Distribution Technology

Changes in catalyst polymerization technology have been the objective of ongoing studies. Results have allowed materials suppliers to better control the molecular weight and molecular weight distribution, and this led to development of newer grades of polyethylene having narrower and more defined molecular weight distributions (and various low density grades). Polymerization processes have been referred to as "single site catalysis," and metallocene catalyst represented one methodology focused on in the past years [7].

This area of activity for improved control of polymer properties has received much attention. Fundamental properties such as molecular weight were described in Section 5.2. It was noted there that the polyethylene or EPR (polyethylene copolymer) used for cable manufacturing does not have a single uniform molecular weight (all the molecules do not have the same length) but possesses a distribution of molecular weights. This is because the catalyst technology used to manufacture conventional high-, medium- or low-density polyethylene cannot provide such an exact control of the polymerization process. This distribution of molecular weights (and branching) normally attained by the use of conventional catalysts influences the crystallinity and therefore the properties. Improved control of the molecular weight by using different catalyst technology has created much interest in the polymer industry, as better control of molecular weight distribution means better control of properties.

What is relevant for insulated cable applications is that materials suppliers can attain greater control over the polymerization process to produce polymers that are more uniform in nature. The term "metallocene" was used initially to describe these modified materials; the term is based on the nature of the catalyst, which was a metallic compound that incorporated a special chemical structure called "cyclopentadienyl." More recently, other catalysts have been developed and the general term "single site catalysis" is more technically appropriate. What this means is that the ethylene is polymerized at one single site on the catalyst. Further details on catalyst technology are beyond the scope of this discussion. As might be expected, much of this new technology is proprietary and patented.

The ability to control the molecular structure means that the materials supplier can apply fundamental knowledge of structure–property relationships to develop products geared for a specific end-use application, in this case, wire and cable. From a property perspective, the product would be fine-tuned for mechanical, physical, and electrical properties. From an application perspective, these newer materials must also be capable of being processed (extruded) at the same (or faster) rates and with the existing equipment employed for cable manufacturing. As a result of these basic material improvements, one should expect equivalent (or better) life characteristics from cables made with these materials. Any commercial application for products developed from this newer technology will be influenced strongly by the processing and lifetime characteristics.

As with any new technology, advantages are balanced by "trade-offs." In this case, not only were older metallocene catalysts more expensive (leading to higher finished product costs), but polyethylenes produced in this manner were more difficult to process. The narrow molecular weight distribution of the metallocene-based resins modified the flow properties (rheology) during processing. This experience provides a clear practical application of the need to understand molecular weight distribution aspects, as discussed in Sections 5.2.2 and 5.2.3. (Appendix B reviews catalyst technology in greater detail.)

5.4 CROSS-LINKED POLYETHYLENE

5.4.1 FUNDAMENTALS

Until now, all the polyethylene chains discussed have been separated to various extents. Cross-linking is the process of joining different polyethylene chains together by chemical reaction. It is the term used to describe the conversion of the polymer chains from two dimensions into a three-dimensional network. Cross-linking is also referred to as vulcanization or curing, and the polymer so obtained is often described as being "thermoset."

This is shown in Figure 5.8. In a sense, XLPE can be considered as a branched polyethylene, where the end of the branch is connected to a different polyethylene chain instead of just "hanging loose." Cross-linking imparts certain desirable properties to the polyethylene; from a cable perspective, it allows the polymer to maintain its form stability at elevated temperatures. Cross-linking can be visualized as preventing the chains from separating "too far" under thermal overload. Other advantages of cross-linked materials include resistance to deformation (i.e., softening) and stress-cracking and improved tensile strength and modulus. It should be noted that the electrical properties of polyethylene are not improved by cross-linking.

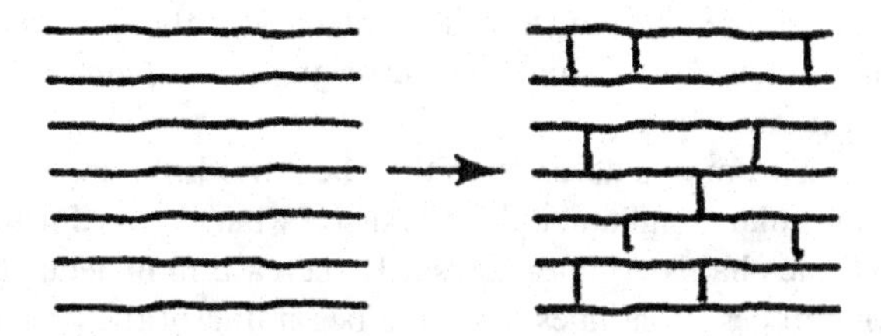

FIGURE 5.8 Depiction of cross-linked network.

As we have seen from the previous discussion, conventional polyethylene is composed of long chain polymers that, in turn, are composed of ethylene groups. The individual molecules are very long. The backbone may contain 10,000 to 60,000 atoms, sometimes more. Further, we have also seen that there are branches, crystalline and amorphous regions and that any additives or impurities must be residing in the amorphous regions—not in the crystalline regions. Cross-linking adds yet another dimension to the complexity of the molecular arrangement.

Figure 5.9 provides a description of how a conventional, non-XLPE "parent" (Figure 5.9a) is converted to the cross-linked "child" (Figure 5.9b through 5.9e). For simplicity, the chains (Figure 5.9a) are all shown adjacent to each other and are not coiled. The linear chains represent a simplified description to fit our purposes here. First, two adjacent chains link together (Figure 5.9b). We immediately see that the molecular weight has increased. The first cross-link leads to two branches. In Figure 5.9c, the first two chains have been simply redrawn from Figure 5.9b in a more familiar way. In Figure 5.9d, three additional cross-links have been (arbitrarily) added, two to different chains. The third shows that the newer (previously cross-linked) higher molecular weight chain is again linked to another chain. In Figure 5.9e, it

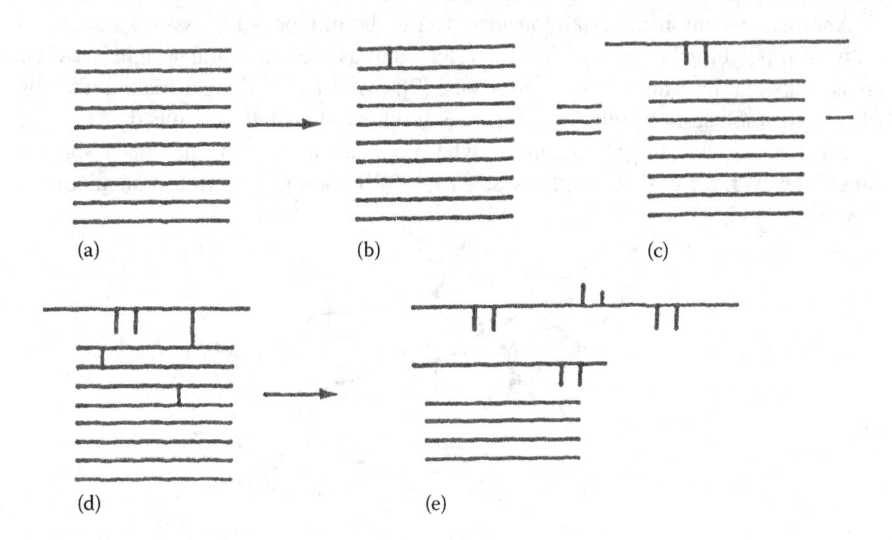

FIGURE 5.9 (a–e) Effect of cross-linking on chain length of polyethylene.

has been redrawn (Figure 5.9d) to show how the cross-linking process looks as the chains are again "stretched out." Note how the original two chains have dramatically increased in molecular weight.

It should be clear from this description that the cross-linking process is a way of increasing the molecular weight, and this is exactly what occurs during this process. Note also that all the chains do not necessarily increase in molecular weight at the same rate. As the process continues (only the beginning of the process is depicted here), the molecular weight increases greatly that the XLPE can be considered to have an "infinite" molecular weight. A depiction of the polymer insulation cross-linked "wavy line" structure is shown in Figure 5.10.

One way of characterizing an extremely high molecular weight polymer as compared with a cross-linked polymer is to determine its solubility in an organic solvent such as toluene, xylene, or decalin [6]. A conventional polyethylene, even one of very high molecular weight, will dissolve in a heated solvent of this type. The solubility results from the chains moving apart in the heated solvent. XLPE will not dissolve. The chains do move farther apart when the cross-linked polymer is immersed in the warm solvent, but not so far apart so that dissolution occurs. What happens instead is that the XLPE merely swells in the solvent and produces a gel. Indeed, this is called the gel fraction. A simpler (qualitative) way to determine whether the polyethylene is cross-linked or not is to subject it to heat by placing the sample in contact with a hot surface. The conventional polyethylene will flow while the XLPE will resist flowing and behave more "rubbery."

Commercial XLPE cable insulations also have a "sol" fraction. This is the portion of the polymer chains that never got incorporated into the "infinite" network. In Figure 5.9, we see some chains in (d) and (e) not incorporated into the network. The gel fraction of a commercial XLPE is about 70%–80%; i.e., about 70%–80% of the polymer chains are incorporated into the three dimensional gel network and the remainder are not and would be soluble in the heated solvent.

Another insulation material consideration is the number of cross-links between individual polyethylene chains. This is referred to as the molecular weight between cross-links and has theoretical significance [9]. (Swelling of the gel fraction diminishes as the molecular weight between cross-links is reduced; this results in increased "toughness" at elevated temperatures where crystalline melting has been significant.) However, for practical purposes, a 70%–80% total gel fraction is an adequate

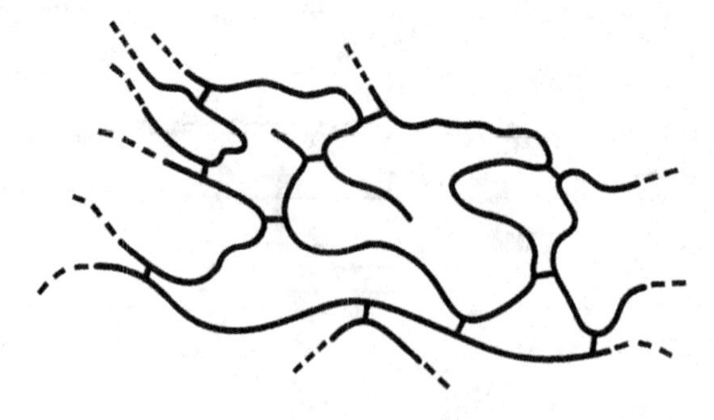

FIGURE 5.10 Cross-linked polymer showing coiled chains.

description. It is also common to refer to the "hot modulus." This is a somewhat easier measurement to make than a sol fraction and does not involve the use of organic solvents. The hot modulus is directly related to the degree of cross-linking, or more correctly to the molecular weight between cross-links. It is greater as the degree of cross-linking increases or as the molecular weight between cross-links decreases. In the case of EPR (and elastomers in general), it is common to refer to "Mooney Viscosity," which is a measure of the hot modulus.

The next issue to consider is just how cross-linking of polyethylene (or copolymers such as EPR) is achieved. Cross-linking of the polyethylene chains can be induced by several different means:

- Use of organic peroxides
- Use of high energy radiation
- Modification of the backbone structure

5.4.2 Peroxide-induced Cross-linking

Polyethylene that is cross-linked by peroxides (the most common method for medium voltage cables) contains a small amount of a cross-linking agent that is dispersed throughout the polymer. This agent is an organic peroxide, the most common being dicumyl peroxide [10]. Organic peroxides are chemicals that are stable at room temperatures, but decompose at elevated temperatures. There are many such peroxides available. Dicumyl peroxide is used commercially for medium and high voltage cables. It has traditionally been incorporated into polyethylene pellets by the material suppliers. When the polyethylene is extruded (conversion of the pellets into cable insulation), the peroxide remains stable due to the fact that its decomposition temperature is higher than the extrusion temperature. After the extrusion process, the polyethylene insulation surrounds the conductor and the conductor shield and is covered by the outer insulation shield; the cable now enters the long curing tube where the temperature is raised above the temperature employed in the extruder. At this higher temperature and pressure, the peroxide now decomposes and induces the cross-linking process. Peroxide-induced cross-linking uses a specific peroxide designed to intentionally decompose at a desired elevated temperature after the conversion of the pellets into cable insulation. The after-extrusion tube is called a curing tube, and the terms "curing" and "cross-linking" are often used synonymously. Note that this process takes place in the molten state of the insulation; i.e., the polymer (polyethylene) is heated to an elevated temperature high enough so that all the crystalline regions are melted while cross-linking is induced. The same process occurs with EPR.

A key component of the overall manufacturing of the cable is that the process must ensure that some degree of cross-linking does not take place prematurely. If this occurs, cluster(s) of oxidized polymer, referred to as "scorch" may form. These components can act as impurities in the sense that they will exhibit poor interfacial contact with the "healthy" polymer, which can therefore facilitate formation of microvoids. The latter in turn can lead to eventual premature failure. Scorch is avoided by proper control of the materials and processing conditions. Cable manufacturing is discussed in more detail in Chapter 11.

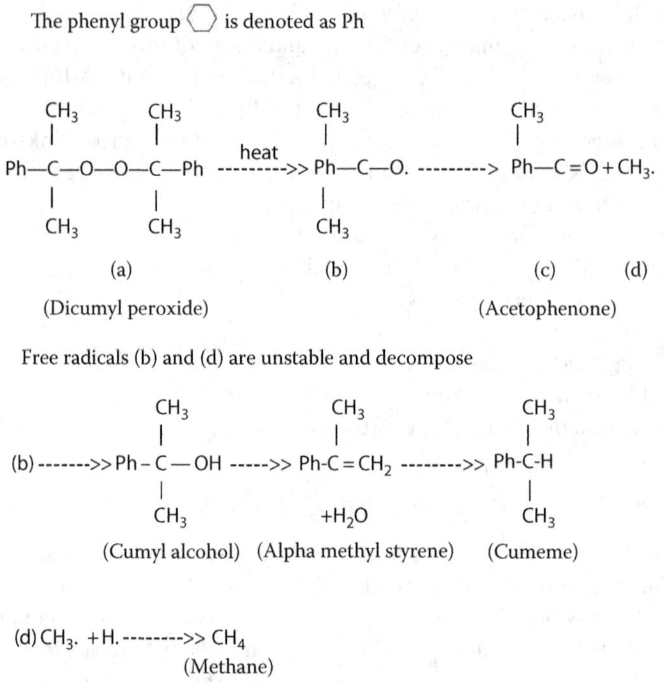

The phenyl group ⬡ is denoted as Ph

```
   CH3         CH3                    CH3                    CH3
   |           |                      |                      |
Ph—C—O—O—C—Ph  ---------heat--->> Ph—C—O.  ---------> Ph—C=O + CH3.
   |           |                      |
   CH3         CH3                    CH3

        (a)                          (b)                   (c)     (d)
  (Dicumyl peroxide)                                   (Acetophenone)
```

Free radicals (b) and (d) are unstable and decompose

```
                  CH3                   CH3                 CH3
                  |                     |                   |
(b) ------->> Ph - C — OH ----->> Ph-C=CH2 -------->> Ph-C-H
                  |                                         |
                  CH3                   +H2O                CH3
         (Cumyl alcohol)  (Alpha methyl styrene)     (Cumeme)
```

(d) $CH_3\cdot + H\cdot$ -------->> CH_4
(Methane)

FIGURE 5.11 Decomposition of dicumyl peroxide leads to formation of volatile by-products.

When the peroxide decomposes during the curing process (Figure 5.11), it forms an active ingredient, called a "free radical," which is unstable. The latter is so active that it interacts with any nearby molecule, which is virtually always the polyethylene chain. This free radical forms when the peroxide "splits" into an active oxygen-containing component that then "pulls" hydrogen atoms off the polymer chains. The polymer chain now becomes the active and unstable component and two such chains immediately combine to "cross-link" and also to stabilize the system once again. During this process, as the peroxide decomposes, and hydrogen atoms are pulled off the polymer chain, several by-products are ultimately formed. The major ones are dimethyl benzyl alcohol, acetophenone, alpha-methyl styrene, and methane.

These by-products form in the following manner. When the free radical (a cumyloxy radical) is generated, it can undergo several different types of reactions in its quest to become stabilized. It can "grab" a hydrogen atom from the polyethylene chain (as described above) and form the relatively stable dimethyl benzyl alcohol molecule. However, the unstable radical may also undergo internal rearrangement and "kick out" a methyl radical and become acetophenone. The unstable and highly reactive methyl radical may also pick off a hydrogen atom from the polymer, hence forming methane gas. Water may also be formed if the dicumyloxy radical expels a hydroxyl and a hydrogen radical, to form the water; it is then converted to alpha methyl styrene in the process.

The first three by-products of cross-linking (dimethyl benzyl alcohol, acetophenone, and methane) are always found in greatest quantities in XLPE. Acetophenone is a solid at temperatures lower than about 20°C. It emits a somewhat sweet odor, is not soluble in water, and is partially soluble in the polyolefin (the extent being dependent on the temperature). Due to its low melting point, acetophenone is a liquid at ambient temperatures. Dimethyl benzyl alcohol is also a liquid at ambient temperatures. These cross-linking agent by-products will remain in the insulation wall initially, but migrate out slowly over time. [There is some evidence that they impart some degree of water tree (and electrical tree) resistance to the insulation on aging.] Their chemical structures are shown in Figure 5.11 along with that of alpha methyl styrene; the figure depicts the chemical changes resulting from the cross-linking process. The methane gas evolved must be allowed to migrate out of newly manufactured XLPE-insulated cable after cross-linking has been accomplished. This is easily induced by allowing the cable to "sit" for a defined time after manufacture. Heating the finished cable shortens the time that is required. Other by-products, such as alpha methyl styrene, may be present in smaller concentrations. It should be noted that, at times, cumene is also found as another by-product; it is believed to develop from the further reaction of alpha methyl styrene.

It should be apparent by now that the peroxide-induced cross-linking process involves rather complex chemical reactions.

To achieve good cable insulation, the peroxide must be uniformly dispersed within the polyethylene. For appropriate uniformity of the cross-linking process to take place in the cable insulation, temperature and pressure must be properly controlled throughout the curing tube, which is quite long, but obviously of finite length. These factors contribute to preventing scorch, referred to previously.

It is important to emphasize that the cross-linking process described here also applies to mineral filled EPR, TR-XLPE, and cable shield materials all of which will also contain a peroxide. The same by-products are produced as long as the same peroxide is employed. It must not be forgotten that the carbon black-containing polymer comprising the inner and outer shields are also being cross-linked concurrently along with the insulation.

Dicumyl peroxide has historically been commercially available in several forms:

- Free flowing powders that contain about 40% active materials; the inert ingredients being calcium carbonate or clay
- As a 94% to 97% active, light yellow, semicrystalline solid
- A slightly more pure 98% active grade

The choice of dicumyl peroxide form is dependent upon requirements of the type of insulation being manufactured.

From what we have learned above, it is clear that peroxide-induced cross-linking takes place in the amorphous regions. Even though the crystalline regions cannot hold the peroxide in the original pellets, this is not a complication during the extrusion and cross-linking process since the crystalline regions necessarily melt during extrusion. By the time the peroxide induced cross-linking takes place in the heated tube after extrusion, the entire polymer is amorphous and the peroxide diffuses and is considered to be relatively uniformly dispersed. It should be emphasized that the complex series of reactions described above all take place within the melted but viscous polyethylene. When the cable is cooled down after extrusion and cross-linking, recrystallization takes place. When this occurs, the newly formed cross-linking agent by-products are "forced" into the newly formed amorphous regions.

XLPE gradually became the preferred insulating material of choice for medium voltage cable starting around the mid- to late 1970s and early 1980s (see Chapter 1). It replaced conventional low-density polyethylene (HMWPE) due to its superior high temperature properties and perceived better resistance to water treeing. Peroxide-induced cross-linking has been the prime method of curing for medium and high voltage cables as the process has been well developed and defined. For 69 kV transmission cables, peroxide-induced XLPE has also been an insulation material of choice. At higher voltages, peroxide XLPE has shared the market with conventional paper-fluid filled cables (These latter cables are not cross-linked.) As might be imagined, the thicker insulation walls of extruded transmission class cables require significant curing process modifications to ensure proper cross-linking.

Other peroxides also have been used; one is bis(tert-butylperoxy)-diisopropylbenzene (known as "Vul-cup"), which has a higher decomposition temperature than dicumyl peroxide. Higher temperature peroxides are of interest where it may be desired to manufacture the cable employing higher than conventional temperatures; the use of a higher decomposition temperature peroxide reduces the possibility of premature decomposition, which could potentially lead to processing problems. It should be noted that not all peroxides will decompose and induce cross-linking over the same temperature ranges. Finally, it should also be noted that most, but not all, of the peroxide necessarily decomposes during the normal curing process.

For low voltage cables (less than 600 V) peroxides may be used to induce cross-linking, but economic factors have allowed both silane and radiation induced cross-linking to share the market (see Sections 5.4.3 and 5.4.4). In this voltage range, it is not uncommon to employ conventional polyethylene since the voltage stresses and temperatures experienced by these cables are generally lower. Numerous additional polymeric insulation types are available, and cross-linking methodology may vary; see Section 5.11.

Again, although polyethylene has been used as the example for this discussion on peroxide-induced cross-linking, the same principles apply to the TR-XLPE and EPR polymers, which are cross-linked by peroxides in the same manner.

Once cross-linking has taken place, the polyethylene structure (which, as we have seen, was complex in nature to begin with) is now even more complex. Cross-linking typically takes place with about 70%–80% of the polymer chains being incorporated into the network, as noted previously. This means that 20%–30% of the remaining insulation is not cross-linked. Typically, this represents the low molecular weight fractions of the initial material (see Figure 5.3). The insulation of such cables that are installed therefore can be viewed as a mixture of LDPE and XLPE (or cross-linked TR-XLPE and a non–cross-linked portion; or cross-linked EPR and a non–cross-linked portion). However, the physical and dielectric properties are clearly dominated and controlled by the cross-linked regions of these insulations. At elevated temperatures, the XLPE cable insulation clearly maintains its form stability and functions as anticipated. For EPR, the low molecular weight sol fraction is the ethylene copolymer.

5.4.3 Radiation-Induced Cross-Linking

It is also possible to cross-link polyethylene using high energy radiation instead of a peroxide. A beam of electrons emanating from special equipment can interact with the polymer chains, causing free radicals to form; a now-reactive polymer chain interacts with another chain (as described previously), hence inducing cross-linking. The electron beam serves the same role as does the catalyst peroxide. Radioactive isotopes such as Cobalt-60 can be used for the same purpose.

In the radiation cross-linking process, energetic electrons come into contact with the polymer chain and break the chemical bonds [8]. A ~C–H or ~C–C~ bond can be cleaved. When a ~C–H bond is broken, a hydrogen atom is released, and the now-highly energized ~C• polymeric free radical seeks to stabilize itself by combining with another like radical. This provides the cross-link. (The hydrogen atom can combine with another hydrogen atom to form a hydrogen molecule.) When a ~C–C~ bond is broken, it is apparent that this can, in principle, lead to a reduction in molecular weight; the shorter free radical chain can combine with another, or with a hydrogen free radical. Hence, the cross-linking and degradation processes compete with each other in radiation cross-linking. In actual practice, for polyethylene it can be generalized that approximately three cross-links form for every polymer chain cleaved, rendering the latter effect of little practical significance.

Radiation cross-linking involves different processing technology as compared with peroxide-induced cross-linking, and is employed primarily for low voltage cables. The radiation process is performed at room temperature, and therefore, for polyethylene, this means (unlike with peroxides) that cross-linking takes place while the polyolefin possesses both crystalline and amorphous regions. However, the insulation temperature can increase during radiation processing (depending on many factors), hence leading to some crystalline melting during the process. This complexity is "controlled" by applying this technology to thin wall insulations at high processing speeds.

The ambient temperature radiation-induced cross-linking process leads to some changes in the insulation not experienced via peroxide-induced cross-linking. First, the distribution of cross-linked regions will differ for the two processes. Second, when melting is not an issue if temperature is properly controlled, the nature of the crystalline regions after radiation-induced cross-linking remains about the same as before, i.e., no crystalline melting has occurred. This is unlike the peroxide cross-linking process, which is performed at elevated temperature, and where recrystallization occurs upon cooling. Finally, the radiation process induces certain chemical changes within the polyethylene not experienced during peroxide-induced cross-linking; this includes as a small amount of degradation products resulting from the radiation breaking a C–C bond, and other small changes on the polymer backbone. Also, no peroxide cross-linking agent by-products are produced. In addition, the use of multifunctional monomers to increase cross-linking efficiency and reduce cost is not uncommon [11].

Whether there are any practical consequences as a result of these differences is not relevant for medium voltage cable, as the radiation process has not been employed commercially for reasons noted. For low voltage wire cross-linking applications where speed of cross-linking is a key issue, radiation technology has been more applicable.

Another potential issue relating to this technology employing electron beams is the nonuniformity of dose (energy) absorbed for thick specimens and therefore the nonuniformity of the degree of cross-linking (gel fraction) as the wall thickness increases. There is an inherent radiation dose-depth relationship that is polymer thickness (film or coated wire) related, and the energy absorbed at different regions of a wire or cable wall will differ as the item being irradiated increases in thickness. The energy absorbed increases at first and then drops after reaching a maximum; the total dose absorbed is dependent upon the electron beam energy. Hence, the degree of cross-linking by electron beam technology is not uniform within the component thickness and much depends on geometry. It is not uncommon to apply a minimum dose in this type of situation or to irradiate from more than one side. This is in contrast to the relatively uniform degree of cross-linking that occurs with peroxides for medium voltage cables. This is not an issue for thin wall cables.

Use of radioactive isotopes, such as Cobalt-60, due to their greater penetrating capability (they emit gamma radiation), does not involve any nonuniform dose-absorbed issues, but Cobalt-60 usage involves a different set of manufacturing concerns, making it less useful for wire and cable cross-linking.

Radiation-induced cross-linking has also been successfully employed to manufacture polyolefin-based heat-shrink polymeric joints, and the crystallinity of the polymer is a key to the concept. In principle, the product is fabricated, cross-linked, heated and deformed (i.e., expanded at high temperature), and then cooled in the expanded state. The cooling process after cross-linking and intentional deformation by expansion causes recrystallization. The cross-linked and now-recrystallized component is indefinitely stable at ambient temperatures; it is provided to the customer in the expanded shape. When the cable joint is later applied in the field, externally applied heat causes the material to shrink by inducing crystalline melting, as it now seeks to regain the shape it had when it was manufactured. The heat shrink component wraps around and hugs the inner joint components tightly (including the connector). Cooling then facilitates the recrystallization process again, as we have seen. The joint now conforms to the shape of the equipment it covers.

The basic unit of radiation dose absorbed is called the rad (which is equal to 100 ergs/gm). Commercial applications require higher doses than rads, and are referred to in terms of grays, with 1 Gray (Gy) being equivalent to 100 rads. Older terminology in the literature refers to Megarads, and 1 Mrad = 10 kilograys. One Gray is equivalent to 1 joule/kilogram.

5.4.4 Silane-Induced Cross-linking

Another method of inducing cross-linking involves "moisture curing." This concept employs organic chemicals called silanes (which are based on silicones) that react with water. In this process, cross-linking occurs at room temperature (but is accelerated by high temperatures). There are several specific approaches that have been applied in the past; this method does not involve the use of a curing tube or radiation equipment. For present-day silane-induced cross-linking technology, the polyethylene insulation has been modified, and is not a homopolymer.

A process called "Sioplas," the first approach to applying this concept, involves grafting a silane monomer onto the polyethylene backbone using a peroxide catalyst (in essence, inducing a special type of branch) and also preparing a separate concentrate (batch) of polyethylene, an antioxidant, and another catalyst (dibutyltin dilaurate). These are then mixed in a specific ratio, extruded, and the completed wire or cable is then immersed in a water tank for a predefined time. Water induces a chemical reaction, leading to cross-linking. The process employs premixed components that require great care in storage. Another process called "Monosil" simplifies the overall procedure by mixing the polyethylene, catalyst, silane, and antioxidant together and then extruding the mix. The curing process is the same. A third silane-based cross-linking process involves the use of a polyethylene–silane copolymer (rather than a grafted material) [12], and allows the wire/cable producer to directly procure the silane-modified polyethylene; this simplifies the handling aspects. Again, the curing step is separate. The chemistry involves formation of ~C–Si–O~ bonds (in contrast to ~C–C~ bonds developed via the other cross-linking methods) [see Figure 5.12.]. The bond strengths are therefore somewhat different for the two types of cross-links, which leads to slightly different physical properties.

The curing process for all the silane technologies described above proceeds at a rate dependent on (a) water diffusion and (b) the insulation wall thickness. The water must penetrate the wall for curing to occur; hours to weeks may be required. Raising the water temperature increases the water diffusion rate into the insulation, and therefore the cross-linking (curing) rate; it is obvious that thinner cable insulation walls will cure more rapidly. However, different wall thicknesses of the same

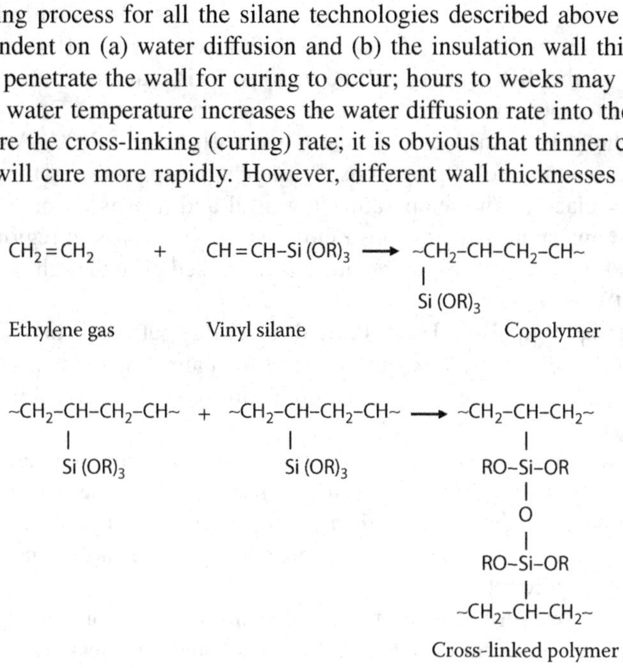

$CH_2=CH_2$ + $CH=CH\text{-}Si(OR)_3$ → $\sim CH_2\text{-}CH(Si(OR)_3)\text{-}CH_2\text{-}CH\sim$

Ethylene gas — Vinyl silane — Copolymer

$\sim CH_2\text{-}CH(Si(OR)_3)\text{-}CH_2\text{-}CH\sim$ + $\sim CH_2\text{-}CH(Si(OR)_3)\text{-}CH_2\text{-}CH\sim$ → $\sim CH_2\text{-}CH\text{-}CH_2\sim$ | RO–Si–OR | O | RO–Si–OR | $\sim CH_2\text{-}CH\text{-}CH_2\sim$

Cross-linked polymer

FIGURE 5.12 Ethylene silane copolymer cross-linking.

insulation material will ultimately cure (cross-link) to the same level, given adequate time. What is unique about the overall silane curing process, and makes it significantly different from peroxide or radiation curing, is that the cross-linking process may continue long after the cable is manufactured.

Due to the wall thickness influence on curing, the silane process has been commonly employed for low voltage (600 V) cables. It has been common to employ an outer layer of "tougher" polymer, such as high-density polyethylene over a silane-cured inner core, for certain applications where outer toughness and abrasion resistance are important.

Silane-induced cross-linking technology has also been applied for the purpose of upgrading aged, water-treed, installed polyethylene or XLPE cables. Here a silicone monomer is incorporated into the aged installed cables system, it migrates through the insulation wall, and then polymerization occurs in-situ (see Section 5.12).

5.4.5 Temperature Influence on Properties

In view of the response of the semicrystalline polymers as a result of temperature changes under normal or overload operation, this subject is treated separately in this section.

Temperature plays a significant role in influencing the properties [11]. One of the properties of semicrystalline polymers that is of relevance for cable applications is that the crystalline regions have a tendency to "separate," or move farther apart, as the temperature is raised. Chain separation converts the crystalline regions into amorphous ones, and is referred to as crystalline melting. Different crystalline regions will melt at different temperatures due to different degrees of "perfection" of the different crystalline regions. This manifests itself as melting over a broad temperature range (starting perhaps at about 60°C), but complete melting of the polyolefin does not take place until about 106°C. At this point, the structure is completely amorphous. While chain separation of the crystalline regions takes place as the temperature is raised and heat is absorbed, the molecules in the amorphous regions also move apart. (Amorphous regions undergo increased chain separation as temperature is increased even though no crystallinity is present.)

These principles apply to HMWPE and to XLPE, but recall that XLPE has a large gel fraction that cannot expand in the same manner as the non–cross-linked (sol) fraction, so thermal expansion is more limited and form stability is better retained.

Clearly, the ratio of crystalline to amorphous regions will change as a cable is thermally load cycled in service. Such chain separation also leads to thermal expansion and is more apparent in cables than in thin sheets. This fundamental phenomenon manifests itself in a nonuniform manner for full size cables under operating conditions (see Figure 5.13).

Focusing first on thermal properties of thin films or sheets of the polyolefin, it is easy to visualize a uniform response across the sample thickness. Here the change in crystallinity as a function of temperature would be uniform at any elevated

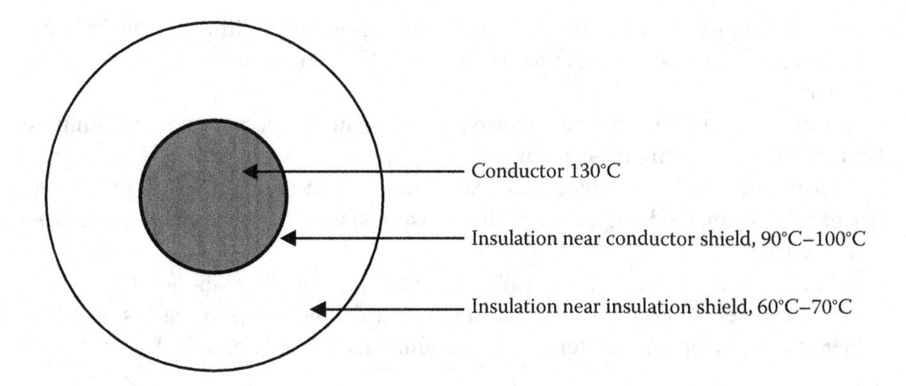

FIGURE 5.13 During thermal loading, cables exhibit thermal gradient across the wall.

temperature. However, such is not the case for cables that have thick walls relative to polymer films or sheets. Due to poor thermal conductivity of polyolefins, the effect of heating (from the conductor outward) on crystallinity is complex. The degree of crystalline melting varies, being greatest closer to the conductor and lesser as one moves toward the outer insulation shield. This is a result of the fact that the change in temperature is time-dependent and a thermal gradient exists across the cable wall. This manifests itself as a gradient of residual crystallinity, which will be dependent on cable wall thickness and varies as a function of time. Given enough time, an equilibrium will be established, but that does not mean that the temperature across the cable wall will be uniform.

Therefore, cables operating continuously at, say, 60°C will have a larger proportion of amorphous regions than a cable operating continuously at 30°C. The thermal gradient across the cable wall means that different amorphous/crystalline ratios will exist at different regions radially away from the conductor. Figure 5.13 demonstrates this principle. In this example, the temperature gradient results from the conductor temperature reaching 130°C. (This "overload" temperature is applicable for XLPE, not HMWPE; however, the objective here is to demonstrate the influence of the thermal gradient on crystallinity rather than the industry specification requirements.) Note that the temperature drops as the distance from the conductor increases.

Therefore, while the fundamental properties of semicrystalline polymers under thermal stress are significant and must be understood, practical application of such principles must be considered within the framework of real-world operating parameters.

This chain separation process leads to property changes such as reduction in physical properties (tensile strength, elongation, and modulus) and also a reduction in dielectric strength. When a cable that has been subjected to thermal overload (heated to elevated temperatures, defined in industry specifications as 130°C–140°C or greater) is later cooled down, the crystalline regions will reform, and the physical properties will now return to being closer to what they were originally. There are fine differences in the nature of the newly formed crystalline regions relative to the original structure, but the nature of these differences is beyond the scope of this chapter.

It is now easy to visualize that thermal load cycling induces even more complex dynamic changes within the cable wall.

Thermal overload is a subject that was of concern in the past; issues relating to thermal expansion (and interaction with accessories) were of concern, but these have been resolved.

At very low temperatures, other phenomena regarding chain motion take place. These phenomena are not as relevant to real-world functioning of cables as is the influence of elevated temperatures, and are summarized in Appendix A.

5.4.6 Role of Antioxidants

Another thermal issue relating to the response of polymeric materials employed as cable insulation (whether or not they are semicrystalline) refers to the role of antioxidants. This is not related to crystalline melting, but to assisting in maintaining properties by interfering with the thermally induced degradation process.

When the cable insulation temperature is raised during manufacturing, it will be susceptible to oxidative degradation. Under these conditions, the polymeric insulation is subjected to temperatures significantly higher than it will ever see in service. As a result, there is virtually always an extremely small amount of oxidation product (carbon–oxygen bonds) in the polyethylene structure that cannot be prevented during the extrusion process. Oxidative degradation, if significant, may be particularly harmful as it can lead to chemical changes within the insulation that introduce more polar materials, which may, in turn, introduce changes in the electrical properties and make the cable more prone to failure during aging. To inhibit this potential degradation mechanism, small amounts of another material called an antioxidant are incorporated into the polymer pellets [13,14]. For medium voltage cables, the common types have historically been either organic amine based or phenolic compounds. Other types, such as phosphites, have also been used. The antioxidant preferentially decomposes in the extruder under the thermal environment and inhibits or prevents decomposition of the polymer. The antioxidant can be considered as a sacrificial component that facilitates high quality product during cable manufacturing.

The antioxidant also resides in the amorphous regions of the polymer at the beginning of the extrusion process, and when crystalline melting occurs as the temperature increases, it can migrate throughout the wall. Upon cooling of the XLPE after manufacture, any unreacted antioxidant would reside in the amorphous regions. Also residing in the amorphous regions will be any antioxidant degradation by-products that are not volatile. This is common for all extruded cross-linked medium voltage insulations.

There are many antioxidants available commercially. In the past, polyethylene-based cables were categorized as "staining" or "nonstaining;" amine types are yellowish in color and phenolics are white. Analytical chemistry techniques can be employed to evaluate antioxidants. Any specific antioxidant employed can generally be determined by obtaining an infrared spectrum of a thin film of the polymer; the antioxidant efficiency, a measure of the amount of "activity" of the antioxidant, can be estimated by an oxidation induction time measurement via thermal analysis. Formation of minute amounts of oxidation in polyolefin insulation during extrusion is not completely preventable, but is kept to an absolute minimum by appropriate processing conditions.

For relatively new or unaged cable systems, very long aging times at high temperatures are required to induce thermal degradation [14]. When changes do occur under dry aging, the property loss is in the direction of: (a) oxidation resistance, (b) physical property change in elongation, and finally (c) electrical properties such as dielectric strength. These changes did not affect the reliability of XLPE cables under thermal overload. [For aged cables that possess water trees, the direction is first (a) electrical properties such as dielectric strength, and then (b) elongation and (c) oxidation resistance.]

5.4.7 Review of Fundamentals

By now, it should be clear that polyethylene is a very complex material. Its apparent simplicity, a composition consisting solely of repeating $–CH_2–$ functional groups, belies the fact that the actual polymer is composed of segments imparting quite different properties. The alignment of some of the chains imparts crystallinity; the nonaligned fractions can coil and are called the amorphous regions; chain branches influence crystallinity; different polyethylenes have different degrees of crystallinity and also different molecular weight distributions. The polymer itself is therefore a "mixture" of different physical segments. That is why it is referred to as "semicrystalline."

The amorphous regions, having "large" distances (on a molecular level) between the polymer chains relative to the crystalline regions, are sites where foreign ingredients can reside. Such foreign contaminants may include ions and cross-linking agent by-products (see Section 5.4). The crystalline regions, where aligned chains are closer together than in the amorphous regions, are the regions that resist foreign ingredient "settlement" and also penetration of most gases. The crystalline regions provide the toughness and resistance to environmental influences. However, without the presence of the amorphous regions, it would not be possible to extrude the polymer into a functional insulation.

One question that arises is what causes different polyethylenes to have different ratios of crystalline to amorphous regions. Any component present in the polymer chain (backbone) that interferes with chain alignment will decrease the degree of crystallinity. Hence, a copolymer of ethylene with propylene, EEA or EVA (as examples), will decrease the number of consecutive methylene links in the chain and increase the tendency for the chains to be more amorphous. This suggests that EPR would be less crystalline than polyethylene and that is exactly the case. The extent to which this occurs is dependent upon the E/P ratio present. As the amount of comonomer increases in the polymer, there is a progressive decrease in crystallinity. One may wonder, therefore, how the "lack" of crystallinity (which imparts "toughness") is compensated for in a completely or almost completely amorphous polymer. The answer is that inorganic fillers need to be added in order to provide the required "toughness" in amorphous insulations (see Section 5.6).

A second factor contributing to influencing the degree of crystallinity is, as noted earlier, the tendency for the chains in homopolymers to have branches. The older conventional high-pressure process of manufacturing polyethylene (from ethylene monomer) facilitates the formation of numerous branches on the backbone. The branches can have different chain lengths themselves. This is depicted in Figure 5.4. It is the number of branches and their lengths in conventionally manufactured polyethylene that influence the tendency to align and, in turn, influence the density and crystallinity. It is for this reason that there are such a large variety of different densities of polyethylene available. It is apparent now that the propylene portion of EPR can be considered to represent very short length branches.

Until the mid-1980s, high molecular weight low-density polyethylene was the commonly used medium voltage insulation material for many users. This polymer has been superseded for new installations, first by XLPE and then by other materials such as EPR and tree-resistant XLPE. Medium- and high-density polyethylenes have traditionally been used as components for cable jackets in low and medium voltage cables.

5.5 TREE-RETARDANT CROSS-LINKED POLYETHYLENE

Over the years, numerous attempts have been made to improve the performance of conventional polyethylene and XLPE in order to attain increased life. In the past, when HMWPE was the insulation of choice, dodecyl alcohol was employed as a tree-retardant additive to HMWPE. When XLPE was first developed, it was reported that one of the cross-linking agent by-products imparted tree-resistant properties to the insulation (acetophenone, which evolves from the decomposition of dicumyl peroxide, see Figure 5.11). Later work focused on improving the tree resistance of XLPE by employing technology not related to the cross-linking agent.

There are several fundamental approaches that are applicable, the basic concepts being either altering of the polymer structure itself, or incorporating additives to a nonaltered structure (or both). Since these approaches would render the polymer more polar (nonpolarity is a desired property of cable insulation, see Chapter 6), a delicate balance of property changes would be required. Such changes would therefore need to first be considered from the framework of the physicochemical goals being sought and what such changes would produce. Hence the following considerations are necessary (as elegantly summarized by Jow and Mendelsohn) [15]:

1. Reducing localized region stresses
2. Introducing a barrier to water tree growth
3. Imparting some level of elasticity to the polymer insulation

Modification of the polymer structure or incorporating superior additives would involve applying these concepts. From this perspective, one could seek:

1. A more polar "backbone" or polar additives that could absorb water and prevent migration toward high stress sites, hence inhibiting water tree growth
2. Employing low molecular weight additives (superior to acetophenone) to prevent treeing
3. Seeking a polymer blend that could facilitate increased elasticity

While specific methods for achieving tree-retardant cable materials is a proprietary arena, one can assume that the principles described here are what has been/is being employed in order to achieve the desired goal [3].

The modified XLPE is usually tested first as a pressed slab (or as extruded miniature thin wall wires) to ensure that improvements have indeed occurred. However, it should be noted that the key method for ensuring that the modified insulation material is indeed tree retardant is to perform tests on completed cable constructions; the accelerated cable life test (ACLT) or accelerated cable water treeing test (AWTT). This means that one ultimate tree-retardant material test is evaluation in conjunction with the shields (and possibly jackets).

The first commercial TR-XLPE material was made available from Union Carbide (now Dow Chemical Company) in the early 1980s. [The original patent literature discloses that a mixture of additives is likely to be present.] Historical information from field aging combined with laboratory data has clearly demonstrated the superiority of TR-XLPE over conventional XLPE.

Table 5.1 provides a simplified summary of comparative material components of conventional and tree-retardant XLPE. The major difference is in the initial ingredients and also potential additional by-products of the cross-linking reaction.

TABLE 5.1
Comparison of XLPE and TR-XLPE Cable Insulation Materials

XLPE Insulation	Tree Retardant XLPE Insulation
XLPE	XLPE
Intentional tree-retardant additive(s)—No	Intentional tree-retardant additive(s)—Yes
Residual amounts of dicumyl peroxide	Residual amounts of dicumyl peroxide
Cross-linking agent by-products	Cross-linking agent by-products
Acetophenone, cumyl alcohol, and alpha-methylstyrene	Acetophenone, cumyl alcohol, and alpha-methylstyrene
Other additives—No	Other additives—may be
Antioxidant	Antioxidant
Antioxidant degradation by-products	Antioxidant degradation by-products

5.6 ETHYLENE COPOLYMER INSULATIONS (EPR)

5.6.1 FUNDAMENTALS

EPR is a copolymer (see Section 5.2.5) composed of ethylene and propylene [2]. The ratio of E/P can vary over a wide range, but in practice, commercial cable insulation is composed of different EPR copolymers having ~50%–80% ethylene. This copolymer has significantly different properties as compared to polyethylene, XLPE, or TR-XLPE, while providing the dielectric characteristics required for low and medium voltage cables. Perhaps most significant is the fact that the propylene segments in the polymer chain interfere with the natural tendency of the polyethylene chains to align and crystallize (see Section 5.2.4). While EPR copolymers are composed of polymer chains of varying molecular weights, and have branches (much like as described for polyethylene), the lack of crystallinity renders it relatively soft and extremely flexible (unlike polyethylene). As with polyethylene, EPR may have "high" or "low" molecular weight molecules and broad or narrow molecular weight distributions. Hence, there are many grades available, each having different properties.

Figure 5.14 depicts one molecule each of ethylene and propylene. A copolymer of this composition would yield a polymer that contains 50% of each monomer. The polymer molecular structure shown in the lower portion of the figure represents an approximation of what the branched polymer structure would look like, with the propylene (methyl groups) branching out from the backbone. These short chain branches are sufficient to prevent crystallization and this EPR copolymer would be soft, flexible, and lacking in stiffness.

It can be visualized that if the ratio of E/P was increased, there would be greater distances between the small branches due to the propylene presence. This is what happens when the ethylene ratio is increased to the ~70%–80% range. In those cases, it is easier for segments of the polymer chains to align and crystallize, and those EPRs can be referred to as being "semicrystalline." Indeed, the higher the ratio, the greater the small segment chain alignment and the greater the crystallinity of the

Copolymer of ethylene and propylene (~50:50)

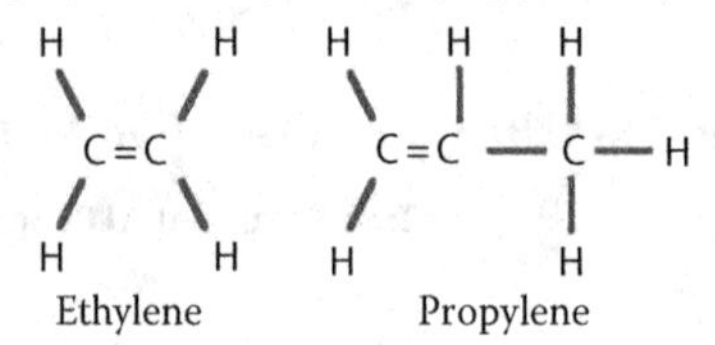

Single molecule structure

- Short chain branches
- Noncrystalline-amorphous structure
- Needs reinforcing fillers (~50%)
- Needs to be crosslinked

FIGURE 5.14 Copolymer of ethylene and propylene (~50/50 E/P ratio).

EPR copolymer. In actual fact, however, even the highest level of crystallinity in practical EPR insulations is at most only about 10% of that which is present in conventional low density polyethylene (LDPE). As a result, the level of "stiffness" in any EPR does not compare with that of any conventional LDPE (or XLPE or TR-XLPE). (Comparison with LLDPE or HDPE is therefore irrelevant.)

EPR materials are known as elastomers. Elastomers are soft polymers and as a class also include natural rubber, synthetic rubber, and butyl rubber (all of which were employed as cable insulation prior to the development of EPRs; see Chapter 1). This class also includes Neoprene, Hypalon, and PVC. Conventional elastomers have significantly different properties than semicrystalline polymers. Reduced crystallinity means that the desirable properties imparted by the crystallinity are minimized or missing altogether: modulus (or "toughness") and high tensile strength of the polymer are examples of properties that have been reduced. This means that a functional insulation that is analogous to HMWPE cannot be produced with a pure, ethylene–propylene copolymer (even if it were cross-linked). A pure EPR copolymer neither has the physical properties required to perform as a functional insulation, nor it could be extruded into cable form. While the electrical properties (e.g., dielectric losses) are low and desired (as with polyethylene), the ability to apply the good properties requires that the EPR copolymer be modified with additives. These additives improve the physical and mechanical properties but are not helpful to electrical properties (such as dielectric losses).

It is important to note that the absence or minimization of crystallinity produces a beneficial side effect; there can be no thermal expansion on heating due to crystalline melting. There will be minimal thermal expansion due to chain motion at elevated temperatures as occurs with polyethylene above the melting range (>106°C).

The required inorganic mineral filler additives in an EPR compound serve various roles; some serve to toughen the system by improving the physical and mechanical properties (and are referred to as reinforcing), while others serve to facilitate processing [16]. Silane-coated calcined clay is the common reinforcing filler employed in EPRs. Also the EPR resin must be cross-linked to render it suitable as an insulation material. If it were not cross-linked, even mineral-filled EPR would not be capable of serving as cable insulation.

In addition to inorganic mineral fillers, many additional components are also required to allow EPR to be useful as insulation. These are mixed with the polymer, and the blend is now generally referred to as a "formulation" or "compound." All the additives (there may be 10 to perhaps 20) are incorporated using special mixing equipment. Therefore, the EPR polymeric material is the only component in the eventual cable insulation. The method of mixing the ingredients together is referred to as a compounding process. It must be noted that many of the ingredients present in EPR compounds that are used as insulations for cables are considered proprietary by some organizations and the exact formulations are not published. However, that is not universal, and Table 5.2 shows typical EP formulations supplied commercially in the past, up until around the late 1990s. The components, approximate amounts, and their purpose are described.

TABLE 5.2
Typical EP Insulation Compounds Employed Until the Late 1990s (in Parts per Hundred)

Ingredient	Amorphous	Semicrystalline
Nordel 1040 (amorphous)	100.0	—
Nordel 2722 (semicrystalline)	—	100.0
LDPE	—	5.0
Zinc oxide	5.0	5.0
Red lead (90% dispersion)	5.0	5.0
Silane treated Kaolin	120.0	60.0
Vinyl silane A-172	1.0	1.0
Process oil	15.0	
Paraffin wax	5.0	5.0
Antioxidant	1.5	1.5
Dicumyl peroxide	3.5	2.6

- *EPR:* The base material that forms a continuous phase in which all the ingredients noted here are uniformly dispersed. The polymer provides flexibility and good electrical properties. All amounts shown are based on 100 parts of the polymer. The two EPR polymers noted have been described as being either amorphous or semicrystalline.
- *Low density polyethylene:* A small amount is added to a formulation of EPR compound employed in "semicrystalline" EPR.
- *Zinc oxide:* A traditional component in EPR and EPDM compounds that improves thermal stability. ZnO was initially incorporated into cable insulation in the past, as it was employed in EPR compounds for automobile tire applications.
- *Red lead:* Lead oxide (Pb_3O_4) serves as an ion scavenger. It improves electrical properties under wet aging. It was incorporated into EPR compounds many years ago to meet low voltage cable test requirements.
- *Silane-treated Kaolin (clay):* This inorganic mineral is a coated calcined (heat treated) clay that serves to improve the mechanical properties of the formulation. Since the rubber material has little or no crystallinity, the filler imparts mechanical strength. The clay is coated with a silane to improve polymer–filler interaction at the interface of the particles with polymer. If the clay was not coated, "gaps" could develop at the interfaces, leading to poor properties. On a microscopic level, the filler particles are large compared to the polymer chain.
- *Vinyl silane:* Additional silane is often incorporated to ensure adequate polymer–filler interfacial contact.
- *Processing aids:* A wax or oil that serves as a lubricant since the inorganic additives are "abrasive" in nature.
- *Antioxidant:* This serves the same role as it does in polyethylene or XLPE; to prevent polymer decomposition during extrusion.

- *Dicumyl peroxide:* This is the cross-linking agent and serves the same role as it does in XLPE. All EPR compounds must be cross-linked to be useful as insulation. The peroxide-induced cross-linking process is exactly the same as for medium voltage XLPE cables (although the extrusion rates may differ).

As noted previously, individual compound manufacturers, or cable manufacturers who perform compounding, may incorporate additional proprietary ingredients in their commercial EPR formulations. The exact nature of these ingredients may vary. Hence, different antioxidants, process oils, or waxes may be used. Some additives can serve multiple purposes such as to further enhance processing, enhance aging properties, or modify dielectric properties. The entire compounding process becomes more complex when one considers other factors, one being the particle size and shape of the clay component.

Mixing of the ingredients in an EPR formulation is an art as well as a science. It is more complex than the technology for mixing ingredients into polyethylene, and the history of mixing ingredients into elastomers predates the use of EPR. The key for achieving satisfactory EPR compounds suitable for later extrusion (along with proper components) is uniformity and control of mixing parameters. Mixing technology employs either a batch process employing a Banbury mixer, or a continuous process employing a Buss kneader. Both technologies facilitate proper mixing of ingredients without degrading the polymer, but the technologies differ.

The Banbury process involves the use of a conveyer belt to carry preweighed ingredients into the mixer and has been described as a high intensity batch mixing process. It is an internal mixer that employs spiral shaped blades encased in segments of cylindrical housings. These intersect to leave a ridge between the blades, which may be cored to provide heating or cooling capability. Under elevated temperature, the heated elastomer is allowed to mix with the additives under controlled conditions. Along with inducing dispersion and preventing agglomeration of the organic and inorganic components, it is equally important to prevent premature cross-linking (commonly referred to as "scorch") and therefore parameters under control include mixing time, temperature, as well as the order of additive incorporation. Rotor design induces satisfactory mixing by employing high shear followed by intimate mixing. The Banbury process has a long successful history of usage with elastomers [17].

The Buss kneader is a continuous single step (not batch) mixing process; additives are incorporated directly into the molten polymer at controlled temperatures. Injection of liquids directly into the molten polymer can be performed. While the batch process predates the continuous mixing process by many years, continuous mixers have been claimed to provide certain advantages, uninterrupted operation being the most obvious: claims include: (a) facilitating improved heat transfer (due to greater surface to volume ratio); (b) availability of interchangeable parts (facilitating increased versatility); (c) holding a portion of the load of a batch mixer at any one time, while producing product at the same rate; and (d) precise temperature control [18].

Regardless of the mixing technology employed, the key involves attaining proper dispersion at acceptable rates without inducing degradation. [It can also be noted that these principles apply to all cable systems requiring mixing; other examples include peroxide addition to XLPE or TR-XLPE, semiconducting carbon black in polyethylene, and also flame retardant additives into low and medium voltage formulations.]

5.6.2 Additional EPR Considerations

1. Sometimes, it is desired to add a third monomer to the ethylene–propylene monomer blend prior to polymerization. These polymeric materials are called EPDM and have been used to facilitate certain (nonperoxide) cross-linking processes. The E refers to ethylene, the P refers to propylene, and the D refers to diene. The reference to M, however, has changed over the years. EPDM has previously been referred to as ethylene–propylene–diene–monomer, and also as ethylene–propylene–diene–methane. Today, M refers to its classification in ASTM standard D-1418. The "M" class includes rubbers having a saturated chain of the polymethylene type.
2. The dienes possibly used in the manufacture of EPDM rubbers are dicyclopentadiene (DCPD), ethylidene norbornene (ENB), and vinyl norbornene (VNB). These dienes typically comprise 2.5 wt% up to 12 wt% of the composition.
3. EPRs that have the higher percentage of ethylene are pelletized; those that have lower levels of ethylene are provided as strips and fed into the extruder. Again, due to their overall lack of crystallinity, cables employing any EP compound will have reduced thermal expansion as well as lower heat capacity and higher thermal conductivity as compared with XLPE (or TR-XLPE). Processing of pelletized (i.e., higher E/P ratio) medium voltage EPR insulated cables is performed on the same equipment used for XLPE or TR-XLPE. Steam or dry curing may be employed, although steam curing has traditionally been more common for EPR.
4. Compounding of EPR (i.e., mixing of all the required ingredients) may be performed by the cable manufacturer, or may be performed by a separate organization that provides the finished compound to the manufacturer. In the past, components in an EPR that were compounded by a commercial supplier had been made public; cable manufacturers may keep their compositions proprietary.
5. While the added ingredients impart the desired mechanical and physical properties to the EP compound, they do have a detrimental effect on initial dielectric properties. The dielectric losses (dissipation factor and power factor) and dielectric constant all increase. This magnitude of the increase will differ as the nature of the EP compound changes (see Section 5.6.3 and Chapter 6).
6. In the past, some medium voltage EPR compounds were provided with carbon black mixed as one of the ingredients in the insulation; this is not the present-day practice.
7. Inorganic clay fillers are an integral part of the EPR formulation. The handling and treatment of the clays have been modified and improved over the years. Clay can be described as being hydrous (possessing moisture, as received), calcined (heat treated), or coated (heat treated and modified

by an organosilane). Present-day EPR compounds employ the latter, while older insulations such as butyl rubber employed calcined clay. Factors influencing clay effectiveness include particle size, (i.e., surface area exposed to the polymer), the degree of interfacial contact with the polymer, and the clay structure itself.

5.6.3 All Ethylene–Propylene Rubbers Are Not Alike

As we have noted, different EPR compounds employed as cable insulation are composed of ethylene and propylene having different ratios in the chain length (also referred to as "backbone"); these differences influence the flexibility. In addition, the compounds contain a plethora of different ingredients varying in nature and concentration, and which are required in order to render the soft EPR polymer useful as insulation. It is not surprising, therefore, to expect that different EPRs would respond differently upon aging, even assuming identical or optimum mixing and extrusion conditions. Indeed, the differences in additives (nature and concentration) along with the EP ratio are what impart different characteristics to EPRs suitable at different voltage levels (say secondary vs. primary or feeder cable). These changes are what render different EPRs suitable to meet different industry specifications.

However, what is being referred to here are differences upon field or laboratory aging for different EPRs employed for medium voltage cables that have met industry specifications. The literature contains many reports of the differences in response on aging. Unfortunately, the literature does not always report the ingredients present in the EPRs studied.

It is worth making a comparison with XLPE, which has only three ingredients: polyethylene, cross-linking agent, and antioxidant. No host of additives to impart "toughness" or compatibility or complex mixing technology is required. One might anticipate that any difference in response on aging of a single compound might be related to the extrusion processing, rather than the nature and mixing of the ingredients (all other things being equal).

A few examples are noted as follows:

Bartnikas [19] refers to different dielectric loss characteristics for three different EPRs as a function of frequency. (Electrical properties of polymeric materials are discussed in Chapter 6.)

The elastic modulus and thermal expansion of different EPRs have been demonstrated to differ as the temperature increases. Differences between four different EPRs in thermal conductivity (resistivity), heat capacity, thermal expansion, and weight loss (on heating) have been reported [20].

A full size cable study by Cable Technology Laboratories provided conclusive evidence; in that study, five different commercially available full size EPR cables were aged both in the field and in the laboratory under both normal operating stress and under accelerated voltage stress [21].

Significant differences were observed. To demonstrate the point, Figures 5.15 and 5.16 show the responses of the different EPRs. In Figure 5.15, the EPRs aged for 24 to 48 months under normal operating stress show that the retained dielectric strength varies extremely widely; from 40% to 100% between the different EPRs.

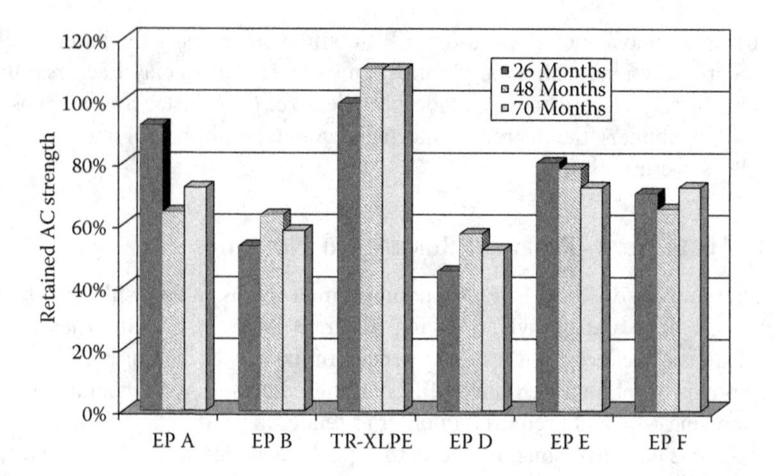

FIGURE 5.15 AC breakdown stress retention of EPR cables laboratory aged at 1.0 Vo.

(Changes in aging response up to 70 months are also shown.) Similar differences are shown in Figure 5.16 for the same cables aged under accelerated voltage stress in the laboratory.

5.6.4 Discharge Free versus Discharge Resistant

A good example illustrating the fact that all EPRs are not alike is represented in the manner by which long life is achieved. Up until now, we have focused on technology designed to prevent partial discharges (see Chapter 6) in the insulation wall and maintain low dielectric losses. This is achieved by ensuring that the insulation is clean (no foreign contaminants), does not possess voids, and is of low polarity. Another goal

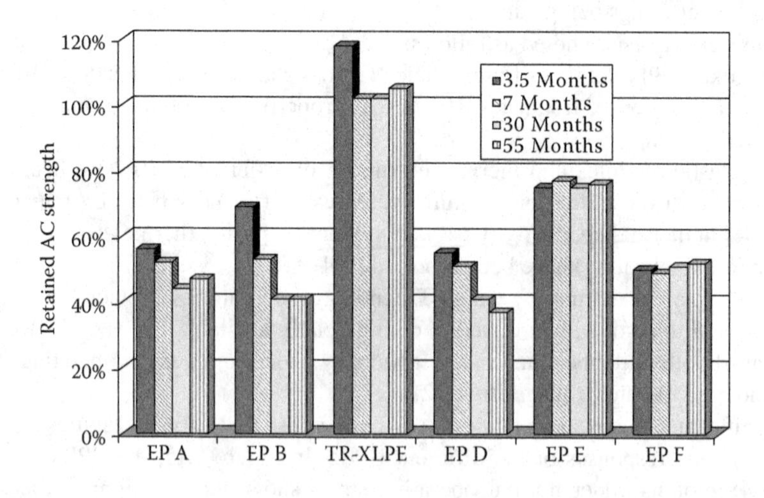

FIGURE 5.16 AC breakdown stress retention of EPR cables laboratory aged at 2.5 Vo.

is to ensure that these properties are maintained upon aging. This philosophy applies to all polyolefins (HMWPE, XLPE, TR-XLPE, as well as EPR). In the case of EPR, where we have seen that there are many required components that are absent in the semicrystalline polymeric insulations, the use of additives that would increase the dielectric losses is intended to be minimized, as some increase in losses is unavoidable. The goal of this approach is to achieve EPRs that are discharge free.

Another approach in developing long-life EPR formulations turns the methodology around. The alternate approach assumes that it is not consistently possible to achieve the goals described above after significant aging, and seeks to provide an insulation that operates reliably under partial discharge conditions (i.e., is not degraded on aging); hence it is considered to be discharge-resistant rather than discharge-free. Therefore, the additives in the EPR polymers that follow this approach are different in nature.

Although the specific additives employed to achieve this goal are proprietary, the principles involved can be noted. Such additives can serve the role of: (a) imparting high dielectric constant (see Chapter 6) to the insulation; and where the additives can also (b) migrate toward any voids that develop upon aging and can increase the surface conductivity of the voids; and (c) trap high energy electrons that could cause partial discharge and thus dissipate the energy as heat. This philosophy intentionally increases losses to improve service aging performance. The choice of additives to the EPR and the method of compounding (to achieve this goal) is made by the cable manufacturer, and the insulation has been described as a "proprietary EPR/EPDM based compound" [22].

Clearly, the ability to incorporate a wide variety of additives into an elastomer like EPR increases the capability for designing the compound to meet specific goals. Ultimately, a goal is to compare the semicrystalline XLPEs and the EPRs; a fundamental paper in this regard has been provided by Eichhorn [23].

5.7 SHIELD MATERIALS

5.7.1 Overview and Role of the Polymer

Polymers employed in formulations used as cable shields are based on ethylene copolymers. They may be copolymers with propylene or with other monomers such as EVA, EEA, or butyl acrylate (EBA). In essence then, they are "elastomers." Each of these individual comonomers imparts different properties to the copolymer employed in the shield. As with insulation, one must consider molecular weight and molecular weight distribution issues, extrusion characteristics, and with cross-linked insulations the shields must also be cross-linked; however, crystallinity is not an issue. Carbon black is present (see Section 5.7.2) and influences both processing and properties.

Strand shields are employed over the conductor and under the insulation, and insulation shields are extruded over the insulation itself and interface with the concentric neutral or metallic tape and (depending upon construction) any jacket that may be present. These shields must be semiconducting in nature, as they provide a "stress gradient" between the insulating and conducting layers, and the intention is to achieve a discharge-free system. While the semiconducting (electrical) properties of materials employed for either strand or insulation shields are similar, their physicomechanical property requirements differ. Hence, the insulation shield must be strippable (for ease of removal at the region where the joint or termination must be installed), yet the strand shield must exhibit the opposite property; it must adhere strongly to the insulation surface at the conductor region (to prevent void formation and water penetration). The modulus and tear strength of the insulation shield are critical parameters, as well as the cohesive strength so as to not allow "pick-off" during stripping, and this is controlled by the copolymer nature, additives, and processing.

5.7.2 Role of Carbon Black

Semiconducting properties are achieved by incorporating a conducting carbon black into the copolymer. By employing the appropriate carbon black and proper handling, the carbon black–filled polymer is rendered semiconducting in nature. Carbon black terminology refers to "furnace black" or "acetylene black;" furnace black is prepared by partial oxidation of oil or gas, while acetylene black is prepared by decomposition of acetylene at elevated temperatures [24]. Along with conductivity, the required characteristics of the carbon black include surface smoothness, particle size control, proper dispersion in the polymer, and purity.

It has been known for a long time that the properties of carbon black–filled elastomers are strongly influenced by numerous factors: (a) the nature of the carbon black, (b) the concentration of black employed, (c) the type of base polymer, and (d) the cure (cross-linking) system used [25]. The latter is not a present-day issue, but the other points remain relevant.

Conductivity is achieved when the carbon black forms aggregates within the polymer matrix, leading to a continuous path that allows electron flow. The details of this technology are beyond the scope of this chapter, but it can be noted that electron "tunneling" from one aggregate to another leads to the conducting properties required in the semiconducting layers. The eventual electrical properties of the semiconducting shield material are controlled by the structure of these aggregates and by the size of the carbon black particles comprising the aggregates. Figure 5.17 demonstrates how this relationship influences physical and electrical properties [24].

Carbon blacks with high structure tend to impart greater conductivity and hardness–structure referring to the ability of the carbon black to aggregate into clusters (as shown in Figure 5.17). The structure can be disrupted when incorporating the carbon black into the elastomer, the extent being determined by the type of black and the mixing process, so proper control is a key issue. The carbon black concentration in a semiconducting shield copolymer will be dependent upon the type of carbon black employed; it may be in the 12%–30% range.

Surface smoothness is necessary so that there are no protrusions (due to mechanical imperfections or poor dispersion) at the region that interfaces with the insulation; these can represent high stress sites from which water tree growth can develop in wet environments. When present at the semiconductive shield interface, protrusion effects are influenced by the carbon black geometry (the tip radius or sharpness) and their height. Hence, the degree of dispersion of the carbon black and its "fineness" in the completed cable are key factors in achieving surface uniformity and smoothness.

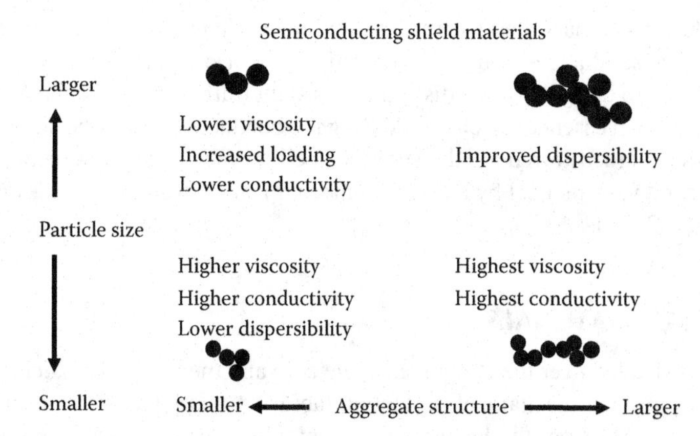

FIGURE 5.17 Effect of carbon black particle size and agglomerate structure on properties.

Proper dispersion of the carbon black in the polymer along with the need for proper mixing and handling is an obvious requirement.

Impurities are a major concern; after preparation, the carbon black may possess small quantities of moisture, sulfur, and inorganic salts. Sulfur can harm the electrical properties, and most of the inorganic salt content comes from the water used for quenching and pelletizing during the manufacturing process [26]. Inorganic ions had been a significant concern in the past. The furnace blacks commonly employed until about the mid-1980s (due to the nature of the processing) had higher levels of inorganic ionic contaminants as compared with acetylene blacks; this led to increased failures due to water treeing. More recently, furnace blacks of increased cleanliness have been employed; this is noted here as many older installed cables possess such older black in their shields.

It is also possible to incorporate additional additives. For example, controlled strippability of insulation shields (for ease of removal) may be achieved with additional additives, but such additives would not be used for conductor shields where strippability is not desired.

Finally, carbon black–containing shield materials must be extrudable and crosslinkable as easily as the insulation, and under such conditions that "true triple" extrusion can be effectively performed with the semiconducting electrical properties remaining in the desired range. Factors that influence the conductivity are the carbon black type and concentration, processing conditions, the cure system, and other components present. The base polymer type is significant; e.g., for strand shields, combinations that have been employed are: (a) furnace black with EVA or EEA copolymers, and (b) acetylene black with EVA or ethylene butyl acrylate (EBA) copolymers.

5.7.3 Nonconducting Shield Materials

It is possible to design shield materials that are nonconducting (rather than semiconducting). The polymer employed is based on EPR and the additives that are incorporated into the EPR are proprietary. The required properties are achieved by ensuring that the additives impart the property of high dielectric constant (see Chapter 6) to the shield. These shield compounds are employed in conjunction with insulation materials that are designed to be discharge-resistant (rather than discharge-free, see Section 5.6.3.1). In essence, when considering the conductor shield, the stress control layer that is designed to reduce the stress at the interface of the conductor and the insulation achieves this goal by a different mechanism than do the shields discussed in Section 5.7.2 [27].

5.8 JACKET MATERIALS

Jackets are extruded over the completed cable core and their metallic shields or neutral wires to provide mechanical protection (abrasion and puncture resistance) and to protect the cable from the local environment. Objectives also include protecting the neutrals from corrosion, and specifically, to prevent moisture penetration. Good environmental stress-crack resistance is also a requirement. Common jackets are insulating, and ideally, they will aid in keeping not only water out of the insulation but also contaminants and foreign ions. Jacketing materials have varying properties that are controlled by their molecular structure and compound ingredients.

Insulating jackets contain small quantities of carbon black that is not conducting (i.e., different in nature from that employed in shields).

For medium voltage cables, several polyethylene types have been commonly employed as jacketing materials:

- LDPE
- MDPE
- HDPE
- LLDPE

It is clear by now that the differences are in the degree of "chain alignment" that takes place, and that HDPE is more crystalline than is LDPE (see Section 5.2.4); the higher the degree of crystallinity, the higher the density. As the name implies, MDPE is "in-between" the other polyethylene grades with reference to the degree of crystallinity. HDPE is manufactured by a different catalyst and pressure control as compared with the LDPEs and MDPEs employed for cable applications. It possesses a higher degree of crystallinity and greater toughness and abrasion resistance, but the extrusion process is more complex. LLDPE is manufactured in such a manner that it approaches HDPE in properties, but the ethylene monomer is polymerized by a different method than conventional LDPE, employing catalysts, and is more easily processed as compared with HDPE. Properties considered during processing and application of jacket materials include abrasion resistance, stress-crack resistance, melt flow rate, and prevention of shrinkage. In practice, LDPE was commonly employed in the past and has gradually been replaced by LLDPE. HDPE has been used in special applications as determined by the user.

It should be noted that VLDPE (density range of 0.880–0.915 g/cc), a linear metallocene-produced copolymer of ethylene and alpha-olefins such as 1-butene,

1-hexene, or 1-octene is commercially available. It possesses short chain branches much as does LLDPE.

More recently, Polypropylene has come into usage as a cable jacket, as a result of new catalyst technology facilitating polymerization. Polypropylene is "tough" but has a significantly different architectural crystalline structure from polyethylene and as a result, possesses lower density. Because of its increased toughness, it is a good jacket material, comparable in some respects to HDPE despite the lower density. Polypropylene extrusion requirements are also different.

Another way of viewing jackets is that they may be thermoplastic or thermoset. The polyethylene and polypropylene jackets discussed above are all thermoplastic in nature. XLPE (thermoset) can be employed as a jacket (but this is more common for low voltage cables). XLPE jackets are based on low density polyethylene; cross-linked HDPE, if desired, is more difficult to extrude, employing peroxide curing due to the higher temperature requirements to induce melting, resulting from the higher level of crystallinity present.

Common thermoplastic jackets based on PVC have been employed for cables; while PVC jackets are relatively poor moisture barriers compared with the polyolefins, they do impart superior flame resistance to the cable construction, a requirement for many low voltage cable constructions. PVC by itself is actually a very rigid material, and flexibility is imparted when plasticizers (softening agents) are added. Typical PVC plasticizers could be based on alkyl phthalates; other additives are also required in a PVC compound. The fact is that plasticized PVC, while providing some degree of protection against dig-ins and corrosion protection, not only has poor resistance to water migration as noted, but also has poorer abrasion and tear resistance compared with the polyolefins. PVC is now less commonly employed as a cable jacket for underground distribution cables for these reasons, even though it is relatively inexpensive compared with other materials. For low voltage cables, it has been common to use other chlorine-containing elastomeric materials such as Neoprene (polychloroprene) or Hypalon (chlorosulphonated polyethylene [CSPE]).

The relative moisture resistance of the different polyethylene types, which vary with the degree of crystallinity, a typical PVC and CSPE, are shown in Table 5.3.

Jackets referred to as "low smoke zero halogen" (LSZH) or "low smoke halogen free" (LSHF) have been available, and newer jackets of this type have been and are

TABLE 5.3
Relative Moisture Vapor Transmission Rates (MVTR) of Jacket Materials

Material	Density (g/cc)	MVTR (ASTM E96 units)
LDPE	0.920	1.16
MDPE	0.930	0.51
HDPE	0.948	0.32
LLDPE	0.920	0.74
PVC	—	10.0
CSPE	—	12.0

being developed; their usage for low voltage cables has been increasing. As the name suggests, their objective is to achieve flame resistance without employing halogens (fluorine, chlorine, or bromine), while maintaining all the other desired properties. Halogen-containing jackets, while preventing burning, decompose during the process to yield toxic gases (hydrogen chloride being an example) and many different formulations of these types of jacket materials are available.

Extruded colored stripes of defined dimensions are employed for ease of identification.

The concentric neutral that is protected by the jacket may be either under the jacket in the cable construction or embedded within the jacket wall.

For some applications, semiconducting jackets have been employed. The purpose of a semiconducting jacket is not only to deter corrosion, but also to ensure adequate grounding of the cable during faults or when lightning occurs. Such jackets reduce the neutral-to-ground impulse voltage (via improved grounding efficiency) and improve the resistance to puncture. These jacket materials, based on polyethylene copolymers, are rendered semiconducting by incorporating conducting carbon black into the polymer. As with any additives to polymers for cable applications, proper dispersion of the carbon black throughout the matrix is vital in order to provide uniform semiconductivity from the neutral. The same particle-to-particle contact technology required for semiconducting shields is applied to semiconducting jackets. However, the carbon black in semiconducting jackets differs in nature and amount from that employed in insulating jackets. (Grounding is discussed in Chapter 16.)

Therefore, numerous options exist for the choice of jacket materials.

5.9 WATER BLOCKING TECHNOLOGY

The discussion so far has focused primarily on keeping water out of the insulation to ensure long cable life. Another approach involving polymer materials technology is designed to protect the insulation if, despite all precautions, water does enter. One method is to incorporate a "filling" material between the strands as part of the cable manufacturing process (at the time of stranding). [The filling compound must not migrate during extrusion, which could cause an uneven surface between the strand and the insulation.] Blocking the strands in this manner of filling the conductor core is intended to prevent longitudinal migration of water that has entered the cable. Several approaches to applying polymer technology have been employed to achieve this objective. As an example, an old patent (1985) refers to "a cable fill composition comprising a paraffin oil, a styrene-ethylene butylene styrene block copolymer and linear polyethylene wax having an average molecular weight of about 1,000–1,500" [28]. This item is cited to demonstrate that the concept has a long history and that the potential polymer technology employed to achieve this objective can be quite complex.

It is also possible to employ water-swellable powders that do more than impede motion of the water. When these powders come in contact with the water, they interact and trap the water, expand, and swell to form a gel; when the water is trapped, migration is prevented. To trap the water, the powder must possess chemical functional groups that will interact with the liquid. These groups must be polar in nature (see Chapter 6). One functional group on the polymer chain that serves this role is referred to as "carboxyl group," but there are many other possibilities. Such powders may be incorporated into yarns or tapes, which are then incorporated into the cable construction during manufacture.

[These components may be applied under the jacket also. One possible approach is to apply both techniques; apply strand filling compound at the conductor interstices and also employ a water-swellable powder or yarn under the outer cable jacket.]

Water blocking technology is suitable, obviously, only for stranded (not solid) conductor cable constructions. While their application in cables certainly must meet the industry standards and specifications, the relatively complex polymer technology employed for this application must consider and meet the criteria noted as follows. In this summary, polymer materials aspects interface with the cable construction parameters.

Properties must be retained under thermal stress, especially thermal overload. This would become a consideration that is related to operating voltage stress, less so for URD as compared with feeder cables.

- Gel viscosity might be anticipated to drop on as temperature increases, and requires consideration regarding functionality.
- Thermal overload requirements that apply to other polymeric cable components must apply to the filling compounds; i.e., there should be no loss of functional capability at a temperature lower than the maximum conditions allowed by industry specifications.
- Components should not interfere with conductor geometry.
- No loss of functional properties should occur under voltage-thermal load cycling condition.
- Additive must not possess contaminants.
- Trapped water must not be released at a later time during cable field service aging.

5.10 PAPER INSULATED CABLES

5.10.1 Fundamentals

The oldest successful type of insulation used for power cables are systems composed of paper and dielectric fluids. Paper insulation is based on the natural polymer cellulose; paper strips composed of cellulose are wound around the conductor and impregnated with the dielectric fluid. The fluids may vary in chemical composition, and both natural and synthetic fluids have also been used. This section reviews the fundamentals of paper-based insulations.

Transmission cables based on paper are called high pressure fluid-filled (HPFF) or low pressure fluid-filled (LPFF). Medium voltage paper-based cables are referred to as PILC cables. Hence, these cables differ in several major ways as compared with extruded cables: they possess numerous layers of paper tapes (both insulating and semiconducting) that are wound around and surround the conductor, are impregnated with the dielectric fluid (as compared with continuous, solid extruded synthetic polymer insulating and semiconducting layers) and are covered by an outer lead sheath.

Paper for electrical cable insulation is derived from wood. Wood consists of essentially three major ingredients; cellulose (about 40%), hemicellulose (about 30%), and lignin (about 30%).

Cellulose is the component of interest as an insulating material and must be separated from the other components; this is performed by bleaching with sulfates or sulfites by the paper industry. The hemicellulose, a nonfibrous material, is more polar than cellulose, more lossy, and is not a useful insulation material; however, a very low level of hemicellulose may be acceptable in the final material, and some small quantity can remain after the bleaching process. (It is notable that cotton is essentially pure cellulose; if cotton were used as a source for paper, no hemicellulose would be present.) Lignin is an amorphous material, serves as a "binder" for the other components in the wood, and is also removed for cable insulation applications. The process of extracting cellulose by converting wood into a useful insulation material is beyond the scope of this chapter, but it should be noted that the paper manufacturing process is mature and has been used for over 100 years. A description is provided in Ref. [29]. The theory and practice of high voltage cables based on paper insulation were elegantly described long ago by Dunsheath [30].

Two methods of depicting the complex chemical structure of cellulose molecules are shown in Figure 5.18a and b [31]. (The latter depicts the folded chain nature of the molecule.) In either depiction, there are repeating chain units, as occurring in polyethylene or EPR; however, the chemical structure of this natural polymer is composed of more complex saturated cyclic components (five carbons and one oxygen in a ring), which are absent in the olefin polymers. The repeating chain units indicate that the cellulose has a high molecular weight. In contrast to polyolefins, it is more common to refer to the molecular weight of cellulose in terms of degree of polymerization, or "DP," rather than weight average or number average (see Section 5.2).

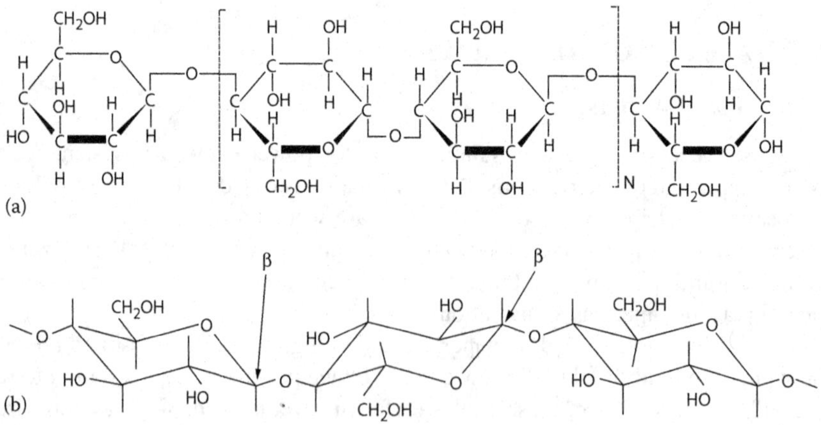

(a)

(b)

FIGURE 5.18 (a) Chemical structure of cellulose: Depiction 1. (b) Chemical structure of cellulose: Depiction 2. β refers to carbon–oxygen bond.

The DP is representative therefore of the number of individual cellulose molecules in a chain and could be considered to be related to Mn (although the industry does not refer to it in that manner). The chain lengths (DP) of cellulose derived from wood varies and may be in the 300–1,700 range; however, the actual DP depends on which part of the wall the cellulose fibrils are located (see following paragraph), and may be significantly greater.

While the chemical structure is understood, the cellulose fine structure architecture in wood layers has still not been unequivocally determined (despite years of study). What is known are the following details: Cellulose does not exhibit branching; the molecules are extended and mostly linear. The hydroxyl groups from one chain form hydrogen bonds with oxygen from the same or a different chain. [Hydrogen bonds are those where the hydrogen molecule is essentially "shared" by two different oxygen molecules.] This interaction imparts a degree of "firmness" to the cellulose structure and allows the formation of clusters called microfibrils; the latter in turn form bundles of parallel microfibrils and these larger structures are referred to as fibrils. Such fibrils (visible under the microscope) display different orientations in different layers of the wood. The most common naturally formed cellulose is referred to as "cellulose I" (of which two forms exist). X-ray diffraction studies have shown that crystalline imperfections do exist.

Cellulose can be swollen by some solvents; a distinction is made between intercrystalline swelling and intracrystalline swelling. The former results from the action of water on the fibers and leads to an increase in the cross-sectional area. Soaking of cellulose in water leads to hydrogen bond formation on the fiber surface. Removal of the water by drying/heating causes the bond strength to increase. This is the physicochemical basis for the conversion of cellulose to paper, hence allowing development of a strong sheet or fiber mesh. Conversion of the fibrils into insulating paper is performed by conventional processes employed in the paper industry and, as noted previously, this natural polymer is manufactured as a tape that is wrapped around the cable core. Therefore, there are butt spaces that overlap in the cable construction. The cellulose in tape form may be employed in various thicknesses (depending on the application) and the fluids employed may vary in nature (chemical structure) as well as molecular weight.

The fluid serves to impregnate the butt spaces and ensure absence of voids. The dielectric fluids themselves are composed of carbon and hydrogen $\sim CH_2 \sim$ groups, much like polyethylene, but their molecular weights are very much lower, so these are liquids that exhibit varying degrees of viscosity. The dielectric fluids are essentially hydrocarbon in nature, but the specific fluids employed have changed over the years. This means that the chemical structure of the fluid employed has been changed (in contrast to the cellulose portion of the paper/fluid combination). Fluids employed have been pine oil, mineral oil, paraffin and naphthenic oils (all of which possess different organic chemical structures), and synthetic butenes (also referred to as polybutene or polyisobutylene). Different polybutene fluids may possess different degrees of branching, hence affecting viscosity. In addition, rosins obtained from trees have mixed composition. The goal of all the fluids is to impart low dielectric losses.

Clearly, the fluid technology for paper cable has varied significantly over the years. Despite the changes, the commonality is that the final fluid remains basically hydrocarbon in nature (even though some of the fluids may contain more polar ingredients than others, and they may possess different antioxidants). The fact is that each modification can be expected to lead to different responses under voltage-thermal stress upon aging (see Section 5.10.2), leading to different long-term aging effects. Polybutene fluids are in common use for present-day PILC cables.

5.10.2 Aging Effects

From an insulation materials perspective, one of the most significant ways to understand the differences between the natural polymer cellulose and the synthetic polyolefins like XLPE or EPR is to compare their behavior upon aging. Aging-induced degradation and loss of property integrity of polyolefins are covered in Chapters 6 and 19. Paper/fluid systems are reviewed here.

5.10.2.1 Cellulose

Cellulose can be considered as a "polyhydric" alcohol. (A simple alcohol would be ethanol.) The oxygen bonds connecting the cyclic structures are referred to as "1,4-glycosidic" bonds (Figure 5.18). These are the regions where the cellulosic molecule is susceptible to cleavage on aging. Water presence can cause breaking of these bonds, leading to a reduction in DP and thus loss of physical, mechanical, and electrical properties (e.g., reduction in tear strength or tensile strength and increase in dissipation factor); water may enter the construction through cracks in the lead sheath that is present over the cable core. Once degradation starts, more water is generated as a result of the chemical degradation of the cellulose molecule. The process is accelerated by elevated temperature. High temperature in the absence of water can also induce such degradation but the temperatures required in that situation are very high.

This degradation mode is significantly different from polyolefins such as XLPE, where the presence of water and voltage stress induce water tree formation and oxidizes the main polymer chain, but does not lead to a significant reduction in molecular weight by chain cleavage.

5.10.2.2 Dielectric Fluid

Under combined thermal/voltage stress, the fluid may undergo degradation. The molecules may be attacked by electrons, leading to free radical formation (see Chapter 6). When this occurs, two events are possible: (a) the fluid molecules may undergo chain cleavage, such degradation leading to a decrease in molecular weight (and viscosity); or (b) the free radicals may combine and lead to an increase in molecular weight. If the latter occurs, the viscosity will increase. As this process continues, and molecular weight continues to increase, the higher viscosity liquid is converted to a wax. Therefore, for our purposes, a wax can be visualized to be a "high molecular weight solidified oil," but not as high in molecular weight as polyethylene. Waxes are intermediate in molecular weight, between those of viscous liquids and those of polymers like polyethylene. It is not uncommon to find wax upon autopsy of failed paper/fluid cables.

5.10.2.3 Paper–Fluid Combination

The degradation of the cellulose and the fluid/wax can occur simultaneously, and it is also possible for free radicals to attack the cellulose also and for synergistic effects to occur. These phenomena render attempts to isolate mechanisms of degradation very difficult on a molecular level. However, on a larger (macro) scale, another effect is very significant regarding the mechanism of degradation involving both components.

As paper-insulated cables age under load cycling and voltage stress, the fluid migrates into and out of the tape butt space regions as the temperature changes. Such migration can lead to voids developing as the fluid migrates. Voids are regions where the fluid is absent and air is present, and therefore are regions susceptible to partial discharges (see Chapter 6), which causes degradation. However, the fluid migration back into the butt space as a result of the load cycling process can render void presence as a transitory phenomenon. If the voids are large enough, even if temporary, they could lead to reliability issues upon continued aging.

The change in the nature of the fluid on aging, as described previously, can influence the migration of the fluid; if the fluid viscosity increases on aging, the ability to migrate under thermal load cycling becomes slower and more difficult. If wax develops, migration can be hindered even further, as waxes can restrict or prevent oil flow during load cycling, allowing detrimental microvoids to remain longer in butt spaces.

It is also possible that (some of) the wax may dissolve in the liquid portion of the fluid that did not degrade, hence increasing the viscosity. Also, wax that exists in the solid state at ambient temperature may soften and melt under high load cycling temperatures. Therefore, what is observed as a wax during an autopsy (always performed after the cable has cooled) may have been a viscous liquid during operating conditions. Another possible consideration is the potential role of water, if present, in displacing the fluid or wax.

This situation is obviously quite complex, and all these factors influence what happens in the butt space region of the paper tape–fluid insulation.

It is clear that paper-insulated cables are more complex than are their synthetic polyolefins counterparts; these issues do not exist for extruded solid dielectric cables. Both the fluid and the cellulose may degrade.

The fact is, however, that paper-insulated cables have an excellent history of reliability, and this is primarily attributed to the presence of an outer metallic lead sheath on these cables which prevents the incursion of air or water as long as the sheath retains its integrity.

5.10.3 Test Methods

A variety of tests are often employed to assess the state of the degraded paper-based cable as a result of aging. Some of these tests are noted here, as examples.

Moisture in tapes: As the paper portion of the cable degrades, moisture is evolved and it may be retained in the tapes. Measurement provides a qualitative indication of degradation.

DP: As noted above, the cellulose chain cleaves on aging and the DP is reduced. The technology involves dissolving the cellulose in a solvent and measuring the viscosity of the solution. Various methods have been applied over the years as cellulose is not easily dissolved and care in interpretation is required, as it has been reported that some measurement methods have caused depolymerization [32].

Fold endurance: Folding endurance represents the ability of paper to withstand multiple folds before it breaks. It is defined as the number of double folds that a strip of specific width and length can withstand under a specified load until breaking occurs. Folding endurance is useful in measuring the deterioration of paper upon aging. Long and flexible fibers provide high folding endurance.

Tensile strength: Tensile strength is a measure of the force required to produce a rupture in a strip and can be measured in the machine and cross machine directions. Tensile strength is indicative of fiber strength, bonding, and length. It can be measured when the paper is either dry or wet; both may drop on aging, the latter more. The wet tensile strength is less than the dry tensile strength.

Dissipation factor of fluid: As the oil degrades, it becomes more polar and the losses increase (see Chapter 6). Also, any evolved moisture that is "held" by the fluid will influence the dissipation factor. This is a common measurement applied to aged paper-insulated cables.

Additional tests on fluids: Diagnostic tests are available that can be applied to assess the specific chemical changes resulting from aging-induced degradation. One useful method is infrared analysis to determine chemical changes resulting from oxidation; degraded, oxidized regions absorb light in the infrared region to reveal their exact nature. Ultraviolet analysis is less common.

One further point is to be noted with reference to paper/fluid filled cables. It is possible that additives may be present that are not known and not addressed by industry specifications, but that may influence aging [33]. When this occurs, projections become difficult.

The discussion on aging reviews the significant differences between paper insulated cables and extruded cables with reference to: (1) construction, and (2) diagnostic testing after aging. One further point is notable to provide emphasis. Because of the significantly different (a) chemical structure of paper (cellulose molecules) as compared with extruded dielectric polymers, and (b) the paper cable construction itself (i.e., the presence of fluid to fill any voids that may develop in the butt gaps), it should not be surprising that each of these fundamentally different constructions will respond differently to external stresses. This is demonstrated by direct current (DC) high potential (HiPot) testing (see Chapter 18) that has traditionally been applied to installed aged paper cable systems to remove the "weak link" at the time of choosing (during the test). PILC cable of high integrity is not affected by application of DC when applied in accordance with industry guidelines, but aged PILC cable is induced to fail; the test can be employed to address cable integrity without harming the cable that has not degraded. The advantage is that the user can remove the weak link at the time of choosing, not during a crisis situation.

However, XLPE cables, due to the development of water trees, reduced dielectric strength and the tendency to "trap charges" (as well as the absence of any low molecular weight fluid to fill the voids), may be susceptible to harmful effects when subjected to DC HiPot testing. Much depends on the degree of degradation that the XLPE has undergone prior to performing the test. The test may remove the weak link as desired, but it may not. The problem is that the DC HiPot test may shorten the life of the XLPE insulated cable that has not failed during the test [34]. This does not happen with PILC cables.

The point here is that the response of electrical insulation materials to outside stresses in their application environment, both natural and imposed by the user, is intimately related to the physicochemical structure of the insulation material and the construction.

5.11 LOW VOLTAGE POLYMERIC INSULATION MATERIALS

5.11.1 Insulation Materials

The discussion to this point has focused primarily on polymeric materials employed for medium (and high) voltage insulation. Such materials must function under conditions where voltage stress is a significant parameter influencing reliability. For insulation materials operating at lower voltage stresses (e.g., 5 kV or less), thermal stress becomes a more significant factor in influencing reliability. Polymers employed in this voltage range include polyethylene and XLPE, but a variety of elastomers beyond EPRs may also be employed. These elastomers are reviewed in this section (the polyolefins having been reviewed earlier). Like all rubbers employed for insulation, these polymeric insulation materials must be compounded with numerous ingredients (as described in Section 5.6 for EPR) and they must be cross-linked.

An advantage of compounding polymers for low voltage applications is that it is possible to incorporate additives that enable a variety of specific characteristics to be met; an example is flame retardancy. It is possible to incorporate additives into the polymer formulation that will render the insulation more resistant to burning, yet still meet the required electrical properties. This is because the electrical property requirements for low voltage cable insulation allows greater flexibility, as operating voltage stresses are lower. This capability leads to a variety of different industry specifications designed to meet specific needs. This capability applies to polyethylene and XLPE, as well as to the elastomers. Another characteristic of this class of materials is that they find application as either insulation or jackets.

Elastomeric materials in this category include Neoprene, Hypalon, SIR, chlorinated polyethylene (CPE), and also PVC. (The historical development of Neoprene and Hypalon is discussed in Chapter 1). Table 5.4 shows the chemical structure of these materials.

Neoprene: Neoprene is also referred to as chloroprene; it is an elastomer derived from polymerization of the monomer 2-chloro-1,3 butadiene (recall that polyethylene is derived from the monomer, ethylene). The chlorine in the polymer backbone imparts the property of flame resistance. Neoprene resembles natural rubber (see Section 5.1.6) in many respects; mechanical properties such as tensile strength, elongation tear, and abrasion resistance, deformation and flex cracking resistance, and retention of properties at low temperatures. It is useful over a wide practical temperature range (–55°C to 90°C) depending on the specific formulation used (there are a number of grades available). This insulation material also possesses heat, ozone, corona, weather, chemical, and oil resistance as well as flame resistance. This polymer is considered a "general purpose" elastomer and these characteristics make it useful for wire insulation and cable jackets.

TABLE 5.4
Chemical Structure of Low Voltage Polymer Insulation Materials

Polymers Plus Monomers of Neoprene/Hypalon	Structure
Neoprene (monomer) 2 chloro-1,3 butadiene	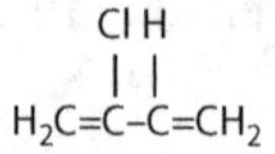 $H_2C{=}C(Cl){-}C(H){=}CH_2$
Hypalon (monomer) Chlorosulfonated polyethylene	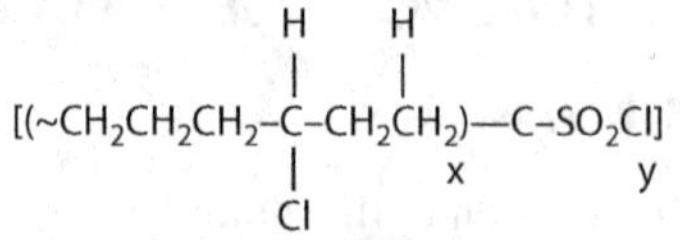 $[(\sim CH_2CH_2CH_2{-}C(H)(Cl){-}CH_2CH_2)_x{-}C(H){-}SO_2Cl]_y$
Chlorinated Polyethylene (CPE) [depiction/ chlorination is random]	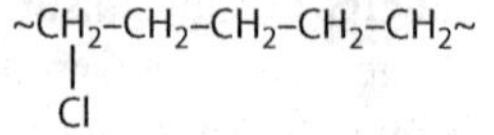 $\sim CH_2(Cl){-}CH_2{-}CH_2{-}CH_2{-}CH_2\sim$
Polyvinyl Chloride (PVC)	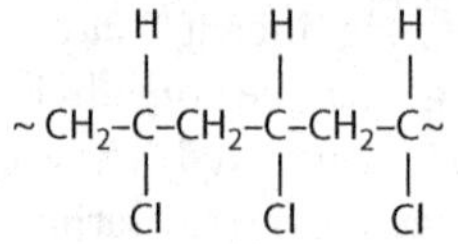 $\sim CH_2{-}C(H)(Cl){-}CH_2{-}C(H)(Cl){-}CH_2{-}C(H)(Cl)\sim$
Silicone Polydimethyl Siloxane	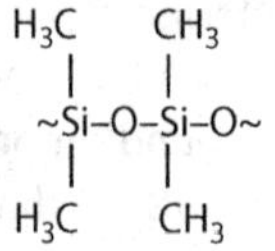 $\sim Si(H_3C)(H_3C){-}O{-}Si(CH_3)(CH_3){-}O\sim$
RTV Silicone Polymer n 200 to 1,000	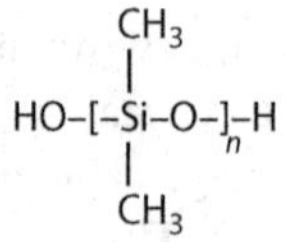 $HO{-}[{-}Si(CH_3)(CH_3){-}O{-}]_n{-}H$
Polyethylene	$\sim CH_2{-}CH_2{-}CH_2\sim$
Ethylene Propylene Copolymer (EPR)	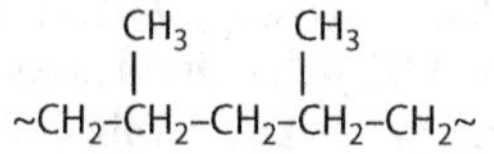 $\sim CH_2{-}CH(CH_3){-}CH_2{-}CH_2(CH_3){-}CH_2\sim$

As with other elastomers, many additives are required to render Neoprene functional. The ingredients in a Neoprene formulation may vary widely; included are conventional fillers, an antioxidant, and a processing aid. Other components may include carbon black, paraffin, and stearic acid. A major difference between Neoprene grades for wire and cable applications is in the curing technology; a combination of magnesium and zinc oxide is employed and an accelerator such as ethylene-thiourea is employed to accelerate the curing process. Different accelerators lead to different properties [34].

Although the curing technology demonstrates significant differences between Neoprene and modern-day EPRs, the objectives of incorporating the other additives are similar. Antioxidants prevent oxidative degradation, fillers serve as reinforcing agents and improve mechanical properties such as toughness, as well as improve high temperature resistance, and carbon black serves to improve tear resistance, ozone and weather resistance.

Hypalon (CSPE): This polymer is an elastomer developed by DuPont (see Section 1.6) that also possesses chlorine in the backbone. There are various grades that impart not only flame resistance, but also resistance to ozone, heat, corrosive chemicals, and weathering, as well as provide superior thermal, mechanical, and electrical characteristics. It has been described as being capable of surviving abrasive, thermally and chemically abusive environments, while maintaining flexibility and inherent UV (sunlight) resistance. It is suitable for insulation and jacketing for low voltage applications.

Not only may the compositions vary, but also the curing systems; cross-linking is induced via different means (different functional portions of the polymer chain may be linked together.) State-of-the-art technology for processing (handling, mixing, and extruding) and various methods of cross-linking of Hypalon have been described [36]. It is noted therein that cross-linking can be induced via various means that include not only peroxides, but also through the highly reactive sulfonyl chloride sites.

To achieve the combination of desired properties, CSPE incorporates components in the formulations that have, more recently, become of concern. One reason is that the gases produced during combustion have varying levels of toxicity; a second reason is that CSPE often contains additives made from lead or lead compounds. Combining environmental issues with high manufacturing and labor costs, manufacturing has been restricted in some geographical areas. Fifty-nine years after first manufacturing CSPE, DuPont discontinued supplying the polymer. (Other overseas sources are available.)

Chlorinated polyethylene: By subjecting HMWPE to chlorination, the polyethylene is converted into an elastomer. (The larger chlorine replaces a portion of the smaller hydrogen in the polymer chain, and prevents chain alignment and crystallization). The modified polymer is flexible, has high tear strength, good chemical, and UV resistance, and possesses superior flame resistance as compared with the parent polyethylene. CPE is available as thermoplastic or thermoset; the technique for cross-linking CPE is the same as for other halogenated polymers. CPE has been suggested as a potential replacement for CSPE.

Polyvinyl chloride: Some basic information on PVC is in Section 5.8, where jackets are discussed. It is inherently a tough brittle polymer with poor heat stability, yet inherent flame resistance. The chlorine presence inhibits flammability (but at the expense of evolving toxic hydrogen chloride gas.) As with elastomers, PVC must be compounded with additives to render it useful. Interestingly, whereas elastomers require additives to impart toughness, PVC requires additives to induce softening. The major additive types that serve this role are called phthalate esters, which are high boiling stable liquids, and there are many (major ones being dioctyl phthalate and diisodecyl phthalate). The nature of this additive type affects low temperature properties, degree of softness, and stability on heating. Plasticizers also reduce the tensile strength (which is regained by use of other additives). Plasticizers are actually classified as two types, primary and secondary; the latter have limited compatibility and when used, the purpose is primarily to reduce cost.

Other additives include phosphates to impart flame resistance (the inherent flame resistance of PVC due to the chlorine is compromised to some extent by the phthalates), stearates to react with any hydrogen chloride gas that evolves on heating (lead, cadmium, barium, and zinc salts have also been used for this purpose), oils (lubricants), mineral fillers like clay or talc to provide "toughness," carbon black, and sometimes organic phosphites have been used to prevent oxidation. Even other polymers have been incorporated to impart impact resistance. The chemistry and technology of PVC are massive and complex. In addition, due to the significant number and quantities of additives, the technology can be considered to be an art as well as a science.

Nevertheless, despite these complexities, studies with PVC have led to formulations successfully applied for numerous cable applications, and have been reliably and safely employed where the voltage stress and temperatures during lifetime usage are not excessive. Such PVC insulations provide: (a) good electrical insulating properties over temperature ranges defined in specifications; (b) ease of processing to meet the required specifications for end-use application; (c) long-life expectancy, as long as operated according to specifications; (d) durability; (e) resistance to ultraviolet light–induced degradation; (f) inherent flame resistance; (g) no need for cross-linking (hence reduced cost); (h) recyclability (as it is not cross-linked); and (i) cost-effectiveness.

Silicone rubbers: Unlike the low voltage polymers described above, silicone polymers are inorganic in nature (although they have also been defined as neither organic nor inorganic). The reason is that the polymers described so far are composed of carbon–carbon bonds in the backbone (organic) while silicones consist of silicon–oxygen or silicon–silicon) linkages. (A silicon-to-silicon linkage is called a silicone.)

Branches off the main chain in the silicone polymer may be methyl or larger molecules. These polymer chains are very flexible and maintain their flexibility over an extremely wide temperature range (−65°C to +315°C), which is a major incentive for its use. The exact useful temperature range depends on the specific nature of the SIR.

As with other elastomers, compounding with additives is required (the silicone itself may make up about 30% of the total composition). A typical compound may consist of the polymer, fillers, processing aids, other additives to facilitate resistance to heat aging, and the curing agent. The fillers have been defined as either reinforcing (which assists in improving properties) or extending (which serves no such role, but reduces cost). The resistance to oils, oxidation, UV (weathering), and abrasion are excellent, as are the electrical properties; however, tensile strength is not as good as that of the other polymers noted above. Nevertheless, the maintenance of properties of SIRs over the wide temperature range provides an advantage in that this property does not change significantly with temperature. Some SIRs swell in some solvents.

SIRs are manufactured by, first, reaction of metallic silicon (obtained by treating sand with carbon) with methyl chloride; then the products (mostly dimethyl dichlorosilane) condense in the presence of water to yield low molecular weight product consisting mostly of polymethyl siloxane. These products are then separated and polymerized further. Vinyl modified polymer is obtained by replacing a very small portion of the methyl chloride with vinyl chloride; the vinyl trimethoxy silane thus produced is much easier to polymerize. Cross-linking of the silicone polymer is performed by use of peroxides.

Room temperature vulcanizing silicone rubber (RTV): The SIRs described above possess nonreactive (e.g., methyl) end groups; hence, the cross-linking process being induced via peroxides. However, it is possible to manufacture low molecular weight silicones with reactive end groups. A typical reactive end group could be hydroxyl (–OH), meaning the polymer end group is a silanol (–Si–OH). This end group can react (link) with another silanol group, or with an alkoxy group (not a silanol), but in the latter case a catalyst is required. It can also react with water. There are two types of room-temperature vulcanizing silicones: RTV-1 is a one component system and cures in the presence of moisture (humidity). RTV-2 is a two-component system that cures when mixed and converts to a solid elastomer or a gel at room-temperature. The compositions are used for electrical insulation due to their dielectric properties.

EAM polymers: Recent developments (compared in time with the polymers described previously) have led to polymer compositions composed of ethylene, methyl acrylate, and an alkenoic acid (e.g., ethoxyethyl acrylate) for low voltage applications. These are referred to as EAMs and are amorphous materials that possess good low temperature resistance combined with resistance to solvents, the extent depending on the ratio of ethylene to methyl acrylate in the chains). They can be readily modified to provide flame resistant compositions. The third monomer (1%–5% level) facilitates cross-linking by use of diamines [36].

This overview summary of various low voltage insulation polymeric material systems demonstrates both the similarities and differences of the different types. Similarities include: (a) the polymers employed are virtually all elastomeric in nature, (b) all require numerous additives to function effectively (some of which are identical), and (c) the ability to "tailor" properties to meet specific needs and specifications exists. Differences include: (a) possession of a wide range of physicochemical properties exhibiting different behavior over a broad temperature range, (b) significant differences in curing methodologies (which alters the properties and behavior of even a single polymer insulation type), and (c) the numerous variety of additives required to toughen some polymers yet soften another.

While the polymer technology in this area is quite mature, the additive technology remains dynamic. As an example, methodology is being sought to eliminate halogens from formulations while maintaining flame resistance (see Section 5.8). Numerous approaches have been explored to develop and apply low smoke, zero halogen or halogen free additives (LSZH or LSHF). These are intended to prevent flammability, smoke, and toxic by-product generation while still maintaining the ability to meet industry specification requirements (in an economical fashion). Conventional additives of this type are aluminum hydroxide and magnesium hydroxide, and their use with nanofillers (see the following paragraph) is an area of ongoing study.

Another dynamic area relates to inorganic fillers and particle size control. The use of fillers, (referred to throughout the discussions on elastomers) is essential to allow rubbery polymers to function effectively. A major role of fillers is to reinforce the polymer properties such as modulus and tensile strength.

Filler effectiveness can be characterized by: (1) the total surface area in contact with the polymer; (2) degree and nature of the interaction with the polymer (i.e., wetting capability); and (3) structure or geometry. Smaller particles clearly allow greater surface contact with the polymer, and typical conventional fillers would be in the millimicron range. Developments in the field of nanotechnology have allowed studies to be performed where fillers having particle sizes in the nanotechnolgy range (or three orders of magnitude smaller than conventional) are being studied. Materials reduced to the nanoscale have been shown to exhibit different properties compared with what they exhibit on a larger scale, thus enabling unique applications.

One attraction in this area of study is that nano-sized particles modify the electrical insulating properties of polymers. It appears that a significant increase in interfacial area contact (item 1 above) leads to far superior properties of the insulation employed for wire and cable applications. The reasons for this are under study. (It has been suggested that electronic properties of solids are altered in transition from micrometer to nanometer dimensions. With reference to this effect on fillers in polymers, considerations include greater electron scattering across the interfaces, increased electron trapping, reduced polymer free volume, or modified local conductivity.) Efforts to apply nanomaterials are proceeding independently of the basic research. Clearly, the factors noted in (2) and (3) above also play a role.

The role of conventional clay fillers in medium voltage EPRs is discussed in Section 5.6.2. Low voltage cables are also discussed in Chapter 9.

5.11.2 Secondary Cables

Insulation materials for low voltage (600 V) find application in secondary cables. The secondary network in urban underground systems is fed by primary feeders, the insulation materials of the latter being what has been discussed earlier. HMWPE, XLPE, and EPRs are employed in secondary networks, as are many of the polymeric insulations discussed in Section 5.11.1. These cables, due to their function, are expected to be able to serve reliably at elevated temperatures. Many are required to possess flame resistance. Materials that are inherently flame-resistant such as Hypalon, Neoprene, and PVC serve this role as low voltage jackets. Polyethylene, as discussed above, can be formulated with appropriate additives to possess this property. This applies to XLPE and EPR also. Application of LSZH insulation systems is applied in this direction. Where abrasion resistance is required, it is common to employ double layer; a tough outer jacket (like HDPE) over the more flexible inner insulation component. It is also possible to design secondary systems that are intended to "self-heal" if the outer jacket is penetrated. Insulation materials technology for secondary cables is very broad [38].

5.12 EXTRUDED CABLE REJUVENATION

5.12.1 Introduction to the Concept

This section is concerned with technology that relates solely to aged extruded cables. As cables age over time, the utility is faced with the fact that such cables are eventually going to fail. Action can be taken to prolong life prior to actual failure. The process involves rejuvenating the cables so that they will remain "healthy" for a prolonged, extended period of time. This rejuvenation process involves injecting a silane monomer into the cable strands from where it diffuses into the insulation; it then polymerizes and also reacts with any remaining water during the rejuvenation process. The technology allows the now polymerized silane to fill any voids that may be present, and leads to an increase in the dielectric strength. This process is also applied after a failure occurs, but in that case the failed region must be removed and a splice installed prior to applying injection.

When cables begin to reach the point of exceeding their useful life, the utility is faced with the decision of whether to replace, repair, or rejuvenate. Cost is one factor, and replacement is often the most expensive path; disruption of service is another consideration. An additional consideration relates to the type of cable insulation involved. This section is concerned with the technical issues.

The history of the development of the rejuvenation process is of interest. The concept evolved from work that was done in the late 1970s–early1980s, after it was observed that many installed polyethylene-insulated cables were failing sooner than anticipated (due to water treeing). One industry response was to focus on XLPE as a HMWPE replacement; experimental evidence suggested that acetophenone, present in XLPE but not in HMWPE, was a tree-retardant additive. As seen in Figure 5.11, acetophenone is formed as a by-product from the decomposition of the cross-linking agent dicumyl peroxide. Hence, it was the cross-linking agent by-product (formed in-situ), not the XLPE itself, that imparted water tree resistance. However, acetophenone diffuses out of the cable over time and imparting of tree-retardant capability by this cross-linking agent by-product is not permanent. At a later time, industry efforts focused on developing permanent tree-retardant grade(s) of XLPE that were inherently superior to conventional XLPE (see Section 5.5).

Concurrent with these efforts to develop new superior cable insulation materials based on XLPE, efforts were also taken to improve the condition of already-installed older HMWPE and XLPE cable. Since it was clear that water presence was central to the premature loss-of-life upon aging, the first approach employed the concept of forcing the moisture from the buried cables by pumping compressed dry nitrogen gas through the interstitial spaces of the cable conductor strands, by employing a constant positive pressure. This approach was successfully applied as most early polyethylene cables possessed a multistrand core conductor, where there are longitudinal open air spaces between the conductor strands (interstices); this is the passageway used for sending gas through the cable. (With a solid conductor, it would not be possible to treat a cable in this manner.) The constant flow served to adequately remove moisture and at a later time, air was employed in place of nitrogen. Although cable life was extended via this methodology, the overall process had several limitations. Special terminations and fittings had to be crafted to allow proper entry and exit of gas flow; cable elbows and stress cones were modified accordingly, and special crimped connectors were required to allow the entering gas access to the stranded conductor. As the process depended on a constant flow of compressed gas, constant maintenance was required. If drying was interrupted, moisture could reenter the cable system.

The gas-based drying procedures were successful in preventing growth of water trees and in extending cable life. The observation that desiccation extended cable life was significant. However, the practical complexities involved were considered too great for ongoing application. As a result, the concept was extended by switching to the use of a compatible insulating liquid; acetophenone was chosen since, as noted above, at the time it had been demonstrated to impart tree resistance to XLPE-insulated cables. Tests on aged impregnated cables demonstrated that acetophenone incorporation led to the desired increase in alternating current (AC) breakdown strength. In practice, employing acetophenone did indeed force out the moisture and impurities in the cable as did compressed nitrogen. [Since acetophenone solidifies at about 20°C (58°F), the temperature control aspects of incorporating this chemical into a completed cable differ from the situation where it is formed in-situ during the cross-linking process.] Regardless, the use of the liquid in place of the gas was plagued by many of the same shortcomings as did the use of dry nitrogen. Acetophenone diffused out of the cable over time and constant maintenance was necessary to ensure constant flow of material, hence increasing costs.

An improved more practical approach that built upon this knowledge learned by working with dry nitrogen and acetophenone was developed in the 1980s. This process employed a special class of silicone materials (called alkyoxy silanes) to overcome the ongoing issue of diffusion of the fluid out of the cable. Here the small liquid molecule (which easily diffuses after being injected) undergoes a chemical reaction that simultaneously removes the water (a goal of the dry nitrogen or acetophenone treatment) and also increases in molecular weight (so the liquid cannot migrate out over time). The process was commercialized and first applied successfully by Tarpey and coworkers [39] at Orange and Rockland Utilities. It solved many of the prior issues and reduced the life extension concept to practice in an economical fashion.

Hence, the silicone fluid is injected into the stranded conductor of the cable, diffuses into the insulation, reacts with water and polymerizes, thereby undergoing an increase in molecular weight after impregnation. The higher molecular weight silicone component (called an oligomer) cannot migrate out the way the lower molecular weight acetophenone could. Since this process leads to permanent conversion of the liquid into a "gel" within the cable insulation, there is no need for ongoing application of external pressure. In addition, the curing process leads not only to filling of the voids, but also resists further moisture migration into the cable. Finally, this process leads to increased dielectric strength.

The conversion of the monomer into a polymer is analogous, in a general sense, to the conversion of ethylene into polyethylene (Section 5.2). In both cases, the low molecular weight component is induced to grow in length and form a polymer chain that possesses the desired properties. However, the analogy stops there, as there are very significant differences in the mechanism of the conversion process as well as the properties of the two polymer types; these differences are beyond the scope of this chapter, but it should be noted that the chain length of the polymerized silicone is relatively short compared to that of polyethylene and is referred to as an "oligomer."

We will now review the chemistry of the rejuvenation process and then the practical application of the technology in the field.

5.12.2 Rejuvenation Chemistry

Several silicone-based rejuvenation liquids may be employed. Phenyldimethyl siloxane (PMDMS) is the monomer that has been employed for many years, and is used as the basis for this discussion. This liquid has low initial viscosity (is therefore easy to inject), diffuses rapidly, reacts with water, and increases in chain length; rapid migration through the cable wall is a significant point. As the molecular weight increases, migration rate is reduced, and the oligomer remains locked in the region previously occupied by the water (i.e., the water tree). Figure 5.19 [40] shows the oligomerization process.

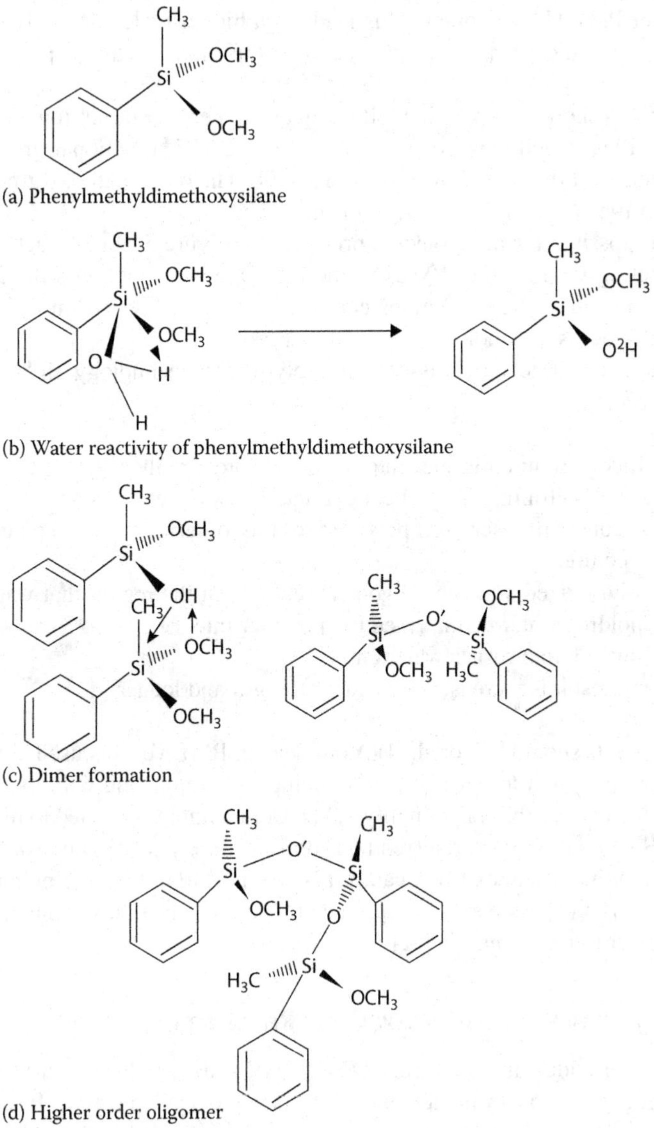

(a) Phenylmethyldimethoxysilane

(b) Water reactivity of phenylmethyldimethoxysilane

(c) Dimer formation

(d) Higher order oligomer

FIGURE 5.19 Polymerization of phenylmethyldimethoxy silane fluid. (The four parts show the reaction with water and depict chain growth.) (a) Phenylmethyldimethoxysilane. (b) Water reactivity of phenylmethyldimethoxysilane. (c) Dimer formation. (d) Higher order oligomer.

Figure 5.19a shows the chemical structure of PMDMS. The slightly polar methoxy functionality (–O–CH_3) allows the molecule to be "attracted" toward the higher stress sites of the degraded polymer (water-treed region). The monomer then reacts with the water in the water tree.

Figure 5.19b shows the initial reaction where the polar hydroxyl functionality of the water (–OH) replaces the methoxy group.

Figure 5.19c shows the newly formed hydroxyl-containing monomer reacting with another PMDMS monomer. This leads to a higher molecular weight silicone, called a dimer, which can be considered to be the first step in the polymerization (oligomerization) process.

Figure 5.19d shows one possible following event; here the dimer (on the right side of Figure 5.19c), which now possesses reactive –O–CH_3 functional groups, reacts with a monomer of the type shown in Figure 5.19b. The other methoxy groups shown in Figures 5.19c and d can also react (not shown).

It is also possible for the molecule produced in Figure 5.19b to react with water (before reacting with another PMDMS monomer), hence yielding a molecule with two hydroxyl groups; these reactions can occur concurrently (also not shown). Via this general process, the molecular weight increases.

Therefore, several technical issues for applying this technology all "move in the right direction."

- The silicone liquid migrates into the cable core rapidly.
- The water is eliminated by chemical reaction with the silicone.
- The silicone monomer itself polymerizes (oligomerizes) and becomes a gel-like structure.
- The polymerized silicone (oligomer) remains in the region that was originally holding water that harmed the polymer integrity.
- The new polymer resists water entry.
- The end result is increased dielectric strength and longer life.

Several points should be noted. (1) As indicated, PMDMS was used as the example; it is now common to employ this monomer in conjunction with other silicone monomers to enhance the functionality. (2) Newer technology has led to replacement of PMDMS by other silanes, referred to as dialkyl alkoxy silanes (an example being dimethyl dimethoxy silane). (3) A catalyst is required to initiate the polymerization process [41]. (4) It is possible to incorporate other chemical additives into the silicone monomer prior to impregnation [42].

5.12.3 In-Service Procedure for Cable Rejuvenation

This section provides an overview of the practical aspects involved in imparting improvements to the performance of installed aged extruded cables. Installation of new cables is covered in Chapter 12; here we are concerned with field procedures that are not related to the original installation, but that are relevant to successful implementation of the chemical reactions. For a discussion of the accessories described, the above chapters should be reviewed. Diagnostic procedures to assess the "state" of the cable that could be applied prior to impregnation are discussed in Chapter 18.

The following description, which clarifies the practical aspects of implementation, applies to procedures for URD cable; for feeder cable, the procedures differ slightly.

1. The cable is de-energized.
2. A diagnostic test (time domain reflectometry or TDR) is performed to determine the cable length and to ensure that the cable concentric neutrals are not corroded and are functioning satisfactorily.
3. Modified accessories are installed: injection elbows or live front injection adaptors are typically retrofit. These remain permanently on the cable (200 amp load break injection elbows are commercial available).
4. The cable may then be re-energized.
5. Flow and pressure tests are performed; nitrogen is injected at one end of the cable and outflow is measured at the other end. The cable is pressurized and flow is monitored. The purpose is to ensure there is good flow, no leaks, and to remove any water that may be in the strands. Splice location is also pinpointed; conventional splices will generally allow flow of fluid, but some premolded splices may require excavation or repair.
6. The cable is then treated (the fluid has been described in Section 5.12.2). A feed tank injects the fluid into the cable and a vacuum tank is placed at the collection end. The process typically takes place overnight.
7. The feed tank is left in place. The time is such as to allow an adequate amount of fluid to diffuse and penetrate into the insulation. Cable insulation wall thickness influences the requirement.

The process described is to ensure that the cable system (including splices and terminations) is suitable for impregnation, that proper and complete impregnation takes place, and that curing takes place when desired (not prematurely).

For further information, the references should be reviewed.

5.13 COMPARISON OF MEDIUM VOLTAGE INSULATING MATERIALS

Table 5.5 provides a general summary comparing the properties of various types of polymer insulation materials for cables. The significant differences between the natural polymer cellulose (paper) based materials and the synthetic polyolefin based materials (XLPE, TR-XLPE and EPR) are summarized for ease of review. This brief table is considered to be general and not definitive. For example, the EPR information provided is based on the general discussion noted in Section 5.6.1. There are caveats about EPRs regarding the composition, dielectric losses, and the general philosophy regarding how to achieve reliability, which affect the limited information caught in short phrases in the entries in the table. Hence, not all EPRs will readily fall into the attempted comparison sought therein. Care must be taken in seeking to compare different EPRs with each other (as well as to XLPE and TR-XLPE and paper systems).

TABLE 5.5
Comparative General Properties of Insulation Materials in Medium Voltage Cables

XLPE/TR-XLPE	EPR	Paper/Cellulose-Fluid
Synthetic polymer: solid dielectric	Synthetic polymer: solid dielectric	Natural polymer; tape construction plus natural or synthetic fluid
Carbon/Hydrogen polymer	Carbon/Hydrogen polymer	Carbon/Hydrogen/Oxygen polymer
Low losses	Filler-induced higher losses	Higher losses
Semicrystalline	Little to no crystallinity: amorphous	Paper: crystalline fibrils fluid: amorphous
Thermal expansion as temperature increases	Slight thermal expansion as temperature increase	Fluid: slight thermal expansion and migration through tapes
Cross-linked polymer structure	Cross-linked polymer structure	Not cross-linked
Water plus voltage stress induces water trees TR-XLPE is water tree resistant	Water plus voltage stress induces water trees; lesser extent than XLPE	Water cleaves C–O linkages. No water trees
Dry aging leads eventually to electric trees	Dry aging leads eventually to electric trees	Dry aging leads eventually to chain cleavage
Low permittivity	Slightly higher permittivity	Highest permittivity
Cable less flexible the EPR	Cable more flexible than XLPE or TR-XLPE	Not applicable due to Pb sheath
Reduced weight vs. PILC	Reduced weight vs. PILC	Greater weight vs. extruded
Easier to repair faults	Easier to repair faults	More difficult to repair faults vs. extruded
Easier to apply accessories	Easier to apply accessories	More complex to apply accessories
Surges, DC HiPot testing may harm aged cable and shorten life, if failure does not occur at time the stress occurs	Surges, DC HiPot testing may harm aged cable and shortened life, if failure does not occur at time the stress occurs	No evidence of shortened life by stress described

There are many different EPR formulations, and all do not necessarily respond in the same manner upon aging. The nature of the nonpolymeric additives, including fillers, plays a major role in influencing the properties as does the nature of the mixing process. Section 5.6.3 should be reviewed for internal EPR comparisons.

The same caveat applies to paper-insulated cables that employ various fluids and a lead sheath.

In comparing polymeric insulation materials, it is common to describe cross-linked (thermoset) polymers as not being reprocessable, reformable, or recyclable. This certainly is true for rigid or "glassy" polymers, one example being epoxies (not employed as cable insulation); once they are cross-linked they cannot be reprocessed and reused, as they cannot be remelted. On the other hand, conventional polyethylene, being thermoplastic, can be remelted (and hence recycled, reprocessed, and reformed into another product, cable or otherwise).

TABLE 5.6
Classification of Polymer Types Regarding Refabrication

Polymer Insulation	Polymer Classification	Cross-linked (Thermoset)	Not Cross-linked (Thermoplastic)	Recyclable into Another Product	Refabricated (Reformable) by Changing Shape
HMWPE	Semicrystalline	No	Yes	Yes	Yes
XLPE	Semicrystalline	Yes	No	No	Yes
Epoxy	Rigid/ amorphous	Yes	No	No	No
EPR insulation	Flexible/ amorphous	Yes	No	No	No
Ethylene (shield) copolymers	Flexible (amorphous)	Yes	No	No	No

Some ambiguity exists in the classification of XLPE. Since it is cross-linked, it cannot be recycled into another product, but it can (and is) be reformed (an option not possible with an epoxy). The reason is due to the semicrystalline nature of the XLPE; the crystalline regions can be remelted at elevated temperatures, even though the cross-linked regions cannot be recycled. Physical modification and cooling then allow the XLPE to assume a reformed shape. Thus, the semicrystalline polymer, even though being cross-linked, can be refabricated. (This has been discussed in the latter portion of Section 5.4.3.) These distinctions are summarized in Table 5.6.

REFERENCES

1. Kressner, T. O., 1969, “Polyolefin Plastics,” Van Nostrand Reinhold Publ. Co.
2. Morton, M. (ed.), 1973, “Rubber Technology,” Second Edition, Van Nostrand Reinhold Publ. Co.
3. Sengupta, S., Person, T. and Caronia, P., June 2010, “A New Generation of Tree Retardant Cross-linked Polyethylene (TR-XLPE),” Conference Record of the 2010 IEEE International Symposium on Electrical Insulation, pp. 1–6, San Diego, CA.
4. Raff, R. and Doak, K., 1965, “Crystalline Olefin Polymers,” Vol. 1 and 2, Interscience Publishers.
5. Billmeyer. F., 1971, “Textbook of Polymer Science,” Chapter 3, pp. 84–90, Wiley Interscience Publisher.
6. Miller, M. L., 1968, “The Structure of Polymers,” Chapter 10, p. 494, Van Nostrand Reinhold Publishing Co.
7. Vasile, C., 2000, “Handbook of Polyolefins (Plastic Engineering),” Second Edition, Marcel Dekker, Inc.
8. Charlesby, A. 1960, “Atomic Radiation and Polymers” Polymers,” Pergamon Press; Chapiro, A., 1962, “Radiation Chemistry of Polymeric Systems,” Interscience Publisher; Odian, G. and Bernstein, B. S., 1964, “Radiation Cross-linking of Polymers via Polyfunctional Monomers,” *Journal of Polymer Science*, Vol. A2, p. 2835.

9. Pauling, L., 1953, "The Nature of the Chemical Bond," Cornell University Press (); Flory, P. J., 1953,"Principals of Polymer Chemistry," Cornell University Press.
10. Winspear, W. G. (ed.) "Vanderbilt Rubber Handbook," Chapter 1, Martens, S.C., "Chemically Crosslinked Polyethylene," R.T. Vanderbilt Company, pp. 226–227; also, 1983, "Hercules Peroxides," Hercules Chemical Company Bulletins ORC, pp. 101E, 104E, 201–208.
11. Bernstein, B. S., Odian, G., Binder, S. and Benderly, A., 1966, "Physical and Electrical Properties of Polyethylene Radiation Cross-linked by Polyfunctional Monomers," *Journal of Applied Polymer Science*, Vol. 10, p. 143.
12. Gross, L., October 1988, "Polyethylene Silane Co-polymers as New Low Voltage Insulation Systems" Paper Presented at Wire Association International Convention, Toronto Canada; see also, *Wire Journal International*, November 1988, pp. 59–66: See also, Caronia, P., April 2005, "Overview of Polyethylene used in 600V Underground Secondary Cables," Paper Presented at Spring 2005 IEEE Insulated Conductors Committee Meeting, pp. 791–806, St.Petersburg, FL.
13. Conley, R. T., 1970, "Thermal Stability of Polymers," Chapter 6, Hansen, R. H., "Thermal and Oxidative Degradation of Polyethylene, Polypropylene and Related Olefin Polymers," Marcel Dekker, Inc.
14. Ezrin, M., Seymour, D., Katz, C., Dima, A. and Bernstein, B. S., June 1986, "Thermal Response of Cable Insulation, Shield and Jacket Materials Aged at 130C and Above," *Proceedings of IEEE International Symposium on Electrical Insulation*, Washington DC, p. 46; also, Katz, C., Dima, A., Zidon, A., Ezrin, M. Zengel, W. and Bernstein, B. S., May 1984, "Emergency Overload Characteristics of Extruded Dielectric Cables Operating at 130C and Above," *Proceedings of IEEE T&D Conference*, Kansas City, MO.
15. Jow, J. and Mendelsohn, A., Spring 1999, "Tree Retardant XLPE Technology Review," *Proceedings of IEEE Insulated Conductors Committee Meeting*, Charlotte, NC.
16. Brown, M., January–February 1994, "Compounding Ethylene Propylene Polymers for Electrical Applications," IEEE Electrical Insulation Magazine, Vol. 10 (No. 1), pp. 16–22.
17. Borzenski, F. J. and Valsamis, L. N., 1 July 2002, "Optimizing Mixing in the Farrell Banbury Mixer" Rubber World.
18. Kalyon, D. M., "Mixing in Continuous Processes," Encyclopedia of Fluid Mechanics, Chapter 28, http://www.hfmi.stevens-tech.edu/publications/174.PDF; also, Buss brochure "Buss Kneader Technology for Cable Compounds," www.busscorp.com.
19. Bartnikas, R. (ed.), 2000, "Power and Communications Cables," (a) Chapter 9, p. 406, (b) Chapter 1, p. 35, McGraw Hill Publ. Co.
20. Qi, X. and Boggs, S., May–June 2006, "Thermal and Mechanical Properties of EPR and XLPE Cable Compounds," IEEE Electrical Insulation Magazine, Vol. 22 (No. 3), pp. 19–24.
21. Banker, W. and Katz. C., 2000, "Update on Field Monitoring and Laboratory Testing of EP & TR-XLPE Distribution Cables," Proceedings of IEEE/PES Insulated Conductors Committee, Fall Meeting, St. Petersburg, FL, pp. 119–124; also, Katz, C., Fryszczyn, B., Bernstein, B. S., Regan, A. M. and Banker, W., July 1999, "Field monitoring of Parameters and Testing or EP and TR-XLPE Distribution Cables," IEEE Transactions on Power Delivery, Vol. 14 (No. 3), pp. 679–684; also, EPRI Report 1009017 "In-Service Performance Evaluation of Underground Distribution Cables," 2003.
22. Smith. R., March 2009, "Kerite Review of Cable Submergence," Paper Presented at Nuclear Regulatory Commission Cable Workshop, Rockville, MD.
23. Eichhorn, R. M., December 1981, "A critical Comparison of XLPE and EPR for use as Electrical Insulation on Underground Power Cables," IEEE Transactions on Electrical Insulation, Vol. EI–16 (No. 6), pp. 469–482.

24. Han, S. J., Mendelsohn, A. and Ramachandran, R., 2006, "Overview of Semiconductive Shield Technology in Power Distribution Cables," *Proceedings of IEEE Power and Energy Society, Transmission and Distribution Conference*, Paper TD2005-000590.
25. Forester, E. O. and Spenadel, L., January 1973, "Black-Filled EP Rubbers," Rubber Age, pp. 39–45.
26. Dannenberg, E. M., 1979, "Carbon Black," Kirk Othmer Encyclopedia of Chemical Technology, Vol. 4, Third Edition, John Wiley and sons.
27. Smith, R. and Hu, H., April 1991, "An Introduction to the Designs and Philosophies of the Kerite Company," Paper Presented at the AEIC Cable Engineering Section, Charleston, SC.
28. Stenger, R. J., 9 April 1985, "Filling materials For Electrical Cable," U.S Patent 4,509,821.
29. Clark, G. L. and Hawley, G. S., 1957, "The Encyclopedia of Chemistry," pp. 696–697.
30. Dunsheath, P., 1929, "High Voltage Cables," Sir Isaac Pitman and Sons Ltd.
31. Brown, T. L., LeMay, Jr., H. E. and Bursten, B. E., 1994, "Chemistry the Central Science," Prentice Hall Publ Co., Englewood Cliffs, New Jersey, Sixth Edition, p. 990.
32. Kaminska, E., 1996, "Determination of Degree of Polymerization of Cellulose in Ligneous Papers," *Materials Research Society Symposium Proceedings Fall 1996*, Vol. 462; Alexander, W. J. and Mitchell, R. L., December 1949, "Rapid Measurement of Cellulose Viscosity by Nitration Methods," *Analytical Chemistry*, Vol. 21 (No. 12), pp. 1497–1500; Lewin, M., (ed.), 2007, "Handbook of Fiber Chemistry," pp. 429, 488–489, CRC Press.
33. El-Sulaimin, A. A. and Uershi, M. I., September 1996, "Effect of Additives on Performance of Cable Oil and Kraft Paper Oil Composite," *Journal of Electric Power Components and Systems*, Vol. 24 (No. 6), pp. 597–608.
34. Srinivas, N. and Bernstein, B. S., 1991, "Effect of DC Testing on Aged XLPE-Insulated Cables with Splices," *Proceedings of JICABLE 91*, Paper B31, Versailles, France.
35. Allinger, G., and Sjothun, L. J. (eds), 1964, "Vulcanization of Elastomers," Stevenson, A.C., Chapter 8, "Neoprene, Hypalon and Fluoroelastomers," pp. 265–285, Reinhold Publ. Co.
36. Klindenger, R. C., 2008, "Handbook of Specialty Elastomers," Chapter 9,"Chlorsulfonated Polyethylene and Alkylated Chlorosulfonated Polyethylene," CRC Press; also, DuPont Dow Brochure H-68574-01"Hypalon Chlorosulfonated Polyethylene," November 1998.
37. White, J. R., 2001, "Rubber Technology Handbook," pp. 66– 67, Smithers Rapra Publ. Co.
38. Doherty, F. and Bernstein B. S., "Self Sealing Cables: State-of- the-Art," IEEE Insulated Conductors Committee Meeting, Paper A-9, pp. 206–219; Szanislo, S., April 2005, "A Review of Secondary Cable Basics," Paper presented at ICC Education Forum, pp. 725–910, St. Petersburg, FL.
39. Nannery, P., Tarpey, J., Lacanere, J., Meyer, D. and Bertini, G., October 1989, "Extending the Service Life of 15kV Polyethylene URD Cable Using Silicone Liquid," IEEE Transactions on Power Delivery, Vol. 4 (No. 4), pp. 1991–1996.
40. Stagi, R. and Chatterton, W., June 2007, "Cable Rejuvenation; Past, Present and Future," *Proceedings JICABLE 2007*, Paper C.7.2.14, pp. 858–861, Versailles, France.
41. Chatterton, W. J., October 2008, "The Chemistry of Cable Rejuvenation," Paper Presented at EUCI Conference, Phoenix, Arizona.
42. Busby, D. and Bertini, G., October 2010, "Cable Rejuvenation Mechanisms: An Update," Paper Presented at CIGRE Canada Conference on Power Systems, Vancouver, Canada.

APPENDIX A: POLYETHYLENE CHAIN MOTION AT VERY LOW TEMPERATURES

At very low temperatures (–100°C to 0°C), brittleness becomes an issue for polyethylene (and EPR also) and cracking can occur. Prior studies of the observed random motion of the polyethylene chains at very low temperatures have been divided into three categories, traditionally referred to as alpha, beta, and gamma. Polyethylene's "simple" $-CH_2-$ repeating unit can behave in a far from simple manner when seeking to understand structure–property relationships.

The first of the three categories has been believed to be due to crystallite chain twisting between amorphous phases; the second is related to chain motion at the boundaries of the crystalline–amorphous interface, and the third is believed to be due to motion of chains of specific lengths. These phenomena influence properties at low temperatures (e.g., low temperature brittleness). These details of the fine molecular structure are noted here to emphasize just how complex is the nature of seemingly simple polyethylene molecule; XLPE, TR-XLPE, and EP are even more complex.

APPENDIX B: SINGLE SITE CATALYST POLYMERIZATION

The advent of newer single site catalysts for producing polyethylenes ultimately raised an older subject to higher visibility. In order for a polyethylene to be adequately extruded for application in cables, it must be melted in the extruder at elevated temperatures and be pushed through a die. A smooth surface is achieved by controlling the extrusion rate. As the extrusion rate increases, potential problems may arise: these are called "sharkskin" and at higher rates "melt fracture." "Sharkskin" refers to a rough surface with repeated patterns of ridges. At conventional extrusion rates, the insulation surface is smooth as it emerges from the extrusion die; at higher rates, flow instability can occur. This phenomenon is related to molecular weight, molecular weight distribution, and surface interfacial phenomena. This is not a problem with polyethylenes produced by the conventional high pressure process, as it is moderated by their relatively wide molecular weight distributions at conventional extrusion rates. Therefore, having a broad molecular weight distribution is not necessarily "bad" from a processing perspective. However, polyethylenes with narrow molecular weight distributions have been more prone to development of "sharkskin," even at relatively moderate extrusion speeds. Research is ongoing to better understand the causes and mitigate the problem. "Melt fracture" only becomes an area for consideration at significantly higher than conventional extrusion rates.

For the record, the newer metallocene catalysts are metal complexes possessing cyclopentadienyl (Cp) or substituted Cp groups. The conventional Ziegler Natta catalysts are typically based on titanium and chlorine. The former technology is used to manufacture LLDPE, used as cable jackets.

Despite the practical processing difficulties, the single site catalyst approach offers much promise, and studies are ongoing. Information provided by materials suppliers suggests the following potential advantages in applying advanced single site catalyst technology:

- Variety of comonomer choices
- Controlled molecular weight distributions
- Better control of crystallinity
- Controlled long chain branching
- Better balance of properties
- Improved physical properties
- Improved ease of installation (due to greater flexibility)
- Improved filled systems
- Superior cleanliness due to better catalyst efficiencies

6 Electrical Properties of Cable Insulating Materials

Bruce S. Bernstein

CONTENTS

6.1 INTRODUCTION

This chapter is concerned with the electrical properties of polymers employed as insulation. The emphasis is on extruded polymers. In Chapter 5, we demonstrated that the physicochemical properties of polymers employed as electrical insulation are determined by the chemical structure of the polymer; the nature of their chemical bonds and their geometry. The fundamental properties of the polymers employed as insulation and as shields were reviewed therein.

With reference to the electrical properties, it is convenient to classify these same polymers according to their response to the voltage stress under which they operate. We will demonstrate that the electrical properties of polymers serving as cable insulation are indeed related to their structure at low or operating stress; however, at high stress (applied or local), other factors become significant and can override this parameter.

Insulation materials, whether they be (for example) cross-linked polyethylene (XLPE), an ethylene propylene rubber (EPR), or polyvinyl chloride (PVC), all serve the role of protecting the conductor from the local environment; they ensure that the current flow path remains constant. They are intended to be poor conductors and are expected to fulfill that function for the projected lifetime of the wire or cable. They are expected to serve that role in dry or wet environments. Key properties that characterize good insulating polymers at operating stress are volume resistivity (VR), dielectric constant (K), and dissipation factor (DF). Any polymer serving this role must possess VR, K, and DF property requirements as defined in industry specifications. Polarization is another key phenomenon to be understood.

VR is a clearly defined parameter for electrical insulation materials; it is the electrical resistance between opposite faces of a cube of the polymer of interest, when measured at a defined temperature and pressure. DF represents a measure of the amount of energy lost as heat rather than transmitted as electrical energy; it is therefore a measure of dielectric losses. K (also referred to as permittivity or specific inductance capacitance) is a measure of how well the insulation holds charge. A good dielectric (insulation) material is one that holds the charge well (low K) combined with very low losses (low DF). Polyolefins such as polyethylene (PE), XLPE, and most EPRs represent examples of polymers that possess excellent combinations of these properties.

All three phenomena are intimately related; however, each will be discussed individually as we review what happens to the insulation itself when it responds to the operating stress. Some overlap in the discussion of the phenomena is inevitable. Polarization, another key property of polymeric insulation, is discussed first (Section 6.3.1), after VR is defined.

At higher voltage stresses, the ability of the insulation to prevent flow of electrons becomes threatened. The property of the insulation that matters here is referred to as dielectric strength, and if this value (which is high initially and often drops upon aging) is such that the insulation can no longer prevent electron flow, failure results. There are several mechanisms by which this can occur, the common one being partial discharge. Operation under high voltage stresses (applied or local) and resultant failure mechanisms are discussed in Section 6.4.

The emphasis shall be on polyolefins employed for medium voltage insulation; polyethylene, XLPE, tree-retardant cross-linked polyethylene (TR-XLPE), and ethylene–propylene polymers. The principles involved with regard to how they, and other copolymers, respond to voltage stress shall be reviewed.

In this chapter, the terms power factor and DF will be used interchangeably. In addition, the terms K, permittivity, and specific inductance capacitance will be used interchangeably.

The reader is also referred to Chapter 4 for discussion of many of these terms with reference to cables.

6.2 VOLUME RESISTIVITY (VR)

VR can be understood by reference to Figure 6.1, which shows a hypothetical cube of polymer insulation material. Since insulating polymers are poor conductors, they have high resistance to movement of electrons. VR is the term commonly applied and represents (under direct current [DC]) the resistance to passage of electrons between opposite faces of the cube, where the cube dimensions, distance between cube faces, and temperature are all defined.

The greater the resistance between two opposite faces, the better the ability of the polymer to serve in electrical insulation applications. The VR units are ohm-cm. The VR is a fundamental material parameter for insulation and semiconducting materials. Polyethylene is an excellent insulation and its VR is 13.

It should be noted that VR is different from insulation resistance, which is another way of describing the quality of the insulation, but in a cable construction. Insulation resistance is a measure of the ohmic resistance of a cable for a given set of conditions. The value depends on the test parameters (e.g., cable length) and will differ as the test parameters change. It is therefore not a basic polymer insulation materials property, but is useful for estimating the condition of cables where the lengths being measured will likely vary. As noted above, the dimensions for determining VR of an insulation are clearly defined. [In addition, VR (ohm-cm) should not be confused with resistance per unit volume (or ohms per cubic centimeter).]

An additional concept employed for studying insulation materials is "surface resistivity," which is represented by the passage of electrons between opposite portions along the surface of a film of defined dimensions (this can be visualized as one surface of the cube described above, Figure 6.1); the units here are ohms (not ohm-cm), as the width and length of the film being measured (one square cm) are chosen to be the same.

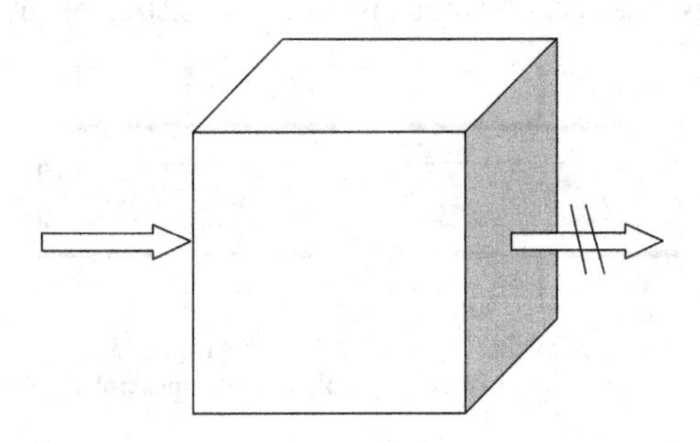

FIGURE 6.1 Volume resistivity.

Hence, VR represents a three dimensional measurement, and surface resistivity represents a two dimensional measurement.

VR, insulation resistance, and the other properties noted above are influenced by moisture content, voltage, time of applied voltage, and temperature. The conditions of measurement are vital to understanding how to compare this property information for different insulations. It is also an important parameter when considering the properties of semiconducting materials.

6.3 INSULATION MATERIALS RESPONSE AT OPERATING STRESS

6.3.1 Polarization

Under an applied field, polymers become polarized. Polarization can be visualized as shown in Figures 6.2 and 6.3. If the polymer is placed between two electrodes and voltage stress is applied, mobile charge carriers will migrate. In simplest terms, anions migrate toward the anode and cations toward the cathode. As seen, there is a charge accumulation at each surface, but no overall change within the polymer. There is migration, but no conduction. Figure 6.3 provides another perspective on charge migration. This is referred to as polarization. Under DC polarization persists, while under AC it is easy to visualize that "cyclic" migration will result. The broader question is, since the polymer is intended to be an insulation, what exactly are these charge carriers?

Polymer molecules employed as insulation can be classified as being polar or nonpolar; polar molecules possess a slight "distortion" in their electronic charge distributions. Even a simple molecule like water is "unbalanced" via this principle, as the oxygen portion of the molecule draws electrons closer, causing the oxygen region of the molecule to have a permanent slightly negative charge (referred to as a dipole). Another way of viewing this is that the electrons spend more time in the vicinity of the oxygen than that of the hydrogen. The overall molecule is neutral but the electronic distortion renders it as being polar in nature, and it is referred to as being a permanent polar dipole.

These principles of orientation apply to polymers that contain atomic constituents that possess polar components. Here we have orientation of the more complex polymer chains, relative to simple polymer molecules. Recall (Chapter 5) that during processing of extruded cables, a small amount of oxidation is virtually unavoidable (see below). This means that instead of the insulation consisting of solely carbon–carbon

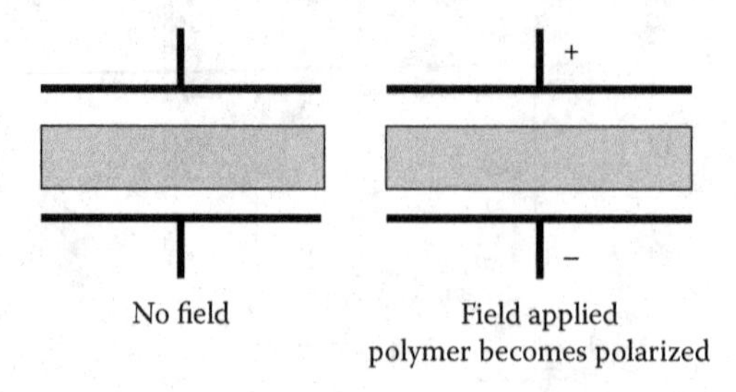

FIGURE 6.2 Polarization of a polymer subjected to an electric field.

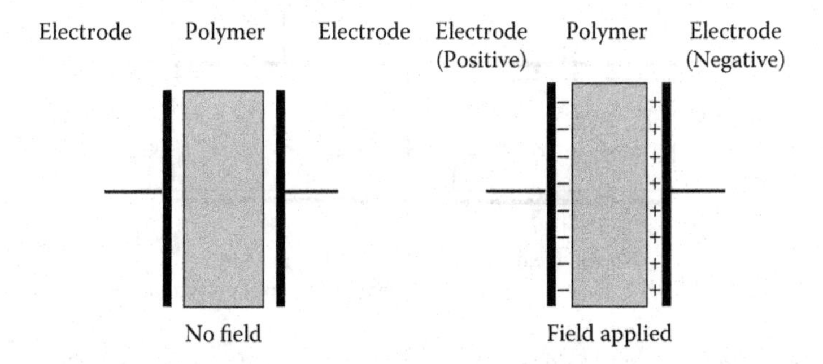

FIGURE 6.3 Charge migration on a polymer subjected to an electric field.

and carbon–hydrogen bonds, there will be a small quantity of carbon–oxygen bonds. In turn, this leads to a permanent "unbalance" in the electronic configuration of that portion of the polymer molecule.

When voltage stress is applied during operation, this situation becomes exacerbated. Under these conditions, the distortion is increased and the extent is related to the magnitude of the applied stress. Ethylene copolymers with vinyl acetate or ethyl acrylate (employed in shields), each possessing oxygen-containing functionality, are more polar than polyethylene homopolymers (or even very slightly oxidized ethylene polymers) and will respond accordingly; however, since these polymers are intended for use in semiconducting layers, their increased polarity involves no practical issue. (If a cable insulation undergoes oxidation during aging [i.e., water tree formation], it is obvious that permanent dipoles will increase in quantity.)

The nonpolar regions (those that contain no permanent dipoles) can also be affected by voltage stress but to a lesser extent. Here, the field can induce a displacement of the electrons (referred to as an induced dipole) and the displacement occurs only while the field exists. Since operating cables are under continuous voltage stress after being energized, from a practical perspective both phenomena become relevant.

One result of these stresses is segmental motion of the polymer chain. In principle, the dipole regions can be on the polymer branch or the main chain. Figure 6.4 is an idealized description of motion of side chains; note the orientation of the side chains in the portion of the energized polymer chain. Figure 6.5 provides an idealized depiction of polarized side chains. Figure 6.6 depicts main chain motion. Note also the segmental motion; the lower half of the upper chain has rotated above the plane of the paper and is now "twisted" over in the same chain as depicted on the right. The other polymer chain behaves in a similar manner. Since there is a tendency for the positive charges on the polymer to move toward the negative electrode, and for the negative charges on the polymer to move toward the positive electrode, the polymer is "pulled" in two directions. This movement is an attempt to align with

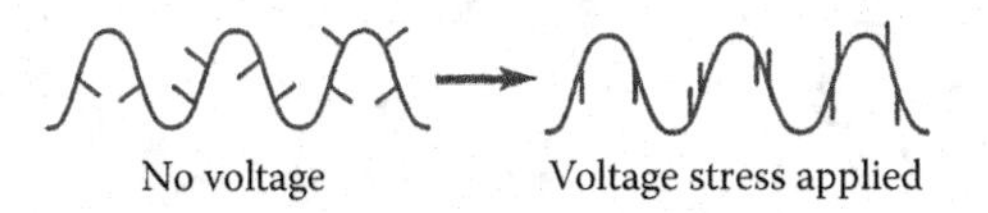

FIGURE 6.4 Orientation of polar functionality on polymer side chains.

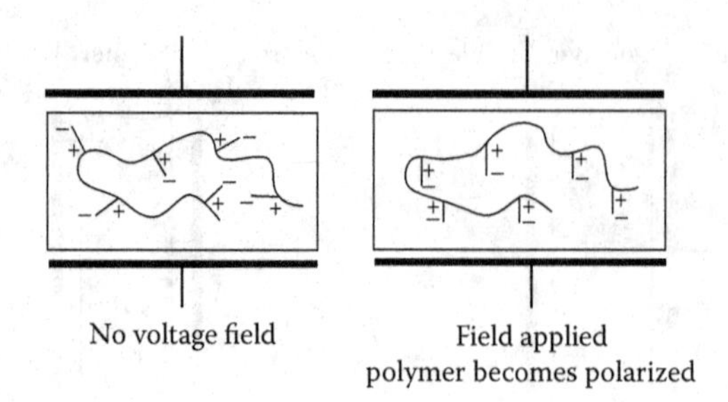

FIGURE 6.5 Polarization of side chains depicted on a coiled polymer.

the field. Hence, the process induces a very slight mechanical stress in the polymer. These figures illustrate the point in a simplified manner. In reality, there will be a distribution of polarities along the chains of different molecular weights.

It must also be noted that any foreign impurities (not part of the polymer structure) can also exhibit migration under the applied stress. Examples include water and ions, as well as cross-linking agent by-products (if present) and antioxidant residues.

It must be emphasized that this generic description focuses on what happens under DC. Consider now what would happen under alternating current (AC); here the polymer chain alignments will be shifting back and forth in accordance with the polarity change, and this realignment will be taking place at a rate controlled by the frequency. In considering these points, it becomes evident that the response of a polyolefin polymer, even a slightly polar one, is quite different under AC than under DC. Such motion may result in increased losses and heat dissipation (Section 6.3.3).

It is worth reviewing some key points established in Chapter 5, but this time from the perspective of electrical properties. Recall that the polymer insulation chains that we have been considering consist of many methylene groups linked together and these are essentially nonpolar in nature. However, during polymer insulation manufacture (polymerization of the monomer or monomer mixtures) and fabrication into cable insulation, these very long chains are always subjected to small chemical changes due to oxidation. More specifically, this may occur during conversion of the monomer to the polymer, during conversion of the polymer to cable insulation over the conductor, or during the high temperature/pressure conditions employed for the chemical cross-linking process. The reason is that when the cable is manufactured, the polymer is heated to very high temperatures in an extruder barrel

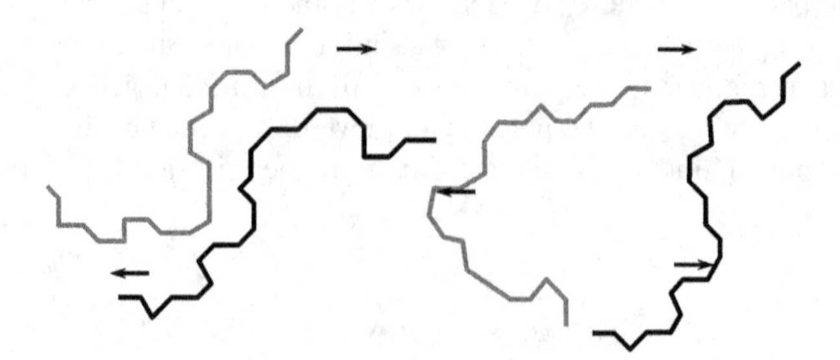

FIGURE 6.6 Schematic of modes of main chain motion of a polymer under voltage stress.

(to melt the crystalline regions and to ensure uniform dispersion of the catalyst), and is subjected to mixing and grinding due to screw motion and frictional contact against the barrel wall. When cross-linking is performed later by using peroxides, the temperature and pressure in the curing tube are raised even higher in order to decompose the peroxide catalyst. As noted earlier, an effort is made to prevent any elevated-temperature-induced degradation (but more realistically, the event is kept to a minimum) by incorporating antioxidant(s) into the polymer. The antioxidant preferentially degrades and protects the polymer insulation. However, a small degree of oxidative degradation, while controlled, is normal. Therefore, there will always be some oxidized functional groups on the polymer chains.

The polymerization process also leads to polymer chains of varying molecular weights and various degrees of branching, so we are working with polymers defined by their molecular weight distributions. To these parameters we must now add another variable; we will have polymer chains having varying degrees of polarity, either on the branch or on the backbone, or both. Recall also that we may have long or short chain branches (only short chain branches are shown in Figure 6.4). These differences mean that the response of the insulation (the polarization phenomena) will be time-dependent. These are important points to keep in mind when reviewing the polymer insulation response to frequency.

The wide variety of molecular weights (broad distribution), variety of short and long chain branches emanating from the backbone, different degrees of oxidation on the branches and backbone, and different nature of polar functionality (carboxyl, ester, carbonyl) possibly present, all ensure that there will be a distribution of relaxation times (even at any single frequency and temperature). The fact is that all the chains cannot respond simultaneously and in the same manner. Also, as the frequency increases, the polar groups have a more difficult time following the field, and at high frequencies (well above 50–60 Hz) the polar groups can no longer follow the field and properties change (e.g., K drops); the phenomenon is referred to as dispersion (see Section 6.3.4). The fundamental principle described here has been referred to as Maxwell–Wagner polarization.

The broad review of polarization of insulation materials operating under AC implies that there will also be dielectric relaxation taking place. The latter also occurs within the polymer molecules during each cycle due to movement under the changing field, and the rate of change is determined by the relaxation time. These phenomena influence permittivity and dielectric losses.

6.3.2 Dielectric Constant/Permittivity/Specific Inductance Capacitance

Dielectric constant (K) can be visualized as what is taking place at the atomic level inside the polymer chain as polarization occurs [polarization being viewed as the observed "outer" response of the chain (i.e., motion or displacement) to voltage stress]. K is not related to motion of the polymer as is polarizability, but instead is related to capacitance, and is a measure of the ability of the insulation material to store electrical energy. The key point is that as a result of motion of the polar functional groups, energy is stored. The ratio of the energy stored in the insulating polymer to that of a vacuum is referred to as the dielectric constant.

Permittivity is intimately related to the polymer structure. A polymer insulation material like polyethylene, which has very few dipole regions, can store very little energy, and has a low K; when polar bonds are present in polymers, K increases. Polymers that have permanent dipoles, such as an ethylene copolymer with vinyl acetate or ethyl acrylate, can store more energy than polyethylene (or XLPE) and they have higher dielectric constants. Very polar polymers have even higher dielectric constants; an example would be a polyamide such as Nylon or Hypalon (chlorosulfonated polyethylene). K varies with frequency and is relatively low at 50–60 Hz. Since the dielectric constant is a ratio (charge stored in the polymer as compared with that in a vacuum), it has no units, and is simply a pure number.

Permittivity can be determined by procedures described in ASTM-D 150 and IEC 60260. In essence, a sample of insulation material is placed between two metallic plates and voltage applied. The sample size is defined in the procedure, and must be flat (and larger than the 50 mm (2 in) circular electrodes used for the measurement). A second voltage run is made without the specimen between the two electrodes. The ratio of these two values is the dielectric constant. Permittivity is the ratio of the capacitance of the dielectric material to that of air. Since an insulation cannot have dielectric constants less than that of a vacuum, K must, obviously, be greater than 1 regardless of the insulating polymer being measured. Dielectric constants of polymers at room temperature are normally in the range of 2 to 10, the lower values also being associated with the lowest dielectric losses. The dielectric constant is dependent upon the temperature and (to a lesser extent) frequency of testing. It is usually reported at 60 Hz or 1,000 Hz. The dielectric constants of various common polymeric materials are shown in the Table 6.1.

[Small differences in K exist between polyethylenes of different densities (crystallinity), K decreasing slightly as density decreases, but these are very minor relative to differences between different polymers.]

TABLE 6.1
Dielectric Constants of Various Polymers Employed as Electrical Insulation

Polymer Type	Dielectric Constant (K)
Polyethylene-Low density	2.25–2.35
Polyethylene-High density	2.30–2.35
Ethylene–Vinyl Acetate Copolymer (EVA)	2.50–3.16
Ethylene–Ethyl Acrylate Copolymer	2.70–2.90
Ethylene–Propylene Copolymer (unfilled)	2.25–2.35
Polypropylene	2.22–2.26
PVC (plasticized)	3.1–10.0
Chlorosulfonated polyethylene (Hypalon)	7.0–10.0
Nylon	3.4–4.0
Polyimide (Nomex)	2.8
Cyanoethyl cellulose	13.3

Based on the principles described, one would expect that as polyolefin cable insulation ages and becomes oxidized (e.g., water trees form), K should increase slightly.

6.3.3 Dielectric Losses

We have seen from the prior discussion that the cable insulation acts as a capacitor. Most capacitors lose part of their energy under AC. The ratio of the energy dissipated through the insulation to the energy stored (in each cycle) is referred to as the dissipation factor (DF), or Tan delta (the tangent of the loss angle when the system is subjected to a sinusoidally varying applied field). This energy is lost as heat, and therefore the DF can be viewed as a measure of the efficiency of the insulation.

The lossyness has also been referred to as power factor, which is the sine of the loss angle. The power factor is not identical to the DF, but the values are nearly identical when the Tan delta is <0.1. This is relevant to low loss insulations, where Tan delta is always less than about 0.1 (and hence the difference is negligible). As losses increase, Tan delta and Sine delta start to diverge.

The losses described result from motion of the polymer as described in Section 6.2.1. As with K, losses are frequency- and temperature-dependent. Dielectric losses are also discussed in Section 6.3.4.

6.3.4 Dispersion

Now that we have discussed polarizability, dielectric constant, and dielectric losses, we will review dispersion in greater detail. The question to consider is what happens if the motion of the chains cannot respond rapidly enough to the changes in frequency, and how this influences permittivity and dielectric losses. Of course, our interest is in the 50–60 Hz range, but to understand the polymer response, it is desirable to review what happens over a very broad frequency range.

Different regions of the polymer chains will be sensitive to and respond differently to voltage stress. Different functional groups will be sensitive to different frequencies. When the "proper" frequency–functional group combination occurs, the chain portion will respond by exhibiting motion (i.e., polarization). Since this phenomenon is frequency-dependent, it should not be surprising that different responses will result from different functional group–frequency combinations. The upper portion of Figure 6.7 depicts the change in dielectric constant with frequency. The lower portion of the figure depicts the change in power factor with frequency. Referring to the top curve in Figure 6.7, we can see that at low frequencies, when stress is applied, the polar region dipoles can respond and "accept" the charge, and realign as previously described. The dielectric constant is relatively high under these conditions. As the frequency increases, no change in this effect will occur as long as the dipoles can respond. At some point as the frequency continues to increase, the chains will respond more slowly as the field is changing more rapidly. When the frequency change occurs at so rapid a rate that no rotation can occur, the charge cannot be held and the dielectric constant drops.

The reason for the change in dielectric constant with frequency is therefore apparent. It should also be noted that other parameters such as temperature will also affect

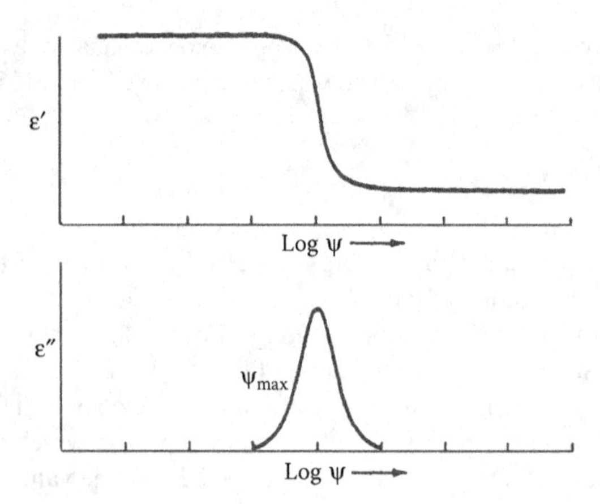

FIGURE 6.7 Permittivity and power factor as a function of frequency.

this response. In essence, any change that affects the motion of the polymer chain will affect the dielectric constant.

The point where the polymer chain segments undergo the rapid change in the rate of rotation is of special interest. The lower curve of Figure 6.7, focusing on losses (e.g., DF), shows a peak at this point. In considering DF, the same explanation applies; changes are affected by frequency and specific polymer nature. At low frequencies, the dipoles on the polymer chains follow variations in the AC field, and the current and voltage are out of phase; hence, the losses are low. At very high frequencies as noted above, the dipoles cannot move rapidly enough to respond, and hence the losses are low here also. But where the change takes place, there the losses are greatest. This can be visualized by thinking in terms of motion causing the energy to be mechanical rather than electrical in nature. As noted above, it is common in technical literature to refer to the dielectric constant and power factor at 50 or 60 Hz and at 1,000 Hz.

We referred earlier to various degrees of oxidation influencing results. For example, as seen in Chapter 5, some insulations possess additional polar functional groups that are a result of additives intentionally incorporated to generate specific properties; chloride, amine, amide, and sulfur are examples. Chloride, for example, a flame retardant additive, is a relatively large molecule, even compared to oxygen, and it is rigidly attached to the backbone in some polymers. It will impede the rotation response of the backbone to the AC stress. Different functional groups will respond differently at the same frequency, whether on the main chain or a branch.

Therefore, chain motion can be hindered due to what is referred to as the polymer viscoelastic nature. If the dipole is rigidly attached on the polymer backbone, then main chain motion is involved. If the dipole is on a branch, it can be considered to be "flexibly attached" in which case the rate of motion of the branch will be expected to differ from that of the main chain, even if the functional group is the same.

The end result of all of this is the phenomenon called dispersion. In response to the applied field, different chains will exhibit motion at different rates at any single frequency and temperature. The polymer chains may exhibit this change over a broad region rather than a sharp or localized region as the frequency and temperature is changed slightly.

For the purposes of understanding power cable insulation response, the main interest is, of course, at 50 or 60 Hz. Also, our interest is in what is intended to be relatively nonpolar systems. It is always necessary to remember that no system is perfect and there will be variations in degrees of polarity not only from one insulation material to another, and not only from one grade of the same material to another, but perhaps also from one batch of supposedly identical material to another. Much depends upon the processing control parameters during manufacture of the materials and during extrusion.

The literature reports dielectric losses of many different types of polyolefins as a function of temperature at controlled frequencies. Hence, it is known that conventional low-density polyethylene undergoes losses at various different temperatures. In addition, antioxidants, antioxidant degradation by-products, and low molecular weight molecules will also respond; this complicates interpretation. With conventional XLPE, the situation is even more complex as there are peroxide residues and cross-linking agent by-products (see Chapter 5). These low molecular weight organic molecules, acetophenone, dimethyl benzyl alcohol, alpha methyl styrene, and smaller quantities of other compounds will gradually migrate out of the insulation over time.

Till now, we have focused on polyethylene and XLPE. TR-XLPE possesses additives or more polar components (see Chapter 5) that lead to slightly increased dielectric constant and DFs.

With reference to EPR, the same principles regarding dispersion apply. In addition, it is necessary to consider the differences between the different EPR polymers themselves, and also distinguish among the filled EPR compounds employed as insulation (see Section 6.3.5). The dielectric properties of unfilled EPR are quite good; low dielectric constant and low DF. But unfilled EPR is not a practical insulation material (see Chapter 5). After the required fillers and other additives are incorporated, the dielectric constant and DF both increase. The reason is due to these additives. Any discussion of EPR must consider that all EPRs are not alike, as discussed in Sections 5.6.3 and 5.6.4.

Contaminants can be defined as "unintended non-polymeric components" that might be present. We have already noted their influence on low operating stress parameters; they can migrate under voltage stress just as do polar portions of polymer insulation chains. [Examples are ions that can enter the cable core from the outside soil or from the shields, and water.] These can also influence other cable properties.

Interpretation of data therefore requires knowledge of the system, and some degree of caution is prudent in the absence of adequate information.

6.3.5 Mineral Filled Systems and Interfacial Polarization

Up to this point in the discussion, we have reviewed what could be viewed as situations solely involving the polymer response to applied stress, and not influenced by other components. However, as seen in Chapter 5 with reference to EPR (and other elastomers employed for low voltage applications), substantial amounts of inorganic fillers and other additives are required to improve the mechanical properties in order to render them useful as cable insulation. Inorganic fillers will affect electrical properties such as dielectric constant and modify the polymer behavior. (They also influence conductivity, as well as other nonelectrical properties.) The nature of the interface where the polymer matrix and the clay additive interact is another key parameter.

An EPR compound can be considered as a heterogeneous dielectric where polarization is generated as a result of the two materials having different dielectric constants; charge is accumulated at the polymer–filler interface. This is referred to as interfacial polarization, and conductivity differences will also exist. Space charge can build up at these interfaces leading to dielectric losses.

Conventional polarization as described in Section 6.3.1 is a result of orientation due to charges that are "bound" in the polymer structure. Charge carriers can also migrate through the dielectric material (hence leakage current results, see Section 6.3.6). However, the motion of these charge carriers can be hindered or even stopped at certain interfaces, and when this occurs, space charge trapping results. This trapping leads to a distortion of the field and an increase in capacitance (but not a change in the inherent dielectric constant of the polymer). The phenomenon is a result of the different nature of the interfaces. Clearly, the nature of the polymer–filler interface is critical, and this means the particle size, shape, and processing history are relevant; it is one reason that silane-coated clays are used in modern EPR insulated cables instead of solely heat-treated (calcined) clays. This phenomenon, in principle, is relevant whenever significantly different surfaces are in contact. In principle, it also applies where an antioxidant is in contact with the polymer matrix (here, however, the concentration is trivial compared with that of mineral fillers). In the case of conductive fillers such as specific types of carbon black, interfacial polarization becomes quite significant and the dielectric constant can be very high (e.g., perhaps ~100 at commercial levels of carbon black present in semiconducting shield materials). It can also apply to water presence and defects, in which cases localized charge will accumulate.

6.3.6 Conduction

The phenomena of polarization, permittivity, and losses all relate to how charge is held in the insulation and the consequences of the polymer response to the applied voltage stress. Despite being a poor conductor, the fact is that the polymer insulation does allow some level of conductivity. A fundamental point relates to the ability of the polymeric insulation to continue to hold charge over time; in this regard, we go beyond charge motion and trapping but to charge release. This can be influenced by morphology changes of the polymer over time (not only to crystalline melting but also to chain motion in the absence of crystallinity); morphology of the clay particles (shape, size, and architecture) in an elastomer, and the nature of the polymer/filler interface (as a subset here, the nature and extent of the coating on the clay). In addition, the numerous other additives can influence events. Not all components present need be trapped initially; foreign ions may facilitate conductivity directly.

6.3.7 Cable Response

To this point, we have examined what happens to the polymer insulation materials when subjected to voltage stress. It is of merit now to compare information generated primarily on cables (rather than solely cable materials) from an engineering perspective.

The classical view of events that occur when applying DC voltage to a cable refer to three components of current flow; capacitive, absorption, and leakage. Capacitive current is seen immediately after application of DC voltage; the cable (i.e., the capacitor) becomes fully charged. There are no chemical reactions, simply charge accumulation. It is likely due to realignment of charged atoms. This phenomenon takes place rapidly and the time constant is less than few seconds. It is not a significant factor in influencing reliability.

Absorption current has also been referred to as polarization current and is the phenomenon we have been discussing in Section 6.2.1 from a materials perspective; temporary realignment of polar molecules induced by the field. This leads to dielectric losses, which are trivial under DC, but significant under AC.

Leakage current has not been discussed earlier in our review of insulation materials (however, see Section 6.3.6). For cables, leakage current is significant, as it provides an indication of the "state" of the insulation. It can be viewed as the inability of the polymer insulation to prevent charge motion; the phenomena of polarization and permittivity to induce trapping are not adequate to keep ions and electrons from passing into and along the insulation.

From a cable reliability perspective, the leakage current is the most significant of the three phenomena discussed here. The leakage current should be proportional to the applied voltage stress; if it is nonlinear, that is an indication of "weakened" insulation.

6.3.8 Paper/Fluid Systems

The significant physicochemical differences in components and construction of an extruded cable as compared with a paper/fluid system has been discussed in Section 5.10. From the perspective of the dielectric phenomena involved, the only difference is that we are now dealing with a solid/liquid system, instead of a solid (e.g., XLPE) or solid/solid (e.g., filled EPR) system. The same principles apply; indeed, they were applied to paper/fluid systems before solid dielectrics were developed.

For example, many principles were established by studying paper/fluid systems. It was known in the 1920s that:

1. The relationship between power factor and temperature was a function of the nature of the paper, as well as the nature of the impregnating compound (and also to the presence of moisture).
2. Additional stored energy (apparent dielectric constant) is due to the presence of polar molecules.
3. Water (which has a high dielectric constant) in a liquid dielectric will be drawn to the strongest part of the field (high stress site).
4. Dielectrics subjected to a potential will experience a "rush of current" (i.e., charging current) that will gradually disappear.

Power factor is, today, considered to be a significant characteristic of interest for paper/fluid systems. Related to this, another measurement referred to as "ionization factor" is relevant; this is the difference in power factor measured at two stresses (one high and one low). Also, the relationship between power factor and temperature is as relevant today as it was in the past (as any increase in losses as one approaches operating temperature could lead to thermal runaway and reliability concerns).

Overall, perhaps the most significant practical technical difference between solid dielectrics and fluid systems is that the latter provide a fluid that can impregnate voids that may develop and prevent them from leading to shortened life (see Section 6.4.4).

6.3.9 Conclusions

We have reviewed the concepts of VR, polarization, dielectric constant, dielectric loss, and interfacial polarization for insulation materials, as well as leakage current regarding cables. We have noted that:

- VR is a clearly defined property that tells how good the polymer is as an insulation material.
- Polarization is a measure of the response of the polymer dipoles to the applied stress and also influences the extent of degradation.
- Dielectric constant provides information on how well the insulation holds charge; nonpolar insulations have lower values than polar ones, and dielectric loss tells us about the energy dissipated as heat as a result of the voltage stress being applied and resulting polarization.
- Interfacial polarization is a phenomenon relevant to mineral filled systems (as well as what happens during aging).

We have also seen that, in a cable construction, the voltage stress may induce a leakage current through the insulation despite it functioning as desired; leakage current provides some practical qualitative indication of the future reliability of the cable system.

It is clear at this point that all these parameters are intimately related and a change in one affects the others.

6.4 INSULATION RESPONSE AT HIGH STRESS

6.4.1 Introduction

High stress is defined here as either higher than normal applied stress (greater than operating stress) or higher than normal localized stress. The latter refers to sites within the insulation wall that have degraded, or sites where voids or foreign contaminants are located. In these latter cases, the local stress may be higher than operating stress, even though the operating stress may be low. This section discusses dielectric strength and breakdown, and encompasses failure of insulation materials and parameters influencing the same, test methods, and partial discharges (the main cause of failure).

In reviewing insulation response at high voltage stress, the discussion will focus on cables as well as the materials. The reason is that we are eventually discussing failure, and there are issues involving cable failure mechanisms that are not relevant for insulation materials alone. An example would be partial discharge (Section 6.4.5), which is a more relevant failure mode for insulation in fabricated cables than for insulations alone; this is due to the greater probability of voids being present in cables as compared with polymer slabs. Hence, we will be reviewing both insulation and cable failure modes in part of the discussion of insulation response to high stress.

6.4.2 Dielectric Strength

When the operating stress is high, the relevant properties discussed in Section 6.3 can be superseded in importance, as the insulation is now more susceptible to failure. This occurs when the voltage can no longer be maintained without physical disruption of the insulation. A stream of electrons is released and the excess current passes through the insulation (see Section 6.4.4). The key parameter influencing the response is the dielectric strength of the insulation. The dielectric strength can be defined as the limiting voltage stress beyond which the dielectric can no longer maintain its integrity; a distinct rupture occurs leaving char and a nonfunctional insulation. The breakdown strength value is high initially, and in order for failure to occur, the applied stress must be very high. As the insulation ages under thermal or thermal/voltage stress (in the presence or absence of water), the dielectric strength drops and the increase in applied voltage stress required to induce failure is reduced. This applies to laboratory testing of insulation materials, full size cables, or cables in service.

The reason for the reduction in dielectric strength upon aging is due to defects developing upon aging, or small defects present initially becoming larger. This applies to all insulation materials (and cables). Water presence accelerates the process, and water tree development upon aging represents one well-known phenomenon leading to reduced dielectric strength of extruded cable materials and cables (see Chapter 19).

The dielectric strength is reported as the applied voltage at which the insulation fails, in volts per mil (or kV/mm) of insulation thickness.

The dielectric strength can be determined via different methods; by applying DC, AC (of different frequencies), or impulse. For cables, AC is most common (as the cables operate at 50–60 Hz) and impulse is of interest to simulate lightning and switching surges. Test methods on cables are discussed in Chapter 10. For materials testing, AC and DC are common. The rate of rise of applied AC voltage on small test samples is often greater than on full size cables (see Section 6.4.4).

Although not a direct cause of failure, mention should be made of water trees as their presence causes a reduction in dielectric strength. Water trees growing under low (normal) operating stress can penetrate the entire insulation wall and yet the cable can continue to function without causing immediate failure. Water trees consist of water-filled small voids separated by "tracks" of oxidized polymer, and water-treed regions possess higher K than the surrounding polymer matrix (at least for XLPE). Failure occurs only after water trees are converted to electrical trees; the two types of treeing defects have different shapes. The differences are summarized as follows:

WATER TREES	ELECTRICAL TREES
Water required	Water not required
Fan or bush shaped	Needle or spindle shaped
Grow for years	Failure shortly after formation
Microvoids connected by tracks of oxidized polymer	Carbonized regions

The next section discusses breakdown measurements and testing in more detail.

6.4.3 Test Methods

When studying insulation materials, several possible methods can be employed. Small samples of the insulation material in slab or block form are placed between electrodes. The electrode shape and size represent a significant parameter for small sample testing (and is always defined). Voltage stress is applied and increased until failure occurs. One common method employs a needle electrode (Figure 6.8). A sharp but controlled radius of curvature exists at the needle tip, and the latter is inserted into the specimen (flat sheet or slab) part way to the ground plane (it may be inserted while warm to avoid inducing "cracks" in semicrystalline polymers). Voltage stress is applied and raised under controlled conditions until failure occurs. Failure of the test specimen is induced at the location of the sharp needle, an intentional high stress site. The dielectric strength is recorded as kilovolts per millimeter (or V/mil). This test methodology simulates the influence of a sharp defect within the insulation, and it is not uncommon to be employed when studying the influence of additives on dielectric strength.

Another common method employs curved electrodes (Figure 6.9) which are inserted into the polymer slab or block; this provides a uniform stress gradient (in contrast to the use of sharp needles) and provides an understanding of the dielectric strength of insulation itself, which can fail anywhere at the interface of the electrode and the polymer. The electrodes are referred to as Rogowski electrodes, and the test conditions are the same.

These test configurations represent two common methods employed to determine the dielectric strength of insulation materials (not cables). In actual practice,

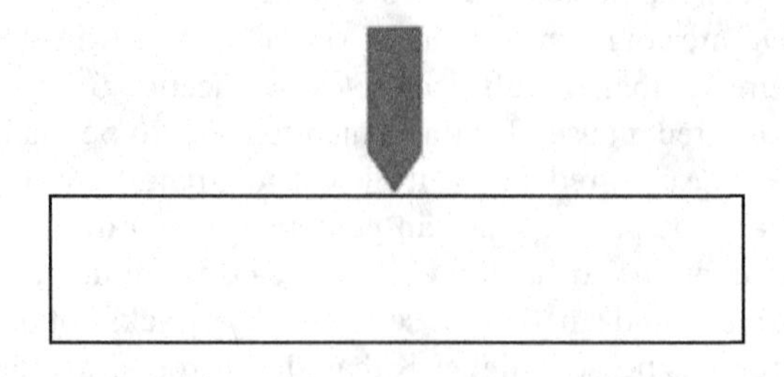

FIGURE 6.8 Depiction of needle electrode for performing breakdown test.

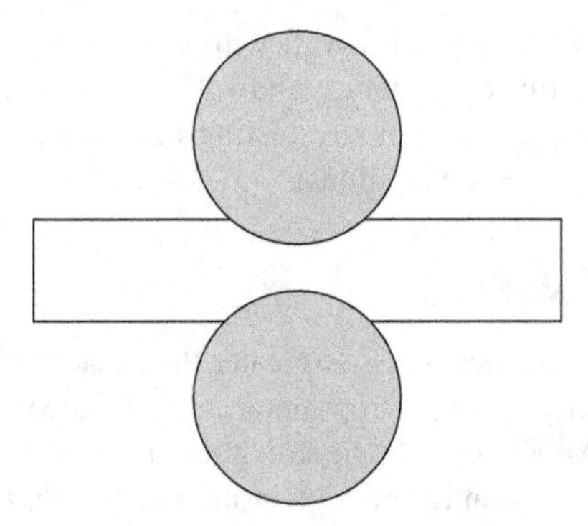

FIGURE 6.9 Curved Rogowski electrodes provide uniform voltage stress gradient during dielectric strength testing.

many variables must be controlled to ensure that reliable data is obtained. These include control of:

- Sample thickness
- Pretreatment of specimens (e.g., annealing)
- Temperature
- Environment (relative humidity)
- Electrode nature (described above)
- Test parameters
    - Rate of rise of applied voltage
    - Test frequency; one can employ DC, 50–60Hz, or impulse

Failure during testing as described must occur at the electrode/polymer interface (see Figure 6.9) and not along the periphery (edge) of the sample; if the latter occurs it is referred to as flashover, which is not a valid indication of the insulation integrity. If foreign contaminants (such as metals) are present within the sample they act as localized high stress sites and will be the region where the failure occurs. Finally, since relatively small specimens are involved (compared to full size cables), a large number of specimens are tested to overcome the inherent variability in results, and the results are analyzed statistically.

One can expect that at the interface of a high surface energy material (metal) and a low surface energy material (nonpolar polymer), there will exist a weak boundary layer.

A materials test method for inducing degradation (ASTM D-6097) that leads to failure requires mention. This test method is intended to study water tree growth, but failure is possible depending on the test conditions and the material. Small molded specimens of insulation (or as a composite with semiconducting shields) are employed. A conical defect is molded into small specimens (using a steel needle). Samples (at least 10) are aged in a defined salt water solution under voltage stress for a controlled time period (30–90 days).

Test methods for full size cables are more complex (and more expensive), as the total construction is involved (conductor, shields, and insulation) and the polymer layers are significantly thicker. For this reason, sometimes model cables are employed;

a model cable might have a ~60 mil wall with strand shield and insulation shield layers significantly less thick. Miniature and full size cable tests are performed for accelerating water tree growth. Full size testing of insulated cables is discussed in Chapter 10 (Standards and Specifications).

6.4.4 Breakdown and Failure

Failure occurs when the applied stress is greater than that the insulation material can withstand. As the dielectric strength drops upon aging, failure is more likely (at any specific operating stress). Although dielectric strength is the most common and most important parameter relative to quantification of insulation failure, there are several issues relative to the technology of measurement and interpretation that require clarification.

6.4.4.1 Intrinsic Strength

Perhaps the most obvious assumption is that the value of the measured dielectric strength would be anticipated to be the intrinsic strength of the insulation. This is defined as the dielectric strength characteristics of the material itself in its pure and defect-free state, and measured under ideal test conditions, and that allows breakdown at the highest possible voltage stress. In practice, however, this is never achieved. One reason is the difficulty in attaining defect-free pure insulation specimens. We have already referred to the small amount of oxidation that normally occurs during manufacture. In addition, the fact is that as the test specimen insulation thickness increases, the chances of extremely small (micro) voids developing as a result of the fabrication process also increases. Voids above a certain size are weak links that lead to failure (see Section 6.4.5). Hence, (for example) a 2 cm thick slab would be expected to possess more small (micro) voids than a 1 cm thick slab (or film). For thicker slabs, the problem increases and this is one of the reasons that the dielectric strength of a full size cable (e.g., 175 mil/445 mm wall thickness) in volts per mil or kilovolts per millimeter is significantly less than that of a slab or thin film of the same insulation material.

The closest one can come to intrinsic dielectric strength is by performing measurements of very thin, carefully prepared films with appropriate electrodes. Under these ideal conditions, the insulation itself would come closer to failing due to its inherent properties.

6.4.4.2 Breakdown Strength Considerations

The following points require consideration when evaluating the significance of breakdown strength information.

1. The field strength at which breakdown occurs depends on the respective geometries of the insulation and the electrodes with which the electric field is applied. For insulation materials tests and cable tests these are defined, but it is obvious that they differ significantly; this is one factor that leads to different values of dielectric strength between the materials and cable tests.
2. We have referred to the fact that the insulation materials will contain minute defects, due to processing-induced localized degradation. This leads to a

reduction of the dielectric strength that may be far below the intrinsic value of the defect-free material. Since this degree of degradation (oxidation) may differ significantly between samples (slabs or cable sections), this may manifest itself as different breakdown values for apparently similar samples. In addition, foreign ions may be nonuniformly dispersed in the insulation wall, contributing to the nonuniformity from one sample to another.

3. The apparent dielectric strength for common polymeric insulation materials at breakdown (kilovolts) increases with an increase in thickness of the test specimen; however, when normalizing the dielectric strength data for thickness, the breakdown strength in terms of volts per mil (kilovolts per millimeter) basis is reduced.
4. The measured dielectric strength is related to the rate of increase at which the voltage stress is applied. Hence, the results will vary as the test conditions are varied. This is another reason why the breakdown values (kilovolts per millimeter) obtained when an insulation fails are not absolute.

   To compensate for this, industry specifications control this parameter. Thus, referring to Figure 6.10 (see Section 6.4.4.3), the AC breakdown strength data shown was obtained employing a 5 min step rise time. What this means is that when starting the test, the voltage (always initially at about 50% of the anticipated breakdown strength) was applied and held for 5 min; the voltage was then increased (by 40 V/mil, or 1.6 V/mm) and held for another 5 min. The process is repeated by increasing the voltage by the same amount and holding for 5 min at each volt per mil (kilovolts per millimeter) value, until failure occurs. By controlling the voltage stress increase and keeping it constant, this variable is eliminated.
5. The apparent dielectric strength decreases with an increase in operating temperature; we have previously seen (Chapter 5) how increasing molecular motion as the temperature is increased influences properties.
6. It is also possible for failure to occur by thermal runaway; this can occur when the rate of temperature increase within the insulation is greater than the rate of heat dissipation. Under voltage stress, some insulation systems

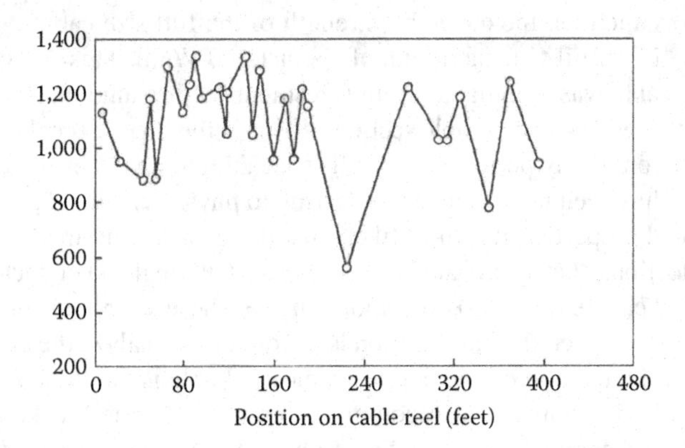

FIGURE 6.10 AC breakdown of XLPE insulated cables.

that possess high losses will generate heat, and if the rate of heating exceeds the rate of cooling (that normally occurs by thermal transfer), then thermal runaway can occur, and the insulation fails by essentially, thermally induced degradation. Several relevant points are as follows: (1) the heat transfer capability of polyolefins is low and heat dissipation is normally not rapid, (2) these events may occur in the presence or absence of discharges (see Section 6.5), and (3) the presence of additives such as inorganic fillers or organic components can contribute to increasing dielectric losses. For direct buried cables, the soil nature plays a role here.

7. For completion, it is noted here that shielded cables can fail by a different mechanism involving corrosion of the concentric neutrals, leading to erosion of the insulation shield and eventual degradation of the outer region of the extruded cable insulation.

The main lesson here is that when discussing dielectric strength, it is essential to understand how the test was performed. In comparing AC dielectric strength values for tests performed at different times and at different locations, it is vital to understand relevant test parameters. This holds true whether one is comparing different grades of the same material (different grades of polyethylene or XLPE) or different insulation materials (Polyethylene versus polypropylene or EPR). It is emphasized that it is particularly important, when comparing the properties of different grades of an insulation, to ensure that the same test conditions are employed.

6.4.4.3 Data Analysis

Measurements involving breakdown testing of polymer insulations provide values that vary among duplicate test samples; the test results are not always identical, and this statistical variation is inherent. As an example, Figure 6.10 shows the AC breakdown strengths of 30 ft lengths of full size XLPE cables obtained from a 5,000 ft reel of a 50,000 foot extrusion run. The cables tested are as identical as possible: same conductor, same strand shield, same insulation shield, same insulation, and same extrusion conditions, as well as same AC breakdown strength test procedure. Hence, despite the fact that sample lengths tested were from the same production run and from the same reel, the dielectric strength of the full size cables varies from a low of about 600 V/mil to a maximum of about 1,350 V/mil. This demonstrates that although the cable was manufactured in (presumably) the same manner, significant differences can exist between cable sections. [This is the type of breakdown strength measurement commonly performed on full size cables as an aid in characterization.]

From what has been noted above in referring to physicochemical phenomena that affect electrical properties, it is most likely that these variations are due to the inevitable imperfections that result during processing (although other factors noted in Section 6.4.4 2 can be involved). Variations such as these are not uncommon.

The manner in which this information is handled is to analyze the data in a statistical manner. The method considered to be most reliable is the Weibull distribution (log normal distributions are also commonly employed). The breakdown strength values are plotted against the cumulative probability of failure. Thus, if there are 10 samples that were tested and failed, the individual breakdown strengths are plotted on the X axis with the first sample representing 10% of the total tested samples on the Y axis, the second value would be 20 %, and so forth. If there were 20 breakdown strength tests performed, the first value plotted would be the 5% value. (If there were only four samples tested, the first value would represent 25%.)

In an ideal situation, a straight line is obtained (commercially available software is employed, although in the past, manual handling of data on probability paper was performed). The least squares fit of this straight line provides the two parameters of interest: (a) the scale parameter, which is the 63.2% probability of failure, and (b) the shape parameter, which is the slope of the line. The scale parameter is what is generally reported in the technical literature; however, one is often interested in the early failures. This is where the shape parameter is important as it provides guidance on the spread of the failure data. A vertical slope indicates a uniform cable, while a slope that trends away from vertical is indicative of nonuniformity and differences between samples (insulation slabs or cable) lengths.

Thus, a breakdown strength value considered to be the "best value" for multiple samples tested in an identical manner requires supplemental information on the spread of the data to be truly meaningful.

6.4.5 Partial Discharge

Partial discharge is due to decomposition of air in the voids within the insulation material wall or at the surface–shield interface. The discharges lead to eventual degradation and failure of the polymer. This process, which involves electron and ion bombardment, is a major cause of insulation failure in cables. Void presence may be due to impurities that lead to poor interfacial contact, aging-induced degradation, or less likely, perhaps, is related to the manufacturing process. This section covers the basic fundamentals from the perspective of electrical properties of polymeric insulation. Aspects of partial discharge events not directly related to insulation materials phenomena are noted here, but not covered in this chapter: these include high frequency signals that travel along the conductor and shield, electromagnetic waves that radiate in the surrounding area, and light and acoustic wave emission. For a discussion of the significance of discharge patterns or pulse magnitudes, references should be consulted.

6.4.5.1 What Happens in Voids

Small voids may be present within the insulation of extruded cables due to aging-induced changes leading to their development. Voids above a certain size are not present in extruded cables that are required to meet industry specification requirements; partial discharge testing is performed on newly manufactured cables precisely to prevent cables with voids of certain sizes (and quantities) to reach the user.

The dielectric strength of air (in the void) must be understood. The greater the dielectric strength of the insulation material, the greater the resistance to degradation under discharge conditions. Thus, while the dielectric strength of the thin films of (polyethylene or XLPE) insulation itself is very high (perhaps as high as 16,000 V/mil), that of air is lower by 2–3 orders of magnitude and it is more susceptible to degradation by the accelerated electrons.

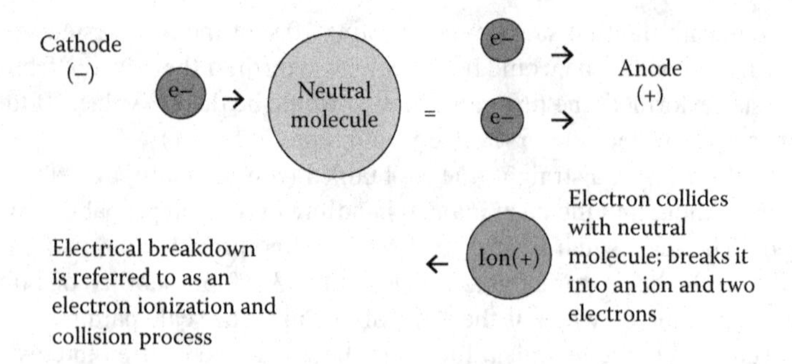

FIGURE 6.11 Electron decomposes air molecule to yield additional electrons and anions.

Void size (diameter), shape, pressure, and temperature all have a significant influence on the discharge events that occur (see below). Different insulation materials can respond differently when subjected to the energized electrons, but the fundamental degradation response of air will not change.

The air degradation process leads to additional electrons and ions being formed. This is depicted in Figure 6.11.

Degradation of the air, which is ~80% nitrogen, can lead to ions and other degradation products that possess nitrogen (e.g., nitrogen oxides). Once the newly formed ions and electrons are generated, they continue the process; attacking the remaining air in the void. This is depicted in Figure 6.12.

At some point a threshold is passed and an avalanche occurs; this avalanche is referred to as the partial discharge inception voltage (PDIV) (see Section 6.5.4.3.1). After the avalanche occurs, the voltage across the void momentarily drops to zero (or almost zero). This is the partial discharge extinction voltage (PDEV). For the discharge process to continue and additional avalanches to occur, the voltage must build up again. It is this repetitive process of degradation of the air in the voids that is so harmful. Hence, discharges lead to avalanches within the voids that cause air degradation.

6.4.5.2 What Happens to the Insulation

As the repetitive avalanche process continues, the insulation wall will eventually be attacked at the solid/gas interfacial region. This leads to polymer chain scission

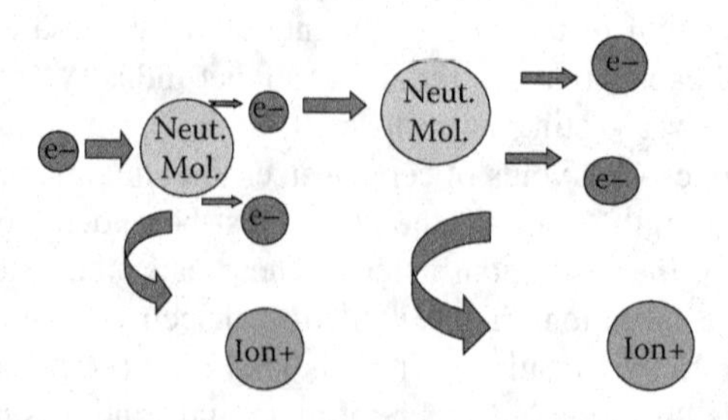

FIGURE 6.12 Generated electrons react with additional neutral molecules.

and additional degradation products; gases such as carbon monoxide and dioxide, methane, and other low molecular weight hydrocarbons are generated. Inorganic carbonates also form; this is in addition to the degradation products caused by the degradation of air. Furthermore, electrons, ions, and other degradation products can be deposited at the surface of the insulation (which is now more highly oxidized and therefore more polar). The trapped charges may remain there (for a finite time) and be released at a later time. As the process continues, the degradation process works its way deeper and into the region away from the original void and from the polymer surface. In place of the original polymer insulation, there now exists degraded polymer, a highly oxidized black char residue that renders the insulation noncontinuous. This is part of an electrical tree, a hollow channel that is filled with air and other gases from the degradation process. As this process continues further, the degraded region will eventually reach the surface of the insulation causing a breach in the insulation continuity. Since this degraded polymer is ruptured, it is no longer insulating and failure occurs.

This degradation mechanism is more likely to occur with cables than with thin films or slabs of the same polymeric insulation material. The processing methodology for preparing films or slabs (e.g., injection molding) is more likely to remove any very small voids that could persist during the extrusion process employed for cables (and grow upon aging), e.g., a conventional medium voltage distribution cable may have an insulation thickness of 175 mils versus 1–5 mils for a slab. However, needle tests on slabs (see Chapter 5) can lead to void development.

For paper/fluid systems, the possibility of void formation at butt spaces during load cycling has been discussed (Section 5.10.2.3).

Although the events in the voids and in the insulation are discussed separately in Section 6.4.5.1 and Section 6.4.5.2, they can occur simultaneously as the process continues.

6.4.5.3 Key Partial Discharge Phenomena Parameters

Some major concepts related to influencing partial discharges that affect degradation rate are introduced here. For further details, references should be consulted.

6.4.5.3.1 Partial Discharge Inception Voltage and Partial Discharge Extinction Voltage

PDIV will be influenced by the following factors:

- *Geometry of the void:* In most cases, the voids are spherical or ellipsoidal. However, a void adjacent to a high energy surface such as a metal particle may be elongated. The "sharper" the tip of the void, the lower the PDIV.
- *Location in the cable wall:* For any void of constant diameter/shape, the closer to the conductor (i.e., the higher stress regions), the lower the PDIV.
- *Pressure of the gas (air) in the void:* Increasing pressure increases PDIV; however, this parameter may vary as a result of the discharge process. Thus, the gases produced by degradation of air or of the insulation wall can lead to dynamic pressure changes.

- *Void size:* Larger voids are more susceptible to discharges (all other factors being equal) as the probability of the air in the larger volume being attacked by an electron is greater.

After the discharge takes place, the voltage drops and at some point, ceases entirely; this is referred to as the PDEV. Then the process starts over again.

6.4.5.3.2 Time Lag

The statistical time lag is the time required for a free electron to initiate the discharge once the voltage across the void reaches the actual breakdown voltage. [The longer this time lag, the greater will be the actual breakdown voltage developed (within the void) before breakdown occurs.] The existence of the time lag can be viewed from the perspective of the major components; what is happening to the air (oxygen and nitrogen) as well as the polymer insulation. Oxygen is an electronegative gas, which means that it will attract electrons to a greater extent than will nitrogen. In the early stages, electrons present are "trapped" by the electronegative gas, resulting in a relatively long waiting time (reduced availability of the initiating electron) for degradation to start. As the oxygen is "consumed," the waiting time is diminished. This waiting time can be, initially, in the order of minutes, and drops later to milliseconds.

This phenomenon is intimately related to what happens not only in the void but at the polymer surface also. As the polymer chains become oxidized, the ability of the modified polymer to trap electrons is increased. Trapped electrons that result early in the process can be released later in the process. [At the beginning, there are very few electrons that can be released in this manner.] This overall process is referred to as the statistical time lag, which is "long" at the beginning but becomes "shorter" as trapped electrons are released.

It should not be forgotten that when discharges occur, ions are deposited at the opposite end of the void.

6.4.5.3.3 Residue Voltage

The collapse of the voltage to zero when PDEV occurs may actually not fully occur. In other words, some "residual" voltage across the void may persist in the absence of discharges. This is not necessarily uncommon, but is generally small. (In the extreme case, not relevant to cables, continuous glow discharges can occur if the residual voltage is large and approaches the actual breakdown voltage.) Residue voltage is related to void size, partial discharge repetition, and conductivity of the degraded polymer insulation wall. The significance of residue voltage is that voids that would not exhibit partial discharge early in the process may do so after the residue voltage builds up due to subsequent discharges.

6.5 SUMMARY

Electrical properties have been reviewed from the perspective of what happens when insulation materials (and cables) are subjected to low and high voltage stress. VR and insulation resistance are defined. Under low stress, the chemical structure of polymeric insulation materials influences the dielectric constant, DF, and polarization of the insulation molecules. The phenomenon of dispersion has been reviewed; the ability of the polymer insulation molecules (of varying molecular weights and degree of polarity) to respond to changes in frequency, and how this influences power factor and permittivity.

These properties are significant at operating stress. Polyolefins such as polyethylene, XLPE, and unfilled ethylene-propylene copolymers have low dielectric constants and low DFs. Low levels of oxidation, generally resulting from processing the polymer or aging-induced changes, lead to slight increases in these properties. Insulation materials with inorganic fillers (such as EPR) or additives that are polar in nature (such as some TR-XLPE insulation materials) have increased dielectric constants and DFs. The phenomenon of interfacial polarization with reference to mineral-filled materials is discussed.

At higher than normal operating voltage stress, the insulation's ability to maintain its continuity and integrity is challenged. Higher than normal stress can be either the applied stress, or even higher than normal localized stress. The key property here is the insulation's dielectric strength, that can drop on aging and is followed by breakdown. Test methods for determining dielectric strength of polymeric insulation materials are discussed and the normal spread of dielectric strength data is analyzed.

Finally, the subject of partial discharge is introduced. Partial discharge is a failure mechanism resulting from decomposition of air in voids within the insulation. The decomposition products attack the air and the insulation itself, eventually causing electric trees and failure. Partial discharge (PD) is a very complex topic and a few key concepts resulting in degradation of the insulation from electron and ion bombardment are reviewed [inception and extinction voltages, time lag, (for PD to initiate) and residue voltage (which influences breakdown on subsequent discharges).

REFERENCES

1. Alger, M., 1997, "Polymer Science Dictionary," Springer Press.
2. Bartnikas, R. (ed.), 2000, "Power and Communication Cables," McGraw Hill and IEEE Press.
3. Dissado, L. A. and Fothergill, J. C., 1992, "Electrical Degradation and Breakdown in Polymers," G. C. Stevens (ed.), Peter Peregrinus Ltd.
4. Dunsheath, P., 1929, "High Voltage Cables," Sir Isaac Pitman and Sons Ltd.
5. Ku, C. C. and Liepens, C. R., 1987, "Electrical Properties of Polymers: Chemical Principles," Hanser Publ.
6. Miller, M. L., "The Structure of Polymers," SPE Polymer Science and Engineering Series, Reinhold Book Corporation.
7. Nobile, P. and LaPlatney, C., September–October 1987, "Field Testing of Cables: Theory and Practice," IEEE Transactions on Industry Applications, Vol. 1A-23 (No. 5).
8. Nelson, W. B., 2005, "Applied Life Data Analysis," John Wiley and Sons.

7 Shielding of Power Cables

Carl C. Landinger

CONTENTS

7.1 GENERAL

Shielding of an electric power cable is accomplished by surrounding the insulation of a single conductor or assembly of insulated conductors with a grounded, conducting medium. This confines the electric field to the inside of this shield. Two distinct types of shields are used: metallic and a combination of nonmetallic and metallic.

The purposes of the insulation shield are to:

1. Obtain symmetrical radial stress distribution within the insulation.
2. Eliminate tangential and longitudinal stresses on the surface of the insulation.
3. Exclude from the electric field those materials such as braids, tapes, and fillers that are not intended as insulation.
4. Protect the cables from induced or direct over voltages. Shields do this by making the surge impedance uniform along the length of the cable and by helping to attenuate surge potentials.

7.2 CONDUCTOR SHIELDING

In cables rated over 2,000 volts, a conductor shield is required by industry standards. The purpose of the conductor shield, also called conductor screen, over the conductor is to provide a smooth cylinder rather than the relatively rough surface of a stranded conductor in order to reduce the stress concentration at the interface with the insulation.

Conductor shielding has been used for cables with both laminar and extruded insulations. The materials used are either semiconducting materials or ones that have a high dielectric constant and are known as stress control materials. Both serve the same function of stress reduction.

Conductor shields for paper-insulated cables are either carbon black tapes or metalized paper tapes.

The conductor shielding materials were originally made of semiconducting tapes that were helically wrapped over the conductor. Present standards still permit such a tape over the conductor. This is done, especially on large conductors, in order to hold the strands together firmly during the application of the extruded semiconducting material that is now required for medium voltage cables. Experience with cables that only had a semiconducting tape was not satisfactory, so the industry changed their requirements to call for an extruded layer over the conductor.

In extruded cables, this layer is now extruded directly over the conductor and, in the case of semiconducting shields, is bonded to the insulation layer that is applied over this stress relief layer. It is extremely important that there are no voids or extraneous material between those two layers since this is the area of maximum voltage stress in a cable.

Present-day semiconducting extruded layers are made of clean materials (a minimum of undesirable impurities) and are extruded to be very smooth and round. This has greatly reduced the formation of water trees that could originate from irregular surfaces (commonly known as protrusions) because of high electrical stress. By extruding the two layers simultaneously, the conductor shield and the insulation are cured at the same time. This provides the inseparable bond that minimizes the chances of the formation of a void at that critical interface.

For compatibility reasons, the extruded shielding layer is usually made from the same or a similar polymer as the insulation. Special carbon black is used to make the layer over the conductor semiconducting to provide the necessary conductivity. Industry standards require that the conductor semiconducting materials have a maximum resistivity of 1,000 ohm-meter. Those standards also require that this material pass a longtime stability test for resistivity at the emergency operating temperature level to ensure that the layer remains conductive and hence provides a long cable life. This procedure is described in Reference [1].

While not widely done, a water-impervious material can be incorporated as part of the conductor shield to prevent radial moisture transmission. This layer consists of a thin layer of aluminum or lead sandwiched between semiconducting materials. A similar laminate may be used for an insulation shield for the same reason.

There is no definitive standard that describes the class of extrudable shielding materials known as "super smooth, super clean." As will be described in Chapter 10 (Standards and Specifications), it is not usually practical to use a manufacturer's trade name or product number to describe any material. The term "super smooth, super clean" is the only way at this time of writing to describe a class of materials that provide a higher quality cable than an earlier version. This is only an academic issue since the older types of material are no longer used for medium voltage cable construction by known suppliers. The point is that these newer materials have tremendously improved cable performance in laboratory evaluations.

7.3 INSULATION SHIELDING FOR MEDIUM VOLTAGE CABLES

The insulation shield for a medium voltage cable is made up of two components:

1. A semiconducting or stress relief layer
2. A metallic layer of tape or tapes, drain wires, concentric neutral wires, or a metal tube

They must function as a unit for a cable to achieve a long service life.

7.3.1 Stress Relief Layer

The polymer layer used with extruded cables has replaced the tape shields that were used many years ago. This extruded layer is called the extruded insulation shield or screen. Its properties and compatibility requirements are similar to that of the conductor shield previously described except that standards require that the volume resistivity of this external layer be limited to 500 ohm-meter. This lower resistivity value, as compared with the conductor shielding value, recognizes that the metallic shield component may not be in continuous contact (such as space between wires) and the fact that it is possible that workers could come in contact with the outer layer while the cable is energized.

The nonmetallic layer is directly over the insulation and the voltage stress at that interface is lower than at the conductor shield interface. Primarily to facilitate splicing and terminating operations, this outer layer is not required to be bonded for cables rated through 46 kV. At voltages above that, it is strongly recommended that this layer be bonded to the insulation.

Since most users want this layer to be easily removable, the Insulated Cable Engineers Association (ICEA) S-94-649-2004 [2] has established strip tension limits. Presently these limits are that a 1/2-inch wide strip cut parallel to the conductor peel off with a minimum of 3 pounds and a maximum of 24 pounds of force that is at a 90° angle to the insulation surface.

7.3.2 Metallic Shield

The metallic portion of the insulation shield or screen is necessary to provide a low resistance path for charging current to flow to ground. It is important to realize that the extruded shield materials will not survive a sustained current flow of more than a few milliamperes. These materials are capable of handling the small amounts of charging current, but cannot tolerate unbalanced or fault currents.

The metallic component of the insulation shield system must be able to accommodate these higher currents. On the other hand, an excessive amount of metal in the shield of a single-conductor cable is costly in two ways. First, additional metal over the amount that is actually required increases the initial cost of the cable. Second, the greater the metal content (conductivity) of the insulation shield, the higher the shield losses that result from the flow of current in the central conductor. This subject is treated completely in Chapter 14 (Ampacity of Cables).

A sufficient amount of metal must be provided in the cable design to ensure that the cable will activate the backup protection in the event of any cable fault over the life of that cable. There is also the concern for shield losses. It therefore becomes essential that:

- The type of circuit interrupting equipment be analyzed. What is the design and operational setting of the fuse, recloser, or circuit breaker?
- What fault current will the cable encounter over its life?
- The level of shield losses that can be tolerated is known. How many times is the shield to be grounded? Will there be shield breaks to prevent circulating currents?

Although there are constructions such as full and one-third neutral listed in ICEA standards for single-conductor, URD and UD cables, these may not be the designs that are the most economical for a particular installation. Studies have been published on the optimum amount of metal to use in the neutral [3,4]. Documents such as these should be reviewed prior to the development of a cable design. In Chapter 14, there is an in-depth discussion of shield losses.

7.3.3 Concentric Neutral Cables

When concentric neutral cables are specified, the concentric neutrals must be manufactured in accordance with applicable ICEA standards. These wires must meet ASTM B3 for uncoated wires or B33 for coated wires. These wires are applied directly over the nonmetallic insulation shield with a lay of not less than six or more than ten times the diameter over the concentric wires. These cables are intended to carry the circuit neutral currents in the shield as well as performing the functions presented for cables in which the metallic shield is not intended to carry neutral current. This may result in the need for additional metal in the shield and acceptance of higher shield losses.

A complicating factor is the growing presence of harmonics in circuits where the sine wave is altered. These may, if sufficient harmonic content exists, require additional neutral capacity.

7.4 SHIELDING FOR LOW VOLTAGE CABLES

Shielding of low voltage cables is generally required where inductive interference can be a problem.

In numerous communication, instrumentation, and control cable applications, small electrical signals may be transmitted through the cable conductor and amplified at the receiving end. Unwanted signals (noise) due to inductive interference can be as large as the desired signal. This can result in false signals or audible noise that can affect voice communications.

Across the entire frequency spectrum, it is necessary to separate disturbances into electric field effects and magnetic field effects.

7.4.1 Electric Fields

Electric field effects are those that are a function of the capacitive coupling or mutual capacitance between the circuits. Shielding can be effected by a continuous metal shield to isolate the disturbed circuit from the disturbing circuit. Even semiconducting extrusions or tapes supplemented by a grounded drain wire can serve some shielding function for electric field effects.

7.4.2 Magnetic Fields

Magnetic field effects are the result of a magnetic field coupling between circuits. This is a bit more complex than for electrical effects.

At relatively high frequencies, the energy emitted from the source is treated as radiation. This increases with the square of the frequency. This electromagnetic radiation can cause disturbances at considerable distance and will penetrate any "openings" in the shielding. This can occur with braid shields or tapes that are not overlapped. The type of metal used in the shield can also affect the amount of disturbance. Any metallic shield material, as opposed to magnetic metals, will provide some shield due to the eddy currents that are set up in the metallic shield by the impinging field. These eddy currents tend to neutralize the disturbing field. Nonmetallic, semiconducting shielding is not effective for magnetic effects. In general, the most effective shielding is a complete steel conduit, but this is not always practical.

The effectiveness of a shield is called the "shielding factor" and is given as:

$$SF = \frac{\text{Induced voltage in shielded circuit}}{\text{Induced voltage in unshielded circuit}} \tag{7.1}$$

Test circuits to measure the effectiveness of various shielding designs against electrical field effects and magnetic field effects have been reported by Gooding and Slade [5].

This subject is also discussed in Chapter 9.

REFERENCES

1. Insulated Cable Engineers Association Publication T-25-425-2003, "Guide for Establishing Stability of Volume Resistivity for Conducting Polymeric Compounds of Power Cables," 1981, Global Engineering Documents, 15 Inverness Way East, Englewood, CO 80112, USA.

2. Insulated Cable Engineers Association Standard S-94-649-2004, "Standard for Concentric Neutral Cables Rated 5,000-46,000 Volts," Global Engineering Documents, 15 Inverness Way East, Englewood, CO 80112, USA.
3. EPRI EL-3014 and EL-3102, RP-1286-2: "Optimization of the Design of Metallic Shield/Concentric Neutral Conductors of Extruded Dielectric Cables Under Fault Conditions," EPRI, P.O. Box 10412, Palo Alto, CA 94303-0813.
4. EPRI EL-5478, RP-2839-1: "Shield Circulating Current Losses in Concentric Neutral Cables," EPRI, Palo Alto, CA.
5. Gooding, F. and Slade, H., July 1955, "Shielding of Communication Cables," *AIEE Transactions*, Vol. 74(Part I), pp. 532–579.

8 Sheaths, Jackets, and Armors

Carl C. Landinger

CONTENTS

8.1 SHEATHS

The terms "sheaths" and "jackets" are frequently used as though they mean the same portion of a cable. In this chapter, sheath is the term that applies to a metallic component over the insulation of a cable. An example is the lead sheath of a paper-insulated, lead-covered cable.

Various metals may be used as the sheath of a cable such as lead, copper, aluminum, bronze, steel, etc. A sheath provides a barrier to moisture vapor or water ingress into the cable insulation. It is necessary to use such a sheath over paper insulation, but it also has a value over extruded materials because of water ingress.

The thickness of the metal sheath is covered by ICEA and AEIC standards and specifications, but there are some constructions that are not covered. The thickness is dependent on the forces that can be anticipated during the installation and operation of the cable. Designs range from a standard tube to the ones that are longitudinally corrugated. The bending radius of the finished cable is dependent on such configurations.

To fully utilize the metal chosen, one should consider first cost, ampacity requirements—especially during fault conditions, and corrosion [1,2].

8.2 THERMOPLASTIC JACKETS

The term jacket should be used for nonmetallic coverings on the outer portions of a cable. They serve as electrical and mechanical protection for the underlying cable materials.

There are many materials that may be used for cable jackets. The two broad categories are thermoplastic and thermosetting. For each application, the operating temperature and environment are important factors that must be considered.

8.2.1 Polyvinyl Chloride

Polyvinyl chloride (PVC) is the most widely used nonmetallic jacketing material in the wire and cable industry. Starting in 1935, when it first became available, the use of PVC grew rapidly because of its low cost, its easy processing, and its excellent combination of overall properties including fire and chemical resistance.

PVC belongs to a group of polymers referred to as vinyls. The unmodified polymer contains approximately 55% chlorine. It is fairly linear in structure (few side chains) with approximately 5%–10% crystallinity. The material must be compounded with additives such as fillers, plasticizers, and stabilizers to attain flexibility, heat resistance, and low temperature properties. General purpose jacketing materials normally possess good physical strength, moisture resistance (but not as compared to polyethylene), adequate oil resistance, good flame resistance, and excellent resistance to weathering and to soil environments. Both flame resistance and low temperature flexibility can be improved (within limits) by the use of additives.

General purpose PVC compounds are recommended for installation at temperatures above −10°C, but specially formulated compounds may be used as low as −40°C.

One of the limitations of PVC jacketed cable is its tendency to creep under continuous pressure. For this reason, cables that are to be supported vertically with grips should not have PVC jackets. Hypalon or Neoprene is recommended for such use.

In the low voltage field, PVC is widely used as a single layer of material where it functions both as insulation and as a jacket. Since PVC is usually a thermoplastic material, it cannot withstand high temperatures. Under high fault current conditions, the insulation can be permanently damaged by melting or can emit plasticizers and become stiff and brittle over a period of time. For this reason, it is not used for utility secondary network cable. Similarly, in industries that handle large amounts of heated material, or where there is the possibility of excessive heat, the use of PVC is avoided because of its tendency to melt or deform when heated to a high temperature. Under continuous direct current (DC) voltage in wet locations, as in battery operated control circuits, single-conductor PVC-insulated cables have frequently failed due to electroendosmosis (water vapor ingress created by voltage stress).

A large percentage of chlorine can be released during a fire. When combined with moisture, hydrochloric acid may be produced. It may leach into concrete structural members and attack and weaken the steel. This situation highlights one of the major problems that can result from the use of PVC. PVC has fallen out of favor for electric utility underground medium voltage power cables because the moisture resistance is not as good as polyethylene in the common case.

8.2.2 Polyethylene (Insulating)

Polyethylene has been widely used as a jacket for underground cables since it became commercially available in large quantities in about 1950. For use as a jacket, polyethylene may be compounded with carbon black or coloring material, and with stabilizers. Carbon black gives the material the necessary sunlight protection for outdoor use.

Polyethylene for jacketing is categorized under three different densities:

- Low density: 0.910–0.925 grams per cm^3
- Medium density: 0.926–0.940 grams per cm^3
- High density: 0.941–0.965 grams per cm^3

Density generally affects the crystallinity, hardness, melting point, and general physical strength of the jacketing material. In addition to density, molecular weight distribution is important since it influences the processing and properties of the polymer.

Polyethylene jackets are an excellent choice where moisture resistance is a prime design criterion since it has the best moisture resistance of any nonmetallic jacket material. When polyethylene is used as a jacket material, it should be compounded with enough carbon black to prevent ultraviolet degradation. Linear, low density, high molecular weight (LLDPE) is the most popular jacket material since it has better stress-crack resistance than the high-density materials. High density provides the best mechanical properties, but may be very difficult to remove from the cable.

In evaluating fillers, both black and nonblack, it has been found that although many of these materials improve the aging characteristics, carbon black is by far the best. It has also been found that the aging resistance increases with carbon black loading from 2% to 5%. Normally, a 2.5%–3.0% loading is used.

Although polyethylene has good moisture resistance and good aging properties in its temperature limits, it has poor flame resistance. This discourages using it as a jacket in many circumstances. Polyethylene jackets have good cold bend properties since they will pass a cold bend test at about –55°C. They are extremely difficult to bend at low temperature because of their stiffness. Like PVC, polyethylene generally is a thermoplastic material and melts at elevated temperatures. This temperature will vary slightly with molecular weight and density, but melt occurs at about 105°C–125°C largely depending on density.

High-density polyethylene (HDPE) has been used extensively as the second (outer) layer for "ruggedized" thermoplastic in secondary and low voltage street light cables because of its toughness.

While black polyethylene for jacketing is most often an insulating material, with higher loadings of carbon black it can be a semiconducting material.

8.2.3 Semiconducting Jackets

This material has been used for over 40 years in direct-buried applications to improve the grounding of the concentric neutral.

The semiconducting material used as a jacket for medium voltage underground residential distribution (URD) cable is a resistive thermoplastic composition. This requirement is met by a polymer containing a suitable conductive material as a filler (commonly carbon black). When semiconducting jackets were first employed in the 1970s, the only available thermoplastic semiconducting materials were the compositions being used for conductor and insulation shields. Several materials were available that were based on thermoplastic ethylene/ethyl acrylate (EEA) or ethylene/vinyl acetate (EVA) copolymers and carbon black. These polymers are relatively amorphous and can tolerate the high filler loading required for conductivity while providing reasonably good physical properties. Other polymers are used in special circumstances. ANSI/ICEA S-94-649 recognizes Type I and Type II semiconducting jackets. The basic difference being that the Type II jacket has a higher use temperature. This is the well-known deformation resistant thermoplastic (DRTP) shield that was used as a strippable insulation shield on URD cable prior to the introduction of the triple extruded cross-linked strippable insulation shield in the 1980s.

It is important that the neutral wires be encapsulated by the jacket. When encapsulation is complete, there is good surface contact and conductivity to ground is ensured. Without this contact, the grounding of the cable through the jacket is compromised. Also if a pin hole in the jacket should develop with well-encapsulated neutrals, there is no opportunity for ground moisture to travel longitudinally along the cable and pool at low spots. There is additional discussion of semiconducting compounds in Chapter 5.

The use of the available thermoplastic semiconducting materials for URD cable jackets in the 1970s and 1980s has been reasonably successful and many miles of cable with these semiconducting jackets are in service. There were several shortcomings that became apparent over time: some of the early semiconducting materials would become less conductive at elevated temperatures. ICEA Publication T-25-425 was developed to address this issue and improved, more conductive compositions were developed. When AEIC CS5 introduced the "Thermo-mechanical bending test," these jackets often cracked and they were occasionally found cracked in the field as well. Finally, the moisture transmission through these jackets was in the order of 10 times greater than the LLDPE insulating jackets that reduce the progress of moisture-induced treeing in the insulation. To address this issue, a new class of thermoplastic semiconductive material was developed in the early 1990s. This

TABLE 8.1
Properties of Jacketing Materials

	Semiconducting	Insulating
Radial res., ohm-m	40	na
Tensile, psi	1,700	2,200
Elongation, %	400	600
Brittleness, °C	–50	–60
Moisture vapor transmission	1.5	0.8

material performs very nearly up to the values attained by LLDPE insulating jackets. Moisture transmission is nearly as low as with the insulating jacket and the new material consistently passes the thermomechanical test in all cable sizes and constructions. The new jacket material also does not adhere to the cross-linked insulation shield and no separator is required (Table 8.1).

Over the years, the availability of high quality thermoplastic semiconducting materials for jackets on URD cables has evolved. The current materials perform well in all respects.

Cables with semiconducting jackets have been successfully used with no reports of cable failures due to corrosion or corrosion failures of any adjacent equipment due to proximity with the semiconducting jacket.

Semiconducting jackets are being used over URD type cables as a deterrent to corrosion and to provide effective grounding to the cable during lightning strikes and faults. Semiconductive jackets reduce the neutral-to-ground impulse voltage by improving the grounding efficiencies.

Another value of a semiconducting jacket is that the jacket does not have to be removed for the installation of more frequent ground connections—only four per mile as required by the National Electrical Safety Code rather than eight. Such grounding points increase the chance of water penetration as well as other construction errors.

8.2.4 Polypropylene

Polypropylene is a very rigid thermoplastic, typically harder than polyethylene and similar in electrical properties to polyethylene. It is commonly thought of as an alternative to HDPE. Polypropylene has a lower specific gravity than HDPE.

8.2.5 Chlorinated Polyethylene

Chlorinated polyethylene (CPE) can be made either as a thermoplastic or as a thermosetting jacket material. As a thermoplastic material, it has properties very similar to PVC, but with better higher temperature properties and better deformation resistance at high temperatures than PVC. CPE jackets also have better low temperature properties than PVC unless the PVC is specifically compounded for this property.

8.2.6 Thermoplastic Elastomer

Thermoplastic elastomer (TPE) is a thermoplastic material with a rubber-like appearance. It is a form of crystalline polyethylene and it comes in various types. It can be compounded for use as an insulation or a jacketing material. By use of compounding techniques, a good electrical insulation can be developed with good moisture resistance properties.

Also, a jacketing material can be compounded to provide flame resistance, low temperature performance, good abrasion resistance, and good physical properties. This material is relatively new as compared with the thermoplastics previously mentioned, but appears to be a very versatile material.

8.2.7 Nylon

Nylon is a thermoplastic with many properties that make it desirable for jacketing of wire and cable. Nylon has relatively high strength, and is tough but rather stiff especially in cold weather. Nylon also has good impact fatigue and, within limitations, good abrasion resistance. A very important feature is the low coefficient of friction in contact with conduit materials. This is an aid in pulling the cables into conduits. Nylon has excellent resistance to hydrocarbon fuels and lubricants as well as organic solvents. However, strong acids and oxidizing agents will attack nylon. The most common use of nylon in cable jacketing is the jacket on THHN and THWN building wire.

8.2.8 Low Smoke Zero Halogen (LSZH) (Also Low Smoke Halogen Free, LSHF) Jackets

LSZH jackets involve a number of polymers formulated to release limited smoke and no halogens when burned. LSZH jackets may be thermoplastic or thermosetting. This increases the escape potential from buildings or other spaces during a fire and reduced soot damage to facilities. The absence of halogens increases human safety (inhalation damage) and reduced corrosion of facilities associated with a fire. At this time of writing, pulling compounds must be selected with care as LSZH is more susceptible to jacket cracking from exposure to such. LSZH jackets commonly have large filler contents and may therefore have poorer chemical, mechanical, moisture, and electrical properties than non-LSZH compounds. LSZH compounds are widely used in Europe and North America and continue to grow in popularity.

8.3 THERMOSETTING JACKET MATERIALS

Thermosetting jackets are not widely used for underground distribution cables except for the special case of medium- or high-density cross-linked polyethylene that is used as the outer layer on two layer, ruggedized secondary cables. Thermosetting jackets are more commonly utilized in industrial and power plant applications.

8.3.1 Cross-linked Polyethylene

Cross-linked polyethylene, with the addition of carbon black to provide sunlight resistance, provides a tough, moisture-, chemical-, and weather-resistant jacket material. The medium- and high-density materials are especially tough and are widely used as the outer layer on two layer ruggedized secondary cables. Only limited use is found for other purposes.

8.3.2 Neoprene

Neoprene has been used as a jacketing material since 1950 for large power cables such as paper-insulated, lead-covered cables and portable cables. Compounds of Neoprene usually contain from 40% to 60% by weight of Neoprene that is compounded with other ingredients to provide the desired properties such as good heat resistance, good flame resistance, resistance to oil and grease, and resistance to sunlight and weathering. Moisture resistance can be compounded into the material when required.

Properties that can be varied by compounding techniques are: improved low temperature characteristics, improved physical strength, and better moisture resistance. Most Neoprene compounds have good low temperature characteristics at −30°C to −40°C. Special compounding can lower this to −60°C, but other properties, such as physical strength, have to be sacrificed.

Because of its ruggedness, tear resistance, abrasion resistance, flame resistance, and heat resistance, Neoprene is a widely used jacketing material in the mining industry. This is probably the most severe application for cables from a physical standpoint. The thermosetting characteristics of Neoprene are desirable in this application since these cables must withstand high temperature while installed on cable reels. Thermoplastic jacketing materials would soften and deform under such environments.

8.3.3 Chlorosulfonated Polyethylene

Chlorosulfonated polyethylene (CSPE) is a thermosetting jacket compound with properties very similar to Neoprene. CSPE is unique in that colored compounds of this material, protected by sunlight stable pigments, have weather-resistant properties similar to black CSPE compounds. Hypalon is the trade name of the most commonly used material.

CSPE compounds are superior to Neoprene compounds in the areas of resistance to heat, oxidizing chemicals, ozone, and moisture. They also have better dielectric properties than Neoprene. The flame resistance of both materials is excellent. The superior heat resistance of CSPE as compared with Neoprene makes it the better choice for cables rated at conductor temperatures of 90°C.

8.3.4 Nitrile Rubber

Nitrile rubber compounds are copolymers of butadiene and acrylonitrile. They provide outstanding resistance to oil at higher temperatures. Since this is their only outstanding feature, they are generally limited to oil well applications where temperatures up to 250°C can be encountered. Their poor oxidation resistance in air limits their use for other applications.

8.3.5 Nitrile–Butadiene/Polyvinyl Chloride

These jacket compounds are blends of nitrile rubber mixed with PVC to provide a thermosetting jacket similar to Neoprene. The advantage of this material over Neoprene is that colored jackets of nitrile–butadiene rubber (NBR)/PVC have properties comparable with that black jackets and can be compounded for physical properties and tear resistance similar to that of Neoprene.

8.3.6 Ethylene–Propylene Rubber

Ethylene–propylene rubber (EPR) is frequently used as an insulating material because of its balance of outstanding electrical properties. It can also be used for jackets, especially in low temperature applications where flexibility is required. These materials can be compounded for −60°C applications with reasonably good physical properties and tear resistance.

EPR is not generally used for a jacketing material in other applications. It is used as jackets in low voltage applications when flame resistance has been compounded into the material.

8.4 ARMOR

Armor is placed (or incorporated) into cables to protect them from physical abuse. Armor design and materials must be carefully selected.

- Armor must be considered when determining ampacity, cable loss, and impedance. High permeability materials such as steel may significantly impact these considerations.
- Armor can greatly add to cable weight.
- The corrosion resistance of the armor is a serious consideration.
- Armor will impact installation requirements such as cable bending radius.

8.4.1 Interlocked Armor

This armor consists of a single metal tape whose turns are shaped to interlock during the manufacturing process. Mechanical protection is therefore provided along the entire cable length.

Galvanized steel is the most common metal provided. Aluminum and bronze are used where magnetic effects or weight must be considered. Other metals, such as stainless steel or copper, are used for special applications.

Interlocked-armor cables are frequently specified for use in cable trays and for aerial applications so that conduit and duct systems can be eliminated. The rounded surface of the armor withstands impact somewhat better than flat steel tapes. The interlocked construction produces a relatively flexible cable that can be moved and repositioned to avoid obstacles during and after installation.

An overall jacket is often specified in industrial and power plants for corrosion protection and circuit identification. Neither flat-taped armor nor interlocked armor is designed to withstand longitudinal stress, so long vertical runs should be avoided.

8.4.2 Round-Wire Armor

This construction consists of one or two layers of round wires applied over a cable core. For submarine cable applications, the wires are usually applied over a bedding of impregnated polypropylene or jute.

Round-wire armor is used where high tensile strength and resistance to abrasion and mechanical damage are desired. Vertical riser cables and borehole cables are made with round-wire armor when end-suspension from the wires is necessary for support for the longitudinal stresses. Round wires have less resistance to piercing than flat-tape armor or interlocked armor, but have superior tensile strength and abrasion resistance.

To calculate the strength of armor wires, the following calculation is suggested:

$$W = wl_n \tag{8.1}$$

where

W = total weight of cable in pounds
w = weight of cable to be suspended in pounds per 1,000 ft
l_n = length of cable in thousands of feet.

Strength of steel armor wires:

$$S = NA \times 50{,}000/f \tag{8.2}$$

where

S = strength in pounds
N = number of armor wires
A = area of one armor wire in square inches
F = factor of safety (usually 5).

For copper or bronze, use tensile strength in ASTM Specifications instead of 50,000. If the strength is found to be less than the total weight W, the next step would be to select the next larger size of armor wires and recalculate the values.

For single-conductor cables, copper or aluminum wires have been used to minimize losses due to circulating currents. Such constructions sacrifice mechanical strength in order to achieve lower losses.

Armor wires can be made with the individual wires coated with polyethylene or other corrosion-resistant coverings. Since there is a portion of the circumference without metal protection, cables with such covered wires are usually made with two layers of armor wires with the second layer in the opposite lay to the first.

For installations in severe rock environments, two layers of steel wires, with no individual coverings, are applied in reverse lay. The outer layer frequently is applied with a very short lay to achieve optimum mechanical protection.

The number of armor wires for a wire-armored cable may be calculated from the following equation:

$$N = \left[\frac{\pi(D+d)}{Fd}\right]^{-2} \tag{8.3}$$

where

N = number or armor wires, nearest whole number
D = core diameter of cable under armor in inches
d = diameter of armor wire in inches
F = lay factor (see Table 8.2)
$D + d$ = pitch diameter or armor wire in inches.

Armor resistance may be calculated from the following equation:

$$R_a = \frac{r_a \bullet F}{1{,}000N} \tag{8.4}$$

where

r_a = DC resistance of one armor wire or tape per 1,000 feet at temperature t in ohms
F = lay factor (see Table 8.3)
N = number or armor wires

Note: For steel wire armor, increase R_a by 50% to obtain approximate alternating current (AC) resistance.

TABLE 8.2
Approximate DC Resistance of Armor Wire

Wire Size, BWG	Diameter, inches	Galvanized Steel, ohms/1,000 ft	Hard Drawn Copper, ohms/1,000 ft	Commercial Bronze, ohms/1,000 ft
12	0.109	7.33	0.895	2.49
10	0.134	4.92	0.592	1.65
8	0.165	3.16	0.391	1.09
6	0.203	2.12	0.258	0.72
4	0.238	1.53	0.188	0.52
Basis:				
Conductivity, temperature	% IACS	12.0	97.5	40.0
Coefficient of resistivity	A	0.0035	0.00383	0.00190

TABLE 8.3
Lay Factor for Round Wire Armor

Ratio of Length of Lay to Pitch Diameter of Armor Wire	Lay Factor
7	1.095
8	1.072
9	1.057
10	1.048
11	1.040
12	1.034

Note: Use 7 as a typical value if the ratio is unknown.

8.4.3 Other Armor Types

Other armor types include flat metal tapes, smooth or corrugated continuous metal sheath or twisted braid (see also Chapter 9).

REFERENCES

1. Landinger, C. C., 2001, adapted from class notes for "Understanding Power Cable Characteristics and Applications," University of Wisconsin–Madison.
2. ICEA S-93-639/NEMA WC 74-2006, "5-46 kV Shielded Power Cable for Use in the Transmission and Distribution of Electric Energy."

9 Low Voltage Cables

Carl C. Landinger

CONTENTS

9.1 INTRODUCTION

While defined differently in Chapter 2, low voltage power cables in this chapter will be those designed to operate at phase-to-phase (V_{L-L}, U) voltages of less than 2 kV. This is done to compensate for some misalignment in standard voltage ratings. In the US standard, rated voltages for low voltage cables are commonly 0 to 600 and 601 to 2,000 volts phase-to-phase (V_{L-L}) based on a grounded "wye" three-phase system (making voltage to ground, $V_g = 0.577V_{L-L}$). Internationally, a common low voltage rating is 0.6/1 kV, where 1 kV is phase-to-phase (U) and 0.6 kV is phase-to-ground voltage (U_0) also based on the same system. There are many exceptions including cables that are rated phase-to-ground. These are generally appropriately marked.

Repeated reference to ICEA [4] and the National Electrical Code (NEC) will be made to US low voltage cables discussed in this chapter. Equivalent (as appropriate) international standards for international designs are given in Chapter 10.

Low voltage power cables might be considered as fully insulated as differentiated from covered cables. They typically have insulation or insulation plus jacket thicknesses (and even armor) that have more to do with mechanical damage and ability to survive the environment in which they will operate than voltage considerations. They operate at voltages low enough that shields (screens) are generally not needed to control voltage stresses but may more likely be used to mitigate the effect of electrostatic and electromagnetic magnetic fields. Despite these common features, they operate in many more applications and environments than higher voltage cables. That results in the need for a very large number of materials and cable designs to meet the needs of the low voltage cable world [7].

One might immediately think of low voltage applications involving electric utility plant or building wires including commercial and industrial applications. However, one quickly must add machine tool wires, cords, appliance wires, oil well cables, transportation and ship board cables, and more, to the subject. The author could not hope to do the entire subject justice. Instead, the chapter will be devoted to presenting some general principles common to designs for numerous applications.

To promote common understanding, it is noted that internationally (and to a declining degree, the US) the word sheath includes both metallic and nonmetallic coverings. In this chapter, nonmetallic coverings will be called "jackets," and "sheath" will be reserved for metallic coverings.

9.2 DESIGN

9.2.1 Conductor Considerations

Consideration must be given to the required flexibility, conductivity, connectability, and cost of the conductor (Chapter 3), which will result in the selection of conductor material (copper, aluminum, or alloy of either) as well as strand design. Copper conductors are favored where cable diameter, flexibility, connectability of the conductor, and cost of components beyond the conductor are major factors. Aluminum conductors offer cost savings where cable diameter is less of a factor and a weight advantage, which is important for cables suspended in air.

Soft annealed copper is the overwhelming choice when copper conductors are involved. However, aluminum alloys are common when aluminum conductors are involved. As an example, extensive use of screw down connections in indoor plant results in considerable use of 8,000 series aluminum alloy conductors as opposed to 1,350 aluminum commonly used by utilities that favor compression connectors for outdoor plant. This is because 8,000 series alloys have reduced cold flow under pressure. These alloys must also be annealed to achieve an acceptable level of flexibility.

Filled strand conductors are common to block the passage of water longitudinally in navel and/or marine and similar applications or the passage of gasses in special application cables.

9.2.2 Conductor with Single Layer of Insulation

The simplest low voltage cable consists of a single conductor having a single extruded layer of insulation applied over the conductor (Figure 9.1). Insulation material must be sufficiently insulating, sufficiently rugged to withstand the rigors of installation and operation, and have properties that are suitable for the application.

Commonly used single layer insulations include thermoplastic polyvinyl chloride (PVC), polyethylene, and a number of fluoropolymers for special purpose wiring (generally in sizes 4/0 AWG and smaller). Thermoset insulations for single layer applications include cross-linked polyethylene (XLPE) and ethylene–propylene rubber (EPR) with lesser use of materials such as chlorosulfonated polyethylene (CSPE) and irradiated PVC.

Consideration must be given as to whether the insulation material will be strippable from the conductor or if there is a chemical incompatibility with the conductor, which might degrade the insulation.

The solution may involve the use of coated conductors such as tinned copper. A common alternative solution is a tape separator over the conductor. The need for separators is certainly not limited to cables with a single extruded layer of insulation.

If a tape separator is used, common practice dictates that it be of a color different from that of the underlying conductor. This minimizes the possibility that installers install a connector over the separator resulting in a very poor connection. A tape separator is shown in Figure 9.2.

Crystalline insulations (such as, but not limited to, polyethylene) have a tendency to develop internal stress due to extrusion and other processing conditions and then "relax" with time. This causes the insulation to shrink along the conductor exposing additional conductor. This is commonly known as shrink back. The degree to which the insulation fills and grips the interstices of stranded conductors can help reduce shrink back. If a separator is used, any grip on the conductor is lost. Careful processing is required to minimize shrink back in this case. It might be noted that solid conductors are also more prone to shrink back.

Cables with a single layer of insulation are widely used as building wires, utility underground, and many other purposes including machine tool wire and appliance wire. Highlighting the importance of this general design will be accomplished by discussing two "groups" that are widely used.

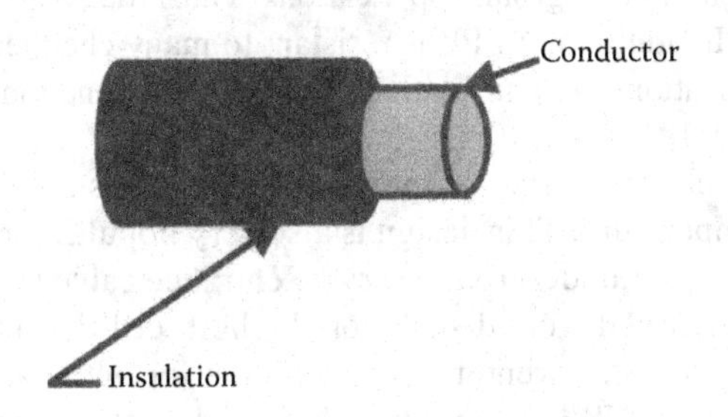

FIGURE 9.1 Conductor with single layer of insulation.

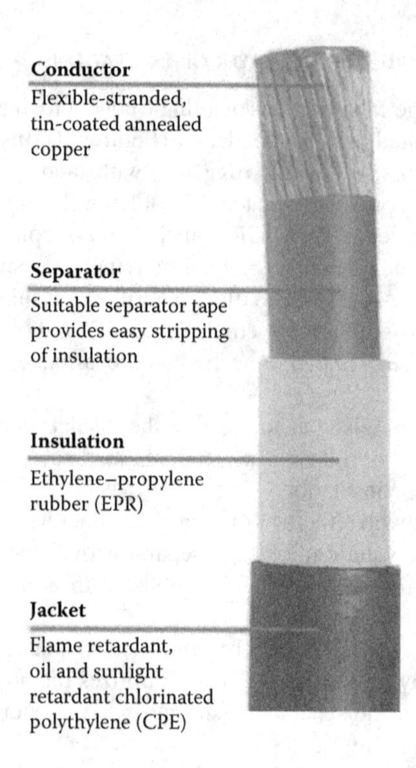

FIGURE 9.2 Typical case of separator. (From Amercable, January 2011, Internet Catalog, with permission. Accessed December 2010.)

9.2.2.1 Underground Residential Distribution Secondary and Similar Designs

Underground residential distribution (URD) electric utility secondary and service cable is most commonly used as an aluminum conductor insulated with a single layer of XLPE. The cable is either directly buried or installed in a duct/conduit. The XLPE is made sunlight-resistant by the incorporation of a low level of carbon black into the XLPE. This low level addition leaves the XLPE with a high voltage breakdown strength greatly exceeding the requirements for a secondary cable and excellent mechanical strength to withstand the rigors of installation and operational conditions common to underground applications. Thus, the XLPE acts as both insulation and jacket. In addition, XLPE is resistant to many chemicals encountered in underground applications. The insulation thickness is a function of conductor size as shown in Table 9.1.

When used for electric utility purposes, the design has a 90°C maximum normal operating temperature. This design is also very popular for applications covered by the NEC [5] for underground service entrance cable (USE). These cables are commonly purchased from distributors by both utilities and contractors. To avoid the need for double inventory, it is common for manufacturers to produce and print the cables as USE even though this would not be a utility requirement for outdoor plant. Indeed, it is not uncommon to see these cables marked USE

TABLE 9.1
Insulation Thickness as a Function of Conductor Size

Conductor Size	USE Insulation Thickness		XHHW Insulation Thicknes	
AWG-kcmils	Mils	mm	Mils	mm
12–9	45	1.14	30	0.76
8–2	60	1.52	45	1.14
1–4/0	80	2.03	55	1.40
213–500	95	2.41	65	1.65
501–1,000	110	2.79	80	2.03

and used by some utilities as well. When used in NEC applications, the designation given to the design must be considered to determine the temperature rating for both wet and dry applications.

If formulated to have sufficient flame resistance to pass the required flame tests, the cable might be designated types RHH–RHW and used for indoor plant in accordance with the NEC. However, in that case, if aluminum conductor, an 8,000 series alloy is required and the design is not commonly used by utilities for URD.

The physical toughness and voltage strength of the single layer XLPE cable have made possible a cable design with a lower insulation thickness. This design is commonly installed in a raceway (as defined by the NEC) in accordance with the NEC requirements and conserves raceway internal area required providing cost savings. The design is Type XHHW (heat and moisture resistant synthetic rubber).

Finally, utilities also use thermoplastic polyethylene insulated cable of the same thicknesses as the URD and/or USE cables for lower temperature service. These cables are normally rated for 75°C maximum normal operation. Thus, they are commonly used on more lightly loaded circuits such as street lighting. However, these cables are not generally suitable for NEC applications because of the relatively poorer flame resistance when compared with PVC, which is also a thermoplastic.

9.2.2.2 Thermoplastic Polyvinyl Chloride Wire Recognized by the NEC

NEC types TW, THW, and MTW with no overall jacket are commonly made with a single layer of PVC. They are differentiated from each other by formulation to provide a level of heat resistance, moisture resistance, and oil resistance required by the NEC for the application. One common application is installation in a raceway, such as a conduit. PVC has excellent flame resistance that has made it suitable for many NEC applications. PVC is relatively low cost, easy to process, and lends itself to formulation to maximize specific properties, such as oil resistance, or different colors required for the application. One disadvantage of PVC is that it becomes brittle and subject to impact fracture at lower temperatures, some $<15°F$ depending on formulation. This requires special care in handling at low temperatures. The author is personally aware of incidents where a contractor threw coils of PVC wire off the truck to the ground in cold weather, resulting in cracks in the insulation. In both utility and NEC applications, careful attention must be paid to the type and application of these PVC insulated wires to determine the maximum allowed operating temperature. This can range from 60°C to 90°C.

There are many applications where a single layer of insulation is not adequate. In such cases, it is common to add a jacket over the insulation layer. The purpose of the jacket is to provide one or a number of features the insulation alone cannot provide.

9.2.3 Conductor with Insulation and Jacket Layer

A jacket over the insulation may be used to provide the following (Figure 9.3):

- Physical toughness
- Sunlight resistance
- Chemical resistance
- Flame resistance
- Low smoke release when burning
- Zero halogen release when burning
- Radiation resistance
- Low coefficient of pulling friction

Some examples are discussed in the following section.

9.2.3.1 Jacket to Add Toughness

9.2.3.1.1 Ruggedized Cables

As discussed in Section 9.2.2, a common underground cable used by utilities and contractors in North America consists of a single layer of polyethylene (75°C normal service) or XLPE (90°C normal service) with carbon black filler to provide sunlight resistance for underground secondary services and services directly buried in earth or in conduit. However, when placed in contact with rubble fill, rocky soils, or soils laden with coral, the damage resistance (impact, cut through, crush and abrasion), is not adequate and the failure rate is elevated. To improve toughness, ruggedized cables are used. The insulation thickness for the single layer cable is partitioned into a thickness of the "original" polyethylene or XLPE insulation overlaid by a thickness of high density polyethylene or medium-to-high density XLPE (insulation/jacket). In the case of the XLPE design, a major challenge is to achieve

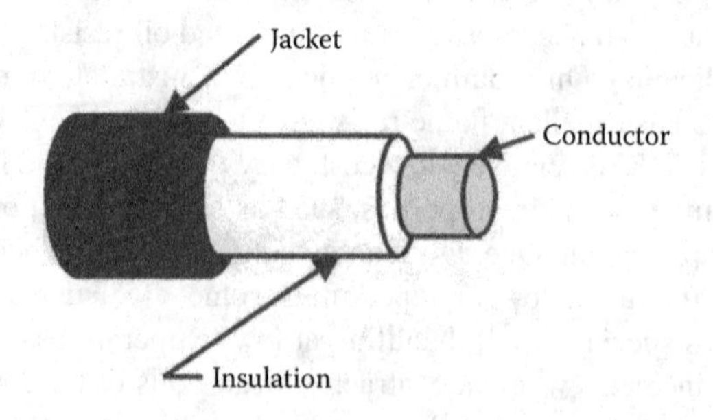

FIGURE 9.3 Single conductor with insulation and jacket.

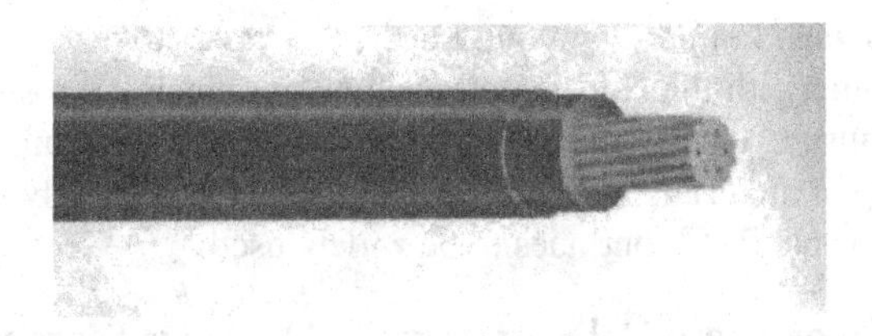

FIGURE 9.4 THHN THWN wire (cable). (From Southwire, January 2011, Internet Catalog, with permission. Accessed December 2010.)

a truly high density cross-linked outer layer. This is because there are fewer cross-linking sites as the density of the XLPE is increased.

9.2.3.1.2 THHN and THWN

In the US, THHN (heat resistant thermoplastic 90°C normal operation in dry and damp locations) and THWN [heat and moisture resistant thermoplastic 75°C normal operation in wet and dry locations (or 90°C with proper rating)] are NEC designations. The dual rated design with a nylon jacket is tough and has a low pulling coefficient of friction. The toughness of the nylon jacket allows for a reduced total (insulation and jacket) wall thickness and pulling at lower tensions than for the single layer equivalent. This design is widely used for rewiring, often allowing larger conductors to be installed in the existing conduit. A THHN THWN wire (cable) is shown in Figure 9.4.

9.2.3.2 Jacket to Add Toughness, Sunlight, and Flame Resistance

9.2.3.2.1 EPR Insulated Cables

EPR is an excellent low voltage insulation offering many desirable properties. While it can be compounded to provide increased toughness and/or flame resistance, this might compromise the properties for which EPR was selected. It is common to add a jacket of polychloroprene (commonly Neoprene), CSPE (commonly Hypalon), or chlorinated polyethylene (CPE). All three of these materials are relatively tougher than the EPR and as the names imply (chlorinated or chlorine), provide flame resistance. These jackets are also flexible so the cables of this design are widely used where flexibility is a requirement. PVC is also commonly used as a jacket over EPR to add sunlight and flame resistance.

9.2.3.3 Polyvinyl Chloride Jacket to Add Sunlight and Flame Resistance

9.2.3.3.1 Polyvinyl Chloride Jacketed Cables

PVC is not considered tough compared with polyethylene or XLPE. However, it is certainly more flame resistant. PVC continues to be used widely over both polyethylene and XLPE as a jacket. The flame resistance and low cost have made PVC a popular jacketing and insulating material. The excellent flame resistance is due to the chlorine in the PVC material. When burned, commonly by an external flame source, the chlorine is released as a gas. Chlorine, when inhaled, can do serious damage to the respiratory system of personnel who are exposed including those escaping a fire or fire fighters engaged in putting it out. Further, methods to put out the fire, such as water, can combine with the chlorine creating an extremely corrosive mixture. The corrosion can cause extensive damage to facilities. This has resulted in increasing use of zero halogen materials for insulation and jackets, especially for indoor plant. However, at this time, PVC continues to be widely used.

9.2.3.4 Low Smoke Zero Halogen Jacket to Maximize Escape Potential

Concern for the epidemiological effects on humans exposed to halogens released by a fire and corrosion consequences has resulted in growing international use of nonhalogenated insulation and jacket materials. However, there has also been concern for the ability of persons involved in, or with a fire, to escape. Smoke release of materials when burning, which might reduce visibility, is now an issue. This has led to increased use of low-smoke zero-halogen (LSZH) materials. LSZH describes a number (rather than single) of families of materials. In addition to the advantages already mentioned, another is reduced soot damage from burning of these materials and there is some reported lower coefficient of friction for LSZH jackets when pulling [2]. Disadvantages reported to date include deterioration of jacket properties when contacted by some pulling compounds and because of high filler loading, poorer mechanical, chemical, and moisture absorption properties may result when compared with non-LSZH jackets.

9.2.4 Plexed Single Conductor Wires and Cables

It is very common to apply two, three, four, or more single conductor cables in the same trench, duct, conduit, tray, or other route. Under such circumstance, it is often convenient to produce an assembly of single conductor cables (also called plexed, cabled, or twisted) with no further covering over all. The configuration is commonly named by the number of conductors in the assembly such as: duplex (two conductors), triplex (three conductors), or quadruplex (four conductors) (Figure 9.5).

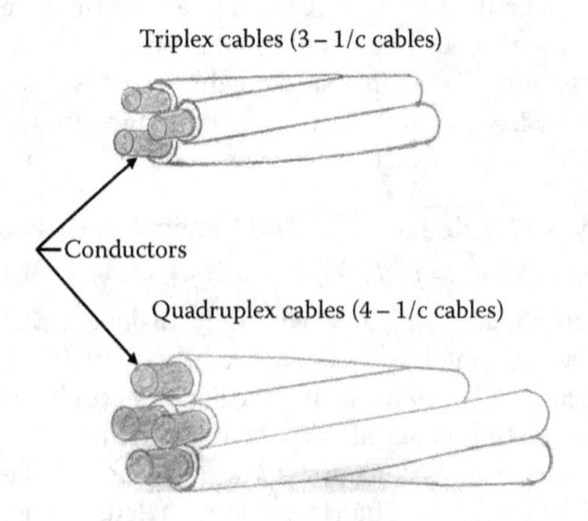

FIGURE 9.5 Triplexed and quadruplexed cables.

Plexing has several advantages. The assembly presents the smallest circumscribing diameter for the cables involved. For single conductor cables of the same diameter, the circumscribing diameter of:

- A duplex assembly is two times the diameter of a single conductor cable
- For a triplex assembly, the multiplier is 2.155
- For a quadruplexed assembly, the multiplier is 2.414

Plexing, which holds the cables together, overcomes some installation issues such as jamming in conduit and ducts, as discussed in Chapter 12. Minimizing the circumscribing diameter also minimizes inductive reactance in AC operation.

Plexing does have a number of disadvantages. Plexing increases the length of the single conductor cable required to make up the plexed assembly length. The increased length results in an increased DC resistance. Maximum cable length on a single reel is reduced, i.e., the length of one single conductor cable on a given reel size is considerably greater than the length of plexed cable of the same size on the same reel.

9.2.5 Multiconductor Cables

Multiconductor cables are extensively used in low voltage applications. They may be roughly differentiated from plexed cables as having additional covering layers (metallic shield or armor, nonmetallic jacket, or both) over the assembly. However, many multiconductor cables do not have the individual conductors twisted together but may be laid in parallel (commonly resulting in a flat configuration). Multiconductor cables commonly have additional conductors or components for purposes such as: grounding, check circuit integrity, signal, and/or communication. They also may have binders to hold individual members together and fillers to achieve and maintain a desired shape (such as round).

9.2.5.1 Flat Configuration Multiconductor Cables

Flat configuration cables may simply result as a matter of practical arrangement of the enclosed conductors. For instance, the NM Type cable shown in Figure 9.6 consisting of phase, neutral, and grounding conductor under a common jacket is widely used in the US for residential (but not limited to) wiring in accordance with the NEC. There is no application reason why the cable is flat but it is convenient for the design involved.

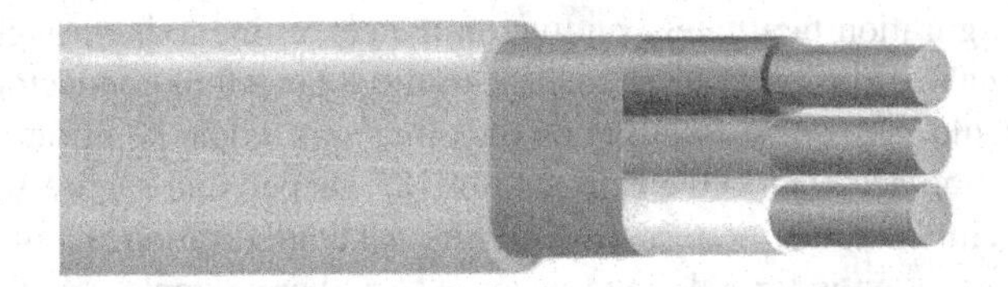

FIGURE 9.6 Type NM cable. (From Southwire, January 2011, Internet Catalog, with permission. Accessed December 2010.)

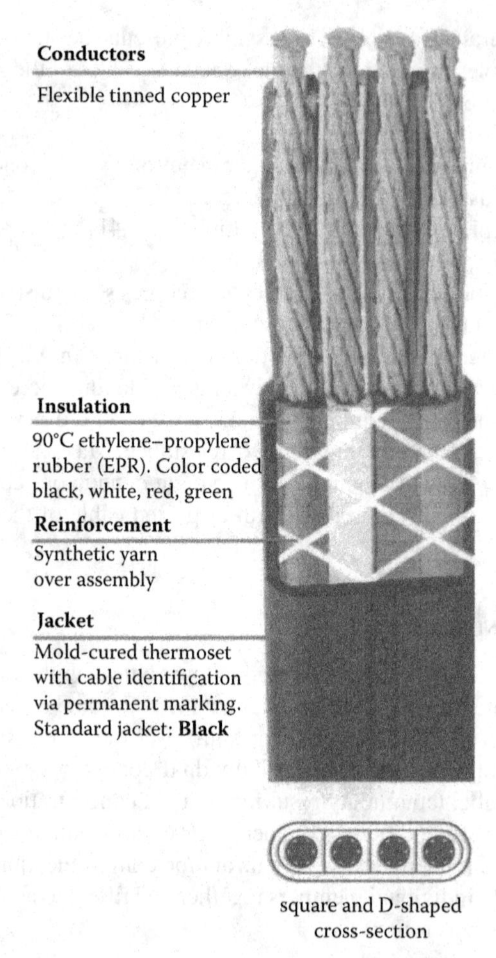

FIGURE 9.7 Type W Flat 4 Conductor Mining Cable, 600/2,000 volt. (From Amercable, January 2011, Internet Catalog, with permission. Accessed December 2010.)

The Type W Flat 4 Conductor Mining Cable shown in Figure 9.7 is subject to reeling and de-reeling while feeding shuttle cars, drills, cutting and loading machines. The flat configuration facilitates bending on the centerline axis through the longer width of the cable. This provides greater flexibility than a four conductor round configuration would. Of course, bending on the other axis would be difficult (and damaging). Attention is called to the extrusion of "D" shaped and square-shaped single conductor members to facilitate a space-saving flat configuration. Thus, in this case, the flat configuration is for a definite purpose. Another example, not shown, would be the Flat Conductor Cable Type FCC designed for installation under carpet squares in accordance with the NEC.

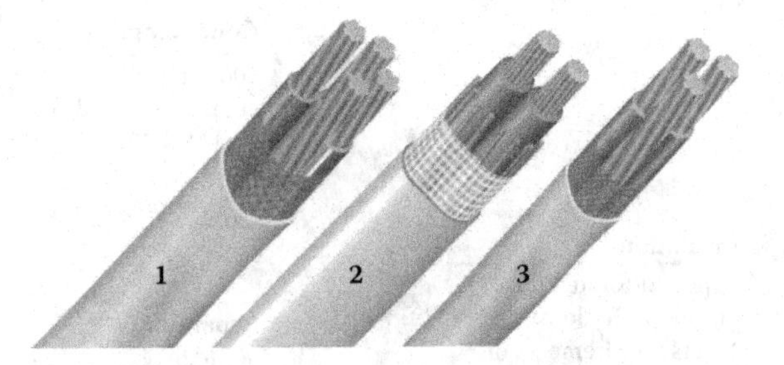

1. **SE Style R** (round) with three insulated conductors and one bare ground, tape, over which a PVC jacket has been extruded.
2. **SE Style U** (flat) with two insulated conductors and concentric bare ground, tape, over which a PVC jacket has been extruded.
3. **SE Style R** (round) with two insulated conductors and one bare ground, tape, over which a PVC jacket has been extruded.

FIGURE 9.8 Service entrance cables. (From Southwire, January 2011, Internet Catalog, with permission. Accessed December 2010.)

Round configuration multiconductor cables are the other (and possibly more) common configuration. Some cable types are made both as flat and round configurations. Service entrance cables are commonly used in the US to bring power from an overhead service drop attached to the building to the meter base and from the meter base to the distribution panel. Service entrance cables may be used in other applications in accordance with the NEC. Figure 9.8 shows a number of available service entrance cables.

9.2.5.2 Round Configuration Multiconductor Cables

There are almost a limitless number of multiconductor cables with a round configuration. With respect to power cables, the most common have three or four insulated power conductors but more are possible for special purpose cables. A simple case might have three insulated conductors and an overall jacket. One such cable typically used for navel shore-to-ship and other pier side power applications is shown in Figure 9.9.

Features not immediately evident from the figure are the red–white–black color coding of the power cables, filling of the interstices at the power cable contacts to minimize water migration, and the reinforcement of the jacket for toughness. In other designs, an interstice might contain a grounding conductor. One, two, or three grounding conductors are common. The number, size, and/or conductivity of the grounding conductors are generally a matter of industry or military specification developed to meet the needs of the application.

ICEA S-95-658/NEMA WC 70 "Standard For Non-shielded Power Cables Rated 2,000 Volts or Less For the Distribution of Electrical Energy" requires that for assemblies of two, three, or four conductor-insulated power cables utilizing a

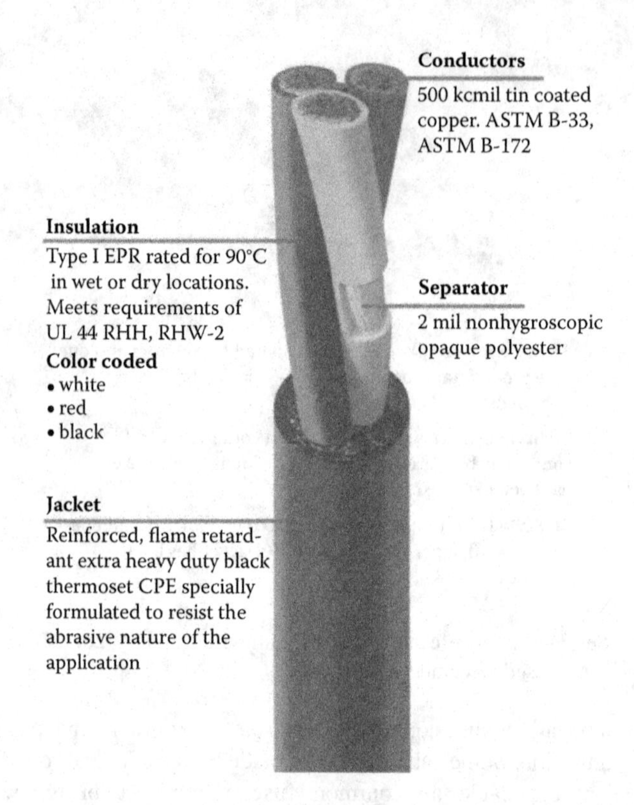

FIGURE 9.9 Three conductor round cable. (From Amercable, January 2011, Internet Catalog, with permission. Accessed December 2010.)

grounding conductor, the minimum grounding conductor size should be as follows unless otherwise specified.

Power Conductor Size (AWG or kcmil)		Grounding Conductor Size (AWG or kcmil)	
Copper	**Aluminum**	**Copper**	**Aluminum**
14	12	14	12
12	10	12	10
10–8	8–6	10	8
6–4	4–2	8	6
3–2/0	1–3/0	6	4
3/0–250	4/0–350	4	2
300–400	400–600	3	1
450–650	700–1,000	2	1/0
700–900	—	1	2/0
1,000	—	1/0	3/0

The grounding conductor size may be subdivided into several wires, but no wire shall be less than #18 AWG in size.

Again, the reader is strongly advised to consult the appropriate standard for the cable and application being considered to meet ground wire requirements as they may differ from the above.

The ICEA/NEMA standard also calls for maximum lengths of lay for multiple conductor cables for:

- Multiple conductor assemblies with no overall covering, cabled with a left hand lay, and a maximum length of lay of 60 times the diameter of the largest insulated conductor.
- Multiple conductor round cables with an overall covering with all conductors having the same direction of lay, the lay shall be left hand with a maximum length of lay as shown.

Number of Conductors in Cable	Maximum Length of Lay
2	30 times individual conductor diameter
3	35 times individual conductor diameter
4	40 times individual conductor diameter
5 or more	15 times assembled diameter

Once again, the governing standard for the cable and application must be consulted as it might not agree with the above.

In Figure 9.10, a design common in international low voltage cables is shown. The underlying insulation may be PVC, XLPE, or EPR overlaid with a PVC jacket. Concern for the safety and health concerns connected with smoke and halogens previously discussed result in increased use of LSZH insulation and jacketing materials internationally. An interesting feature of this design is the use of sector-shaped conductors (120° for 3/C cables and 90° for 4/C cables). Round conductors are also available.

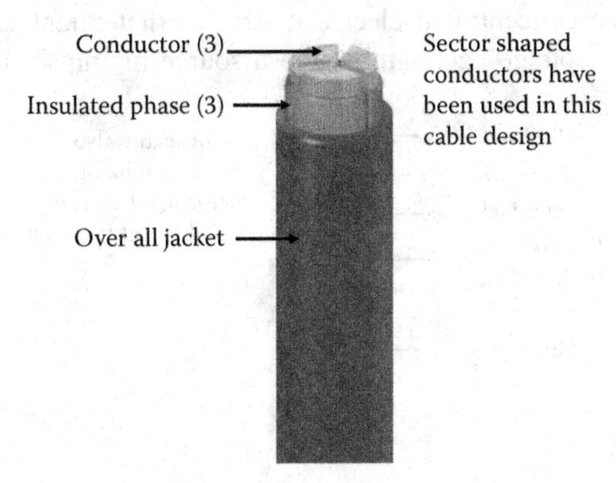

FIGURE 9.10 3/C Low volt cable with sector-shaped conductors. (From General Cable, January 2011, Internet Catalog, with permission. Accessed December 2010.)

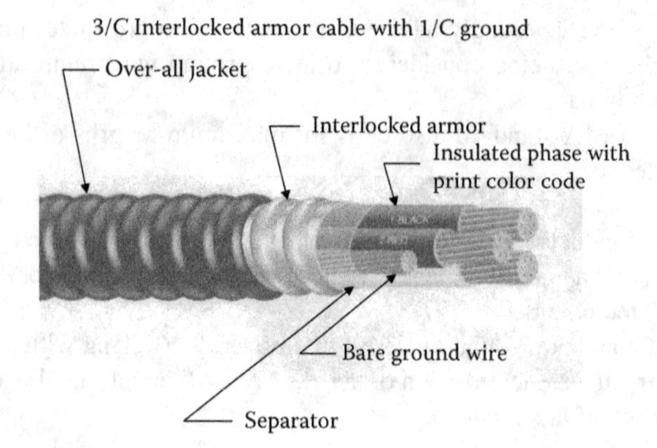

FIGURE 9.11 Interlocked armor low voltage cable. (From Southwire, January 2011, Internet Catalog, with permission. Accessed December 2010.)

9.2.5.3 Low Voltage Armored Cables

Cable armoring is discussed in Chapter 8. Armoring of low voltage cables is very common.

The interlocked armor cable shown in Figure 9.11 is in accordance with the NEC in the US. It is listed type MC (metal clad). The armor may be aluminum or galvanized steel and may be bare or jacketed depending on the application. The Type MC cable shown includes ground wires. The NEC also allows inclusion of optical fiber members.

Armoring is discussed in Chapter 8. There are numerous recognized armors. One that might not be thought of immediately is bronze wire braid armor. In Figure 9.12, the armored version of the low voltage cable common internationally (Figure 9.10) is shown. The stark contrast between these two armoring methods shows the diversity to be found in the low voltage cable world.

9.2.5.4 Shielded Low Voltage Cables

In the introduction, it was stated that low voltage cables did not generally require shielding to reduce concentrated electrical stresses that might lead to cable failure. However, low voltage cables might be a source or impacted by electrostatic

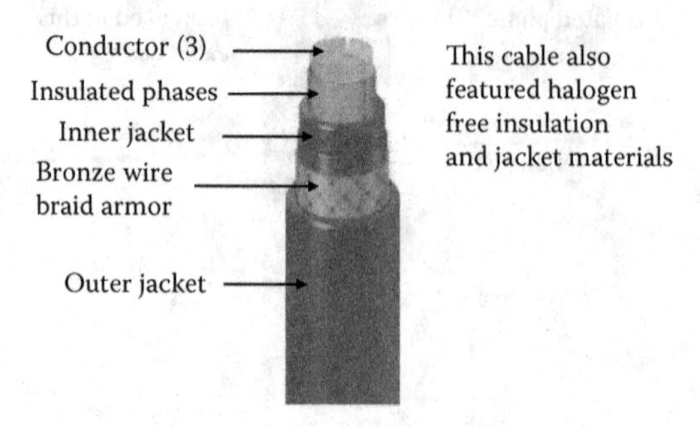

FIGURE 9.12 3/C Armored low volt cable, sector-shaped conductors. (From General Cable, January 2011, Internet Catalog, with permission. Accessed December 2010.)

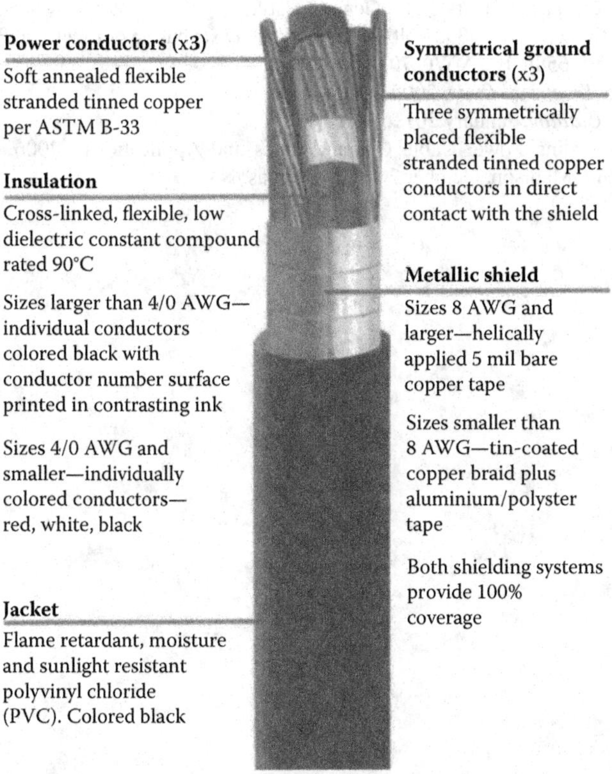

Power conductors (x3)
Soft annealed flexible stranded tinned copper per ASTM B-33

Insulation
Cross-linked, flexible, low dielectric constant compound rated 90°C

Sizes larger than 4/0 AWG—individual conductors colored black with conductor number surface printed in contrasting ink

Sizes 4/0 AWG and smaller—individually colored conductors—red, white, black

Jacket
Flame retardant, moisture and sunlight resistant polyvinyl chloride (PVC). Colored black

Symmetrical ground conductors (x3)
Three symmetrically placed flexible stranded tinned copper conductors in direct contact with the shield

Metallic shield
Sizes 8 AWG and larger—helically applied 5 mil bare copper tape

Sizes smaller than 8 AWG—tin-coated copper braid plus aluminium/polyster tape

Both shielding systems provide 100% coverage

FIGURE 9.13 Shielding low volt cable to contain EMI emissions. (From Amercable, January 2011, Internet Catalog, with permission. Accessed December 2010.)

and electromagnetic field interference (EMI). There are numerous methods to mitigate the interference of such fields. Some may be as simple as to twist the pair of unshielded wires being exposed to the fields. In general, electrostatic fields are more easily shielded against than magnetic fields. The measures needed to deal with magnetic fields are more complex and are frequency dependent. In Figure 9.13, one method of design and shielding to deal with EMI in connection with variable frequency drive applications is shown. Note the emphasis on symmetry of cable components to maximize field cancellation.

This is but one example. As a general rule, the lower the frequency, the more difficult it is to provide shielding for magnetic fields. Thick layers of high magnetic permeability metal may be required at low frequencies (such as 60 Hz) while at higher frequency (>1 KHz) copper braid might provide some measure of shielding. The subject is extremely complex and beyond the scope of this chapter.

REFERENCES

1. *Internet Catalog*, January 2011, AmerCable.
2. *Wire & Cable Technical Information Handbook*, 1996, Anixter.

3. *Internet Catalog,* January 2011, General Cable.
4. "Power Cables Rated 2000 Volts or Less for the Distribution of Electrical Energy," 2009, ICEA S-95-658/NEMA WC 70.
5. *National Electrical Code,* 2005, NFPA 70.
6. *Internet Catalog,* January 2011, Southwire.
7. "Understanding Power Cable Characteristics and Applications," 2006, University of Wisconsin–Madison, October 2006, Las Vegas, NV.

10 Standards and Specifications

Carl C. Landinger and Lauri J. Hiivala

CONTENTS

10.1 INTRODUCTION

Standards and specifications for power, control, signal, instrumentation, and communication cables have been prepared in North America by various industry organizations since the early part of the twentieth century. These are commonly referenced in user specifications and supplemented by the user to include specifics required by the user. Electrical cables are manufactured to these requirements, depending on the application of the particular installation.

The power cables that are covered by these standards and specifications can be classified under three major categories:

- Low voltage cables rated up to 2,000 volts
- Medium voltage cables rated 2,001 to 46,000 volts
- High voltage cables rated 69,000 volts to 765,000 volts

The most widely used documents in North America are those issued by the Aluminum Association (AA), American National Standards Institute (ANSI), American Society for Testing and Materials (ASTM), Canadian Standards Association (CSA), Insulated Cable Engineers Association (ICEA) in conjunction with the National Electrical Manufacturers Association (NEMA), the Association of Edison Illuminating Companies (AEIC), the Rural Utilities Service (RUS), and Underwriter's Laboratories (UL) in conjunction with the National Electrical Code (NEC). These organizations may loosely be categorized into:

1. Manufacturer organizations
2. User organizations
3. Consensus organizations

10.2 MANUFACTURER ORGANIZATIONS

10.2.1 INSULATED CABLE ENGINEERS ASSOCIATION

This group was formerly known as IPCEA. They removed the "Power" from their name to more accurately describe a broader scope of activities. Divisions in ICEA include:

- The Energy Cables Division comprises three Sections: (a) the Power Cable Section, (b) the Control & Instrumentation Cable Section, and (c) the Portable Cable Section.
- The Communication Cables Division comprises two Sections: (a) the Copper Section and (b) the Optical Fiber Section.

Membership is made up of technical employees, who are sponsored by over 30 of North America's leading cable manufacturers. They develop standards, guides, and committee reports on all aspects of insulated cable design, materials, and applications. They work with other organizations toward the development of joint standards. Many of their standards are subject to the approval of NEMA and are published as joint ICEA/NEMA standards. Others are approved as an American National Standard by ANSI.

These standards encompass the entire cable: conductor, shields, insulation, jackets, testing, etc. The only possible omission is packaging. This is considered to be an area that is not allowed by US law.

Both ICEA and NEMA standards may be purchased from Global Engineering Documents, 15 Inverness Way East, Englewood, CO 80112, USA.

10.2.2 National Electrical Manufacturers Association

NEMA is the trade association of choice for the electrical manufacturing industry. NEMA members are from cable manufacturing organizations and generally they are from the commercial side of those companies.

10.2.3 Aluminum Association

The AA tends to impact cable standards and specifications by representation on organizations engaged in these activities. The association has taken the lead in specifying ingredient and impurity levels in aluminum alloys used for conductors, and until recently had published code words for many bare, covered, and insulated aluminum wires and cables. These code words are widely used to specify and purchase the wires and cables coded. The ASTM has now assumed such responsibility for the ones pertaining to bare conductors and the ICEA for the rest.

10.3 USER ORGANIZATIONS

10.3.1 Association of Edison Illuminating Companies

Since November 1, 1924, power cable specifications have been written by the Cable Engineering Committee of the AEIC, a group of investor-owned and municipal utility company engineers. They also prepare guides that pertain to power cables.

Their first specifications were written for paper-insulated cables for medium-voltage applications. Presently, their specifications cover all forms of laminated cables from 1 to 765 kV such as paper-insulated metallic-sheathed, self-contained liquid-filled and high-pressure pipe-type cables. These specifications include conductors, insulations, sheaths, shields, jackets, and testing requirements.

AEIC also prepares specifications for extruded dielectric cables from 5 to 345 kV that build upon the ICEA standards (hence also on the applicable ASTM requirements). AEIC uses the ICEA standards for such items as conductors, shields, jackets, and testing requirements. A unique feature of AEIC's past extruded cable specifications is that they require a qualification test to be performed on a sample of cable that represents the cable to be manufactured. This has now found its way into the ICEA standards for utility cables [14,16].

Another feature of AEIC's specifications for extruded cables is that a checklist of the available options is presented. This can be useful for those users that are in the process of developing a user specification for themselves.

AEIC documents are available from AEIC, 600 N. 18th St., P.O. Box 2641, Birmingham, Alabama 35291-0992, USA.

10.3.2 Rural Utilities Service (RUS) [Previously REA]

This is also a user group of the US Department of Agriculture. Their Electric Program lists borrowers by state; provides guide specifications; building plans and drawings; electrical design and construction policies and procedures; electrical standards for materials and construction for use by the Rural Electric Cooperatives of the US.

10.4 CONSENSUS ORGANIZATIONS

10.4.1 American National Standards Institute (ANSI)

ANSI membership and polling lists involve all interested parties. Thus, ANSI standards are consensus standards. ANSI does not develop standards, but rather elevates standards developed by others to a consensus level by a voting procedure.

10.4.2 American Society for Testing and Materials (ASTM)

ASTM International, formerly known as the American Society for Testing and Materials (ASTM), is a globally recognized leader in the development and delivery of international voluntary consensus standards. Membership includes raw material suppliers, manufacturers, users, and general interest groups. As such, ASTM is a consensus organization. ASTM is organized for "the development of standards on characteristics and performance of materials, products, systems and services."

10.4.3 Canadian Standards Association (CSA)

The CSA is a not-for-profit membership-based association serving business, industry, government, and consumers in Canada and the global marketplace. It functions as a neutral third party, providing a structure and a forum for developing standards.

Their committees are created using a "balanced matrix" approach, which considers the views of all participants and develops the details of the standard by a consensus process. Substantial agreement among committee members, rather than a simple majority of votes, is necessary. When a draft standard has been agreed upon, it is submitted for public review, and amended if necessary.

10.4.4 International Electrotechnical Commission (IEC)

The IEC is the world's leading organization that prepares and publishes International Standards for all electrical, electronic, and related technologies—collectively known as "electrotechnology." Each National Committee of the IEC handles the participation of experts from its country. Technical Committees prepare technical documents on specific subjects within their respective scopes, which are then submitted to the full member National Committees (IEC's members) for voting with a view to their approval as international standards.

10.4.5 National Electrical Code (NEC)

The NEC is written by a number of panels, each addressing specific sections of the NEC. Membership of the panels is carefully controlled to avoid having any single interest group gain control over the panel. The NEC covers wire, cables, and wiring methods in facilities frequented by the general public. The only specific exemption is facilities under the control of electric utilities that are not facilities used to host the general public such as customer service areas. The NEC is adopted by ANSI and is adopted in whole or part (or not at all) by governmental agencies.

10.4.6 Underwriters Laboratories (UL)

UL has published a wide range of standards covering wires and cables specified by and for use in accordance with the NEC. UL both lists and labels acceptable designs and acts as a third party inspector by having UL inspectors visit the manufacturer's facilities to conduct tests to verify that the manufacturer is meeting the standard for the product(s).

10.5 TYPICAL FEATURES OF STANDARDS AND SPECIFICATIONS

10.5.1 Conductors

10.5.1.1 Resistance

CSA, ICEA, and UL power cable standards include both copper and aluminum conductors, which are covered by individual ASTM standards. Since in most instances direct current (DC) resistance is the governing factor for confirming conductor size, these standards specify a maximum resistance for each American Wire Gauge (AWG) and thousands of circular mils (kcmil) size. ICEA conductor diameters are required to meet a nominal diameter ±2%.

10.5.1.2 Compressed Strand

ASTM standards for concentric-lay-stranded conductors give the manufacturer the option of "compressing" Class B and C conductors. This means that the overall diameter of such a conductor may be a maximum of 3% lower than that of a "concentric" conductor. The need and advantages for such compression were presented in Chapter 3. Thus, even if "concentric" stranding is requested, the manufacturer has the option of providing "compressed" strand.

10.5.1.3 Temper

An important decision that must be made involves the temper of the metal. This option should be based on such factors as the anticipated pulling force, required flexibility, and also on the cost.

The harder the temper, the greater will be the pulling force that can be applied to the conductor during installation. A half-hard aluminum conductor will withstand less force than a three-quarter or full-hard conductor. On the other hand, any increase in temper results in a conductor that is harder to bend and is thus less flexible. However, this additional bending force of the bare conductor may be negligible when compared with that of the finished cable. When the wires are drawn for the conductor stranding process, the metal is work hardened and the temper increases. Annealing either during the wire drawing process or after the conductor has been stranded will decrease the temper, but this takes energy so there is an increase in the cost of an annealed conductor. All of these points need to be weighed before a decision is reached.

10.5.1.4 Identification

Cable manufacturers have the capability of indent printing the center strand of a seven-strand or greater Class B or C conductor. If requested at the time of the inquiry, they can also print the year of manufacture and their name at regular intervals on this center strand. This provides a lasting identification of the manufacturer and the year.

10.5.1.5 Blocked Strand

Another consideration is to block, or fill, the strands of a Class B conductor with a compound that eliminates almost all the air from the interstices. This prevents the accumulation of moisture in the air space as well as longitudinal moisture movement along the cable. The elimination of water in the strand reduces the treeing concerns and increases the life of cables in accelerated treeing tests. ICEA standards contain a test for the effectiveness of this "water blocking" [12,14,16].

Another method of keeping water from entering (or leaving) the strand is to install a metal barrier in the semiconducting strand shielding. The layer is a "sandwich" of the semiconducting material with a lead or aluminum overlapped tape in the center.

10.5.2 Conductor Shielding

Conductor shielding (either a semiconducting or nonconducting stress control layer) is required for cables rated 2,000 volts and higher by these standards. Conductor shielding normally consists of an extruded semiconducting layer applied between the conductor and the insulation. For this layer to function properly, it shall be easily removable from the conductor and the outer surface of the extruded shield should be firmly bonded to the overlying insulation to ensure that there are no air voids between the conducting layer and the insulation.

This extruded semiconducting material is made from a polymer that will ensure compatibility with the insulation and a strong bond at the interface. Its conducting properties are obtained by adding particles of special carbon black. The present requirement for the maximum resistivity of this layer is 1,000 ohm-meter at the maximum normal operating temperature and emergency operating temperature. Industry standards require this material to pass a longtime stability test for resistivity at rated emergency overload temperature of the cable. Accelerated tests have shown that the cleanliness of the material can significantly affect the life of the cable when it is in a wet environment. Compounds that are demonstrably smoother and/or cleaner than conventional compounds can improve the life of a cable in an accelerated water treeing test significantly.

The nonconducting rather than semiconducting stress control layer that may be used has properties that are best described as having a high dielectric constant (high K) material. This means that it acts like a rather poor conductor and produces a very low voltage drop between the conductor and the insulation. It does provide the stress control that is needed for smoothing out the conductor surface.

In some applications, a semiconducting tape may be applied over the conductor and under the semiconducting layer. This functions as a binder and is sometimes used for larger conductors.

If a semiconducting conductor stress control layer is used, the resistivity is measured between two painted silver electrodes placed at least 2 inches apart on the conductor stress control layer. If greater accuracy is desired, current electrodes may be placed 1 inch beyond each potential electrode.

The resistance shall be measured between the two potential electrodes. The power of the test circuit shall not exceed 100 milliwatts.

The volume resistivity is then calculated from the following equation:

$$\rho = \frac{R\left(D^2 - d^2\right)}{100L} \tag{10.1}$$

where

ρ = volume resistivity in ohm-meters
R = measured resistance in ohms
D = diameter over conductor stress control layer in inches
d = diameter over conductor in inches
L = distance between electrodes in inches

10.5.3 Insulation

Cross-linked polyethylene (XLPE) (including tree retardant XLPE [TR-XLPE]) and ethylene propylene rubber (EPR) are the dominant materials presently used as the insulation for medium voltage cables.

10.5.3.1 AEIC Specifications

AEIC prepares specifications for extruded dielectric cables from 5 to 345 kV [1] that covers both cross-linked (thermosetting) polyethylene and ethylene–propylene rubber cables. AEIC also cover all forms of laminated cable from 1 to 765 kV. At this time, there is no medium-voltage thermoplastic polyethylene power cable [6] being manufactured in North America.

Both AEIC [1] and ICEA [14,16] require that numerous tests be performed on the material that will be used in the manufacturing process. Applicable tests, with passing requirements, include:

- For filled and unfilled XLPE, including TR-XLPE, and ethylene–propylene rubber Classes I, II, III, and IV, the insulation physical requirements are shown in Table 10.1.
- The insulation electrical property requirements are shown in Table 10.2.

10.5.3.2 Insulation Thickness and Test Voltages

Both XLPE and ethylene–propylene rubber insulated cables have the same insulation thickness requirements and test voltages in accordance with ICEA [14,16] standards. The alternating current (AC) withstand test voltage is approximately 200 volts per mil (7.9 kV/mm) of specified insulation thickness. Two important changes from past ICEA requirements are that a DC test is no longer required for these cables. The

TABLE 10.1
Insulation Physical Requirements

	Insulation Type					
		XLPE Class III and TR-XLPE Class III	EPR Class			
Physical Requirements	**XLPE and TR-XLPE**		I	II	III	IV
Unaged requirements						
Tensile strength, minimum						
psi	1,800	1,800	700	1,200	700	550
MPa	(12.5)	(12.5)	(4.8)	(8.2)	(4.8)	(3.8)
Elongation at rupture, minimum percent	250	250	250	250	250	250
Aging requirements after air oven aging for 168 hours						
Aging temperature, °C ± 1°C	121	136	121	121	136	121
Tensile strength, minimum percentage of unaged value	75	75	75	80	75	75
Elongation						
Minimum percentage of unaged value	75	75	75	80	75	—
Minimum percent at rupture	—	—	—	—	—	175
Hot creep test at 150°C ± 2°C	Unfilled	Filled				
Elongation, maximum percent*	175	100	50	50	50	50
Set, maximum percent*	10	5	5	5	5	5

Source: ANSI/ICEA S-94-649, "Standard for Concentric Neutral Cables Rated 5,000–46,000 Volts", 2004.

* For XLPE and TRXLPE insulations if this value is exceeded, the Solvent Extraction Test may be performed and will serve as a referee method to determine compliance (a maximum of 30 percent weight loss after 20 hour drying time).

TABLE 10.2
Insulation Electrical Requirements

	Insulation Type				
				EPR Class IV	
Property	XLPE and XLPE Class III	TR-XLPE and TR-XLPE Class III	EPR Class I, II and III	28 kV or Less	Above 28 kV
Dielectric Constant	3.5	3.5	4.0	4.0	4.0
Dissipation Factor, %	0.1	0.5	1.5	2.0	1.5
Insulation resistance constant, megohms—1,000 ft at 15.6°C	20,000	20,000	20,000	20,000	20,000

Source: ANSI/ICEA S-94-649, "Standard for Concentric Neutral Cables Rated 5,000–46,000 Volts", 2004.

other is that they provide two insulation thicknesses for each voltage rating, such as 100 Percent Level and 133 Percent Level.

RUS specifications require the use of the 133 Percent Level insulation thickness for cables that are manufactured to their needs unless dispensation is given based on the selective designs.

10.5.4 Extruded Insulation Shields

In addition to the conductor stress control layer, medium voltage, shielded power cables require an insulation shield. The insulation shield consists of a nonmetallic covering directly over the insulation and a nonmagnetic metal component directly over or imbedded in the nonmetallic conducting covering. The insulation shield shall be readily removable in the field at temperatures from −10°C to 40°C. At the option of the purchaser, an insulation shield that is bonded may be supplied. The insulation and the semiconducting material must be compatible since they are in close contact with one another.

10.5.4.1 Strip Tension

The ICEA standard has peel strength limits for the removal of the semiconducting layer for 5 to 46 kV cables. The lower limit is for cable performance, and the upper limit is set to permit removal without damaging the surface of the insulation.

The test calls for a 1/2-inch wide strip to be cut parallel to the center conductor. This cut may be completely through the layer (in contrast to field stripping practices). The 1/2-inch strip is removed by pulling at a 90° angle to the insulation surface at a set rate of speed. The limits are shown in Table 10.3.

TABLE 10.3
ICEA Strip Tension Limits

Material	Lower Limit in Pounds (N)	Upper Limit in Pounds (N)
XLPE and TR-XLPE	3 (13.4)	24 (107)
EPR	3 (13.4)	24 (107)

Source: ANSI/ICEA S-94-649, "Standard for Concentric Neutral Cables Rated 5,000–46,000 Volts", 2004.

10.5.4.2 Resistivity

The volume resistivity of the extruded insulation shield shall not exceed 500 ohm-meter at 90°C and 110°C for 90°C rated cables and 105°C and 125°C for 105°C rated cables, when tested in accordance with ICEA procedures. This layer can only be used as an auxiliary shield and requires a metal shield in contact with it to drain off charging currents and to provide electrostatic shielding.

This volume resistivity requirement is half that for the conductor shield, because this layer is subject to chemical action from the environment. The function of the shielding properties would be acceptable with a higher value, but concerns over longtime stability have influenced this level.

The resistivity of the extruded layer is also measured using silver-painted electrodes. The outer coverings including the metallic shield are removed. Four silver-painted annular-ring electrodes are applied to the outer surface of the insulation shield. The inner two electrodes are for the potential application and are at least 2 inches apart. If a high degree of accuracy is required, a pair of current electrodes are placed at least 1 inch behind each potential electrode. The resistance shall be measured between the two potential electrodes. The power of the test circuit shall exceed 100 milliwatts.

The volume resistivity is then calculated as follows:

$$\rho = \frac{2R\left(D^2 - d^2\right)}{100L} \tag{10.2}$$

where

ρ = volume resistivity in ohm-meters
R = measured resistance in ohms
D = diameter over insulation shield in inches
d = diameter under insulation shield in inches
L = distance between potential electrodes in inches

10.5.4.3 Insulation Shield Thickness

ICEA has established minimum point and maximum point thickness requirements for the extruded insulation shield layer to provide guidance for the manufacturers of molded splices and terminations [14,16].

10.5.5 Metallic Shielding

In addition to the extruded insulation shield previously described, shielded cables must have a nonmagnetic metallic shield over and in contact with the nonmetallic semiconducting layer. The following options are available for the metallic member.

- Helically applied flat tin-coated or uncoated copper tape(s)
- Longitudinally applied corrugated annealed copper tape
- Tin-coated or uncoated copper wire shield (minimum of six #25 AWG or larger)
- Concentric copper wires (#16 AWG or larger to meet neutral cross-sectional area)
- Flat copper straps applied with close coverage to meet neutral cross-sectional area)
- Combination of tape plus wires
- Continuous welded corrugated metal sheath (copper, aluminum, bronze, etc.)
- Extruded lead sheath

Wire shields and flat tapes are popular metallic shields and are almost always copper. A 5-mil copper tape with a minimum 10% overlap is generally used when tapes are specified. For wire shields, #24 to #18 AWG wires are used in proper multiples to provide a minimum of 5,000 circular mils per inch (0.1 mm^2/mm) of insulated core diameter. These types have a limited fault current capacity and are not commonly used by electric utilities for outdoor plant for that reason.

Concentric neutral wires and flat straps are normally specified on underground distribution (UD) and underground residential distribution (URD) cables where the metallic shield functions as both a shield and a neutral. These constructions normally use copper wires with an overall jacket applied over the wires for corrosion protection.

In higher voltage cables such as 35 kV to 138 kV, fault currents often may be greater than the capabilities of wires alone. In those situations, the tape plus wire construction is frequently used.

Where metallic shields must be sized for specific fault-clearing requirements, there are several sources of data such as: ANSI/ICEA P-45-482, "Short Circuit Performance of Metallic Shields and Sheaths on Insulated Cable" and EPRI RP 1286-2, (EL-5478), "Optimization of the Design of Metallic Shield/Concentric Neutral Conductors of Extruded Dielectric Cables under Fault Conditions."

10.5.5.1 Concentric Neutral Cables

ICEA standards cover the number and size of concentric neutral wires for this type of cable. The concentric neutral wires shall be uncoated copper in accordance with ASTM B3 or tin-coated copper in accordance with ASTM B33. The wires of the concentric neutral shall be applied directly over the insulation shield with a lay of not less than six or more than ten times the diameter over the concentric wires.

Although the AEIC specification does not provide information on concentric neutrals, it is important to understand that a full or one-third neutral is not mandated by any standard. Many utilities use fewer neutral wires based on the fact that too much metal leads to increased shield losses. RUS standards do not require even a full neutral for URD cables.

10.5.6 Cable Jackets

Jackets are generally required over metallic shields for mechanical and corrosion protection during cable installation and operation.

There are many possible jacketing materials such as:

- Low density and linear low density polyethylene, black (LDPE/LLDPE)
- Medium density polyethylene, black (MDPE)
- High density polyethylene, black (HDPE)
- Semiconducting jacket Type I
- Semiconducting jacket Type II
- Polyvinyl chloride (PVC)
- Chlorinated polyethylene (CPE)
- Thermoplastic elastomer (TPE)
- Polypropylene, black (PP)

Their attributes are discussed in Chapter 8. ICEA standards cover the thickness of these jackets.

10.5.7 General

Numerous documents are available that provide useful information on standards such as: [2–5,7–11,13,15,17–24].

REFERENCES

1. AEIC CS8-07, "Specification for Extruded Dielectric Shielded Power Cables Rated 5 through 46 kV," Third Edition, 2007, AEIC, 600 North 18th Street, P.O. Box 2641, Birmingham, AL 35291-0992.
2. AEIC CS9-06, "Specification for Extruded Insulation Power Cables and Their Accessories Rated Above 46 kV Through 345 kV AC," Third Edition, 2006, AEIC, 600 North 18th Street, P.O. Box 2641, Birmingham, AL 35291-0992.
3. ICEA T-22-294, "Test Procedures for Extended-Time Testing of Wire and Cable Insulations for Service in Wet Locations," 1983, Withdrawn by ICEA.
4. ANSI/ICEA T-24-380, "Standard for Partial Discharge Test Procedures," 2007, Global Engineering Documents, 15 Inverness Way East, Englewood, CO 80112.
5. ICEA P-32-382, "Short-Circuit Characteristics of Insulated Cable," 2007, Global Engineering Documents, 15 Inverness Way East, Englewood, CO 80112.
6. ICEA S-61-402/NEMA WC 5, "Thermoplastic Insulated Wire and Cable, 0 to 35 kV," Withdrawn by ICEA.
7. ICEA T-25-425, "Guide for Establishing Stability of Volume Resistivity for Conducting Polymeric Components of Power Cables," 2003, Global Engineering Documents, 15 Inverness Way East, Englewood, CO 80112.

8. ANSI/ICEA T-26-465/NEMA WC 54, "Guide for Frequency of Sampling Extruded Dielectric Cables," 2008, Global Engineering Documents, 15 Inverness Way East, Englewood, CO 80112.
9. ANSI/ICEA P-45-482, "Short Circuit Performance of Metallic Shields and Sheaths on Insulated Cable," 2007, Global Engineering Documents, 15 Inverness Way East, Englewood, CO 80112.
10. ICEA T-28-562, "Test Method for Measurement of Hot Creep of Polymeric Insulations," 2003, Global Engineering Documents, 15 Inverness Way East, Englewood, CO 80112.
11. ANSI/ICEA T-27-581/NEMA WC 53, "Test Methods for Extruded Dielectric Cables," 2008, Global Engineering Documents, 15 Inverness Way East, Englewood, CO 80112.
12. ANSI/ICEA T-31-610, "Test Method for Conducting Longitudinal Water Penetration Resistance Tests on Blocked Conductors," 2007, Global Engineering Documents, 15 Inverness Way East, Englewood, CO 80112.
13. ICEA T-32-645, "Guide for Establishing Compatibility of Sealed Conductor Filler Compounds with Conductor Stress Control Materials," 1993, Global Engineering Documents, 15 Inverness Way East, Englewood, CO 80112.
14. ANSI/ICEA S-94-649, "Standard for Concentric Neutral Cables Rated 5,000–46,000 Volts," 2004, Global Engineering Documents, 15 Inverness Way East, Englewood, CO 80112.
15. ANSI/ICEA T-34-664, "Test Method for Conducting Longitudinal Water Penetration Resistance Tests on Longitudinal Water Blocked Cables," 2007, Global Engineering Documents, 15 Inverness Way East, Englewood, CO 80112.
16. ANSI/ICEA S-97-682, "Standard for Utility Shielded Power Cables Rated 5,000–46,000 Volts," 2007, Global Engineering Documents, 15 Inverness Way East, Englewood, CO 80112.
17. ANSI/ICEA S-108-720, "Standard for Extruded Insulation Power Cables Rated Above 46 Through 345 kV," 2004, Global Engineering Documents, 15 Inverness Way East, Englewood, CO 80112.
18. CAN/CSA-C68.5-07, "Primary and Shielded Concentric Neutral Cable for Distribution Utilities," 2007, Canadian Standards Association, 5060 Spectrum Way, Suite 100, Mississauga, Ontario, Canada, L4W 5N6.
19. CSA C68.08, "Shielded Power Cable for Commercial and Industrial Applications, 5–46 kV," 2008, Canadian Standards Association, 5060 Spectrum Way, Suite 100, Mississauga, Ontario, Canada, L4W 5N6.
20. IEC 60502-1, Edition 2.1 2009–11, "Power cables with extruded insulation and their accessories for rated voltages from 1 kV (Um = 1,2 kV) up to 30 kV (Um = 36 kV) — Part 1: Cables for rated voltages of 1 kV (Um = 1,2 kV) and 3 kV (Um = 3,6 kV)," 2009, IEC Central Office, 3 rue de Varembé, P.O. Box 131, CH-1211 Geneva 20, Switzerland.
21. IEC 60502-2, Edition 2.0 2005–03, "Power cables with extruded insulation and their accessories for rated voltages from 1 kV (Um = 1,2 kV) up to 30 kV (Um = 36 kV) — Part 2: Cables for rated voltages from 6 kV (Um = 7,2 kV) up to 30 kV (Um = 36 kV)," IEC Central Office, 3 rue de Varembé, P.O. Box 131, CH-1211 Geneva 20, Switzerland.
22. IEC 60840, Edition 3.0 2004–04, "Power cables with extruded insulation and their accessories for rated voltages above 30 kV (Um = 36 kV) up to 150 kV (Um = 170 kV) — Test methods and requirements," IEC Central Office, 3 rue de Varembé, P.O. Box 131, CH-1211 Geneva 20, Switzerland.
23. IEC 62067, Edition 1.1 2006–03, "Power cables with extruded insulation and their accessories for rated voltages above 150 kV (Um = 170 kV) up to 500 kV (Um = 550 kV) — Test methods and requirements," 2006, IEC Central Office, 3 rue de Varembé, P.O. Box 131, CH-1211 Geneva 20, Switzerland.
24. UL Standard 1072, "MV Cables Rated 2,001 to 35,000 Volts," 2007, Underwriter's Laboratories, 2600 N.W. Lake Rd., Camas, WA 98607-8542.

11 Cable Manufacturing

Carl C. Landinger

CONTENTS

11.1 INTRODUCTION

Insulated electric power cable manufacturing involves a broad range of complexity depending on the cable design to be produced [1,2]. Different cable plants may be capable of a limited or broad range of designs. Those capable of a broad range may limit operations to only a few steps in the manufacturing process.

Despite this large variability in plants, the steps in the manufacturing process remain basically the same, whether done in one facility or a number of facilities. Conductor manufacturing is common to all cables with metallic conductors. The manufacture of extruded dielectric power cables and laminar dielectric power cables follows.

11.2 CONDUCTOR MANUFACTURING

In Chapter 3 (Conductors), it was pointed out that for efficient distribution of electric power, the conductors must be produced from a high conductivity material. It was also shown that copper and aluminum offer the best available combinations of conductivity, workability, strength, and cost to become the most popular power cable conductor materials. From the conductor-manufacturing standpoint (we will not attempt to include mining, refining, and fabricating stages), we will begin at the point where copper and aluminum are received as large coils of round rod. The diameter of aluminum rod for producing conductors is commonly 3/8 inch in diameter (0.375 inches, 9.53 mm) while 5/16 inch (7.94 mm) diameter is common for copper rod. For larger solid conductors; [i.e., #1/0 AWG or larger (0.325 inches diameter, 8.26 mm)], it is common to begin with a larger diameter rod. The larger diameter (often 0.500 inches, 12.7 mm) allows for enough diameter draw-down to remove surface imperfections in the rod, as received, to make a conductor surface without residual surface imperfections.

11.2.1 WIRE DRAWING

In wire drawing, the copper or aluminum rod is drawn through a series of successively smaller dies to reduce the rod to a wire of the desired diameter. The quality of the wire surface depends on sufficient drawing and reduction to eliminate surface defects. Thus, there is the need to utilize a rod or intermediate wire having a diameter somewhat larger than the solid wire to be produced. If fine wire is desired, it is common to utilize a coarse wire-drawing machine followed by a fine wire-drawing machine. The wires are taken up on spools for later stranding operations or on reels for use as a solid conductor.

11.2.2 Annealing

Drawing copper or aluminum wires increases the temper of the metal. That is, a rod of a "softer" temper is "hardened" as the wire is drawn down to the required diameter.

Except for the use of full hard temper aluminum stranded conductors for electric utility outside plant secondary and primary cables, it is generally desirable to use a softer temper. To produce a softer temper, the wire is exposed to elevated temperatures well in advance of emergency operating temperatures of the cable. For many years, this has been accomplished in a large oven. Exposure time using this method is a matter of hours.

It is possible to partially anneal wires "on the fly." This is generally done on a wire before it is used in a stranding operation. The method is not generally suitable for full annealing to a soft temper or to conductors after they have been stranded.

11.2.3 Drawing to Temper

Another method of arriving at wires of the desired temper is to begin with rods of a softer temper and using the work hardening associated with the wire drawing to arrive at the desired temper.

11.2.4 Stranding

The simplest conductor is a solid rod/wire. As the cross-sectional area of the conductor increases, the conductor becomes increasingly stiff. Stranded conductors use a number of smaller wires, the sum of which totals the desired conductor cross-sectional area. The machines used in stranding range from those that apply successive layers of strands over a central core made up of one or a number of strands, to those producing stranded configurations resembling ropes and for very fine wires making bunches of fine strands. See Figure 11.1 for an example of a seven-wire strander in operation.

FIGURE 11.1 Conductor strander in operation.

11.3 MANUFACTURING OF CABLE WITH EXTRUDED INSULATIONS

11.3.1 Insulating and Jacketing Compounds

There are literally thousands of insulating and jacketing compounds used in the cable industry. Many of these compounds are commercially available from compound suppliers. They may also be custom compounded by companies that sell them as finished, "ready-to-extrude" compounds. The cost of "ready-to-extrude" compounds is high enough so there is considerable incentive for the manufacturer to mix many compounds in-house. Low voltage compounds provide special opportunities ranging from the simple addition of one or more ingredients at the extruder to the complete mixing of all the ingredients and the production of strips or pellets suitable for extrusion. The complex subject of compounding is beyond the scope of this chapter. For our purposes, we will assume compounds are complete, "ready-to-extrude." However, it is necessary to recognize that this all-important compounding step is increasingly becoming a part of the manufacturing process.

11.3.2 Extrusion

The extrusion currently in use to produce polymeric layers comprising the cable is similar regardless of the polymer or layer being extruded.

Compound, in the form of pellets or strips, is fed into the back of a screw that rotates in a barrel. The material advances down the screw and is melted during the advance. In general, the barrel is divided into zones that are individually temperature controlled. There are some extruders where the barrel is heated at the start of the extrusion, but as the extrusion continues the mechanical shear and friction result in sufficient heat generation that barrel heating is no longer required. In fact, depending on compound and extrusion parameters, barrel cooling and even screw cooling may be required. Properly executed, the compound is all melted and forced through a diehead arrangement that deposits the melt on the core being passed through the head. This core may be a bare wire or cable in some stage of completion. Figure 11.2 shows

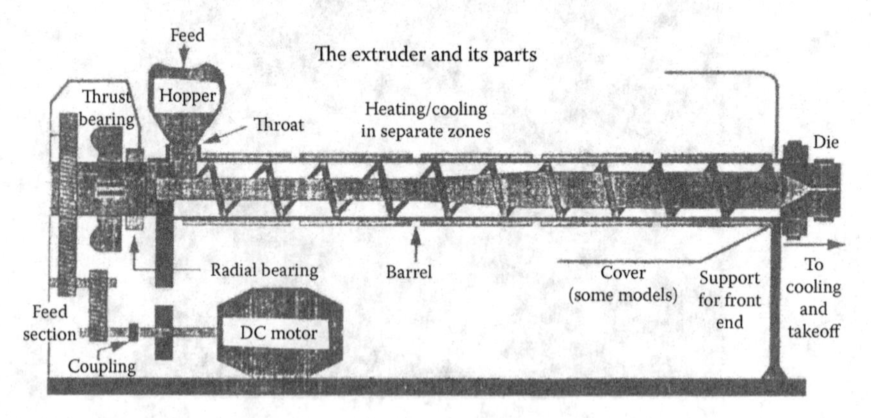

FIGURE 11.2 Extruder.

FIGURE 11.3 Crosshead and extruders.

the common features of an extruder. Figure 11.3 shows the physical arrangement of three extruders and the crosshead at a manufacturing facility.

In many cases, the compound is introduced with all of its final ingredients included. However, variations such as the addition of curing peroxides, color concentrates, or other ingredients at or near the extruder are quite commonly used.

11.3.3 Curing

This term is somewhat of a carryover from the rubber materials that required curing. The cross-linking process for modern thermosetting compounds, such as cross-linked polyethylene and ethylene propylene rubber (EPR), is often referred to as a curing stage. While materials such as polyethylene can be cross-linked by a radiation-induced reaction, the majority of cross-linking for power cables continues to be by the chemical means.

Taking a simple case of polyethylene, the addition of a peroxide agent such as dicumyl peroxide to the polyethylene and supplying heat energy results in a chemical reaction that cross-links the polyethylene turning it from a thermoplastic material into a thermosetting material. Peroxides are also commonly used for cross-linking EPR compounds. The heating period to effect cross-linking is commonly called curing. It is also referred to as vulcanization (a carry over from the tire industry), hence reference to the CV tube is the "continuous vulcanization" tube.

Curing tubes have three distinct configurations. The most commonly used is a curved tube that is in the shape of a catenary as shown in Figure 11.4. The first portion is the curing section and the lower portion is the cooling section. The shape is designed to prevent touch-down of the cable until the cable has cured. The weight of the cable, line speed, and length of the total tube must be considered in this design. Other forms of curing tubes may be horizontal or vertical. Horizontal tubes are used for very small cables or, for large thick insulation walled cables, in special extruders

FIGURE 11.4 Curing section of dry cure CV line.

that employ very long dies. A vertical extruder has the advantage of being able to make very large cables with thick insulation walls, especially transmission cables. They run relatively slowly, but gravity does not work to deform the shape of the cable.

The heat source most commonly used in the past was steam in a tube through which the extruded cable was passed. This continues to be the most popular means for curing secondary cables. When curing relatively heavy walls such as primary cables, the upper limit on temperature that steam can practically impose makes it desirable to use other heat sources. The most popular heat source today for medium and high voltage cables is radiant heating in a nitrogen-filled tube. This is one of a number of dry curing methods. This method allows for much higher curing temperatures and therefore faster line speeds and curing. These curing tubes are divided into a number of zones each of which has its individual temperature controls. This allows for optimum temperature profiling to effect cure.

Because of the high temperature involved, care must be taken to avoid thermal damage of the polymer. More common in Europe, and gaining popularity in North America for cables up to 600 volts, is silane curing. The system is based on the technology of silicones, and "Sioplas" as originally developed is a two-part system of cross-linkable graft polymer and a master batch catalyst.

A further development, "Monosil," introduces ingredients at the extruder and thus eliminates the grafting process. Water is the cross-linking agent in these silane systems and cure rates become very thickness-dependent. One of the advantages of moisture curing is that a thermoplastic extrusion line (no curing tube as required for peroxide curing) can be used to produce the thermosetting cable.

11.3.4 Cooling

Thermoplastic materials, such as polyethylene or polyvinyl chloride, do not require curing. Single layers that are relatively thin, such as in 600 volt building wire, may be cooled in a water trough following extrusion. In the case of polyethylene, care must be taken to avoid too rapid a quench. This rapid cooling can result in locked-in mechanical stresses which will result in shrink-back of the material on the wire.

Heavier thermoplastic layers, such as encountered on primary cables, require gradient cooling to avoid these stresses in the polyethylene.

Following curing, thermosetting materials must also be cooled. When steam is the curing medium, water cooling is universally employed. Cross-linked polyethylene must not be rapidly quenched to avoid "shrink back" that is caused by locked-in stresses. Cooling zones are used to control the cooling process for water-cooled cables.

With the dry cure process, there is the possibility of using nitrogen as the cooling method. This is not frequently used for cables at this time. Cooling is sufficiently gradual that stresses are not locked-in.

11.4 EXTRUSION LINE CONFIGURATIONS

11.4.1 Single Extruder Line

The simplest configuration for an extrusion line is one that can be used for low voltage thermoplastic cables having a single plastic layer over a conductor (Figure 11.5). A few examples of cables that are produced this way are:

- Line wire
- Building wire
- A jacket over other cores

A curing zone may be added just before the cooling zone if curing is needed.

11.4.2 "Two Pass" Extrusion

Thermoplastic primary cables have been produced in a similar line configuration, but two separate extruders were used to apply the conductor shield and the insulation. Another "pass" through the third extruder after the first two layers were applied and cooled became known as "two pass." The figures that are shown here do not

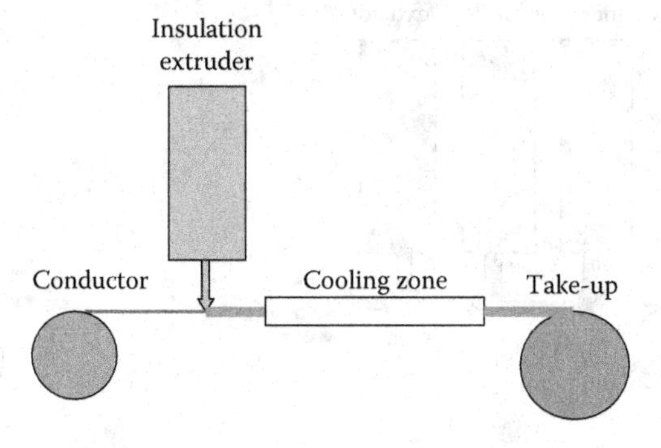

FIGURE 11.5 Single extrusion.

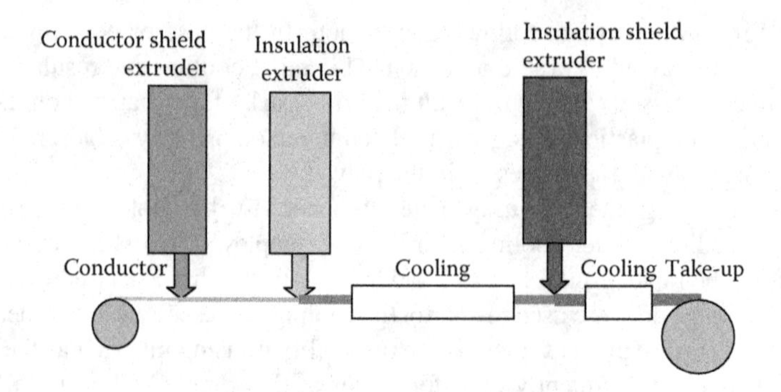

FIGURE 11.6 Triple tandem single pass extrusion.

imply that the curing and cooling tubes are straight. The figures represent all possible configurations.

If the product was to be cured, a curing zone had to be included. Note that the third extruder in Figure 11.6 was placed after the first cooling zone. That made it difficult if not impossible to maintain the desired strippability of thermosetting insulation shields over thermosetting insulations. Thus, it was common to utilize a thermoplastic insulation shield over thermosetting insulation.

11.4.3 "Single Pass" Extrusion

The development of semiconducting thermosetting shield materials that would be readily strippable from thermosetting insulation even though all three layers were cured at the same time led to the development of lines where all three layers of a primary cable could be extruded over the core prior to any curing or cooling.

Figure 11.7 shows a single pass extruder where the conductor goes in at one end and the three layers are applied by three different extruders but in one complete operation.

This was the first time the "triple extrusion" term was applied. While this arrangement was preferable to all previous methods because of minimal exposure

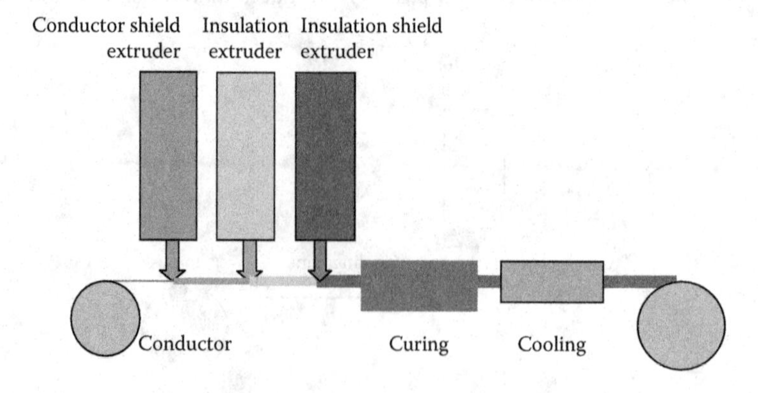

FIGURE 11.7 Single pass with three extruders.

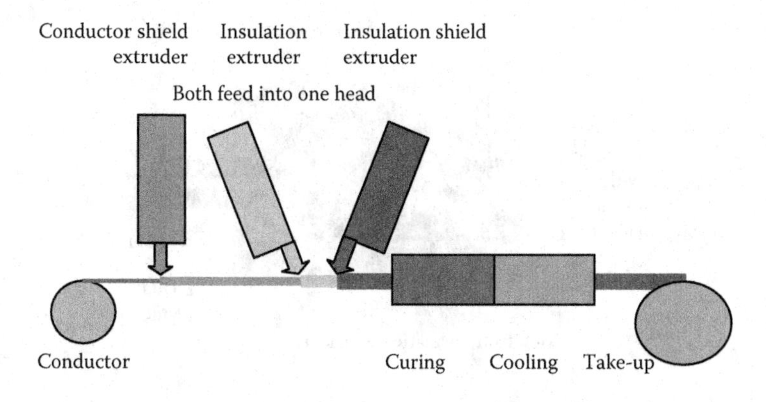

FIGURE 11.8 Single pass with one dual extruder (triple tandem).

of the insulation to possible contamination or abuse, further developments were desired. Dual extrusion of the insulation and insulation shield layers would make for a smoother interface. Thus, the next improvement was single extrusion of the conductor shield and then, a few feet away, the dual extrusion of the insulation and the insulation shield. This was also called "triple extrusion"! About this time, dry curing lines were growing in popularity and many lines of this type were installed.

While the extruders are essentially the same (Figure 11.7), the insulation and insulation shield extruders feed into a single head (Figure 11.8).

Unfortunately, this method continued to allow the conductor shield to be vulnerable to scraping in the next extrusion head, continued to allow build-up on the extruder die face (die drool), and exposed the conductor shield to the environment.

11.4.4 "True Triple" Extrusion

The method now used for the majority of medium voltage cables utilizes a single crosshead where all three layers are applied simultaneously. This is referred to as "true triple" extrusion (Figure 11.9).

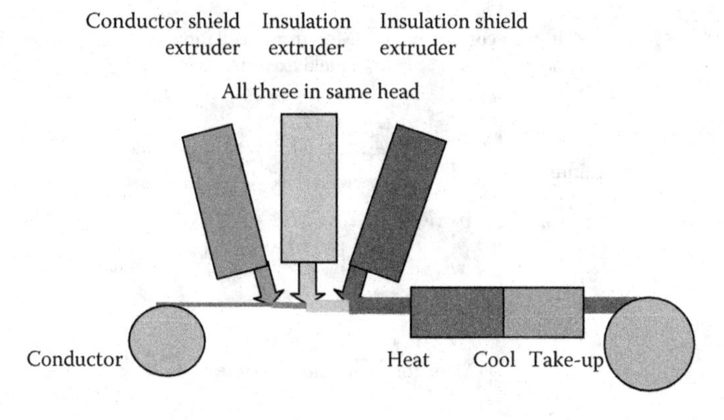

FIGURE 11.9 True triple extrusion.

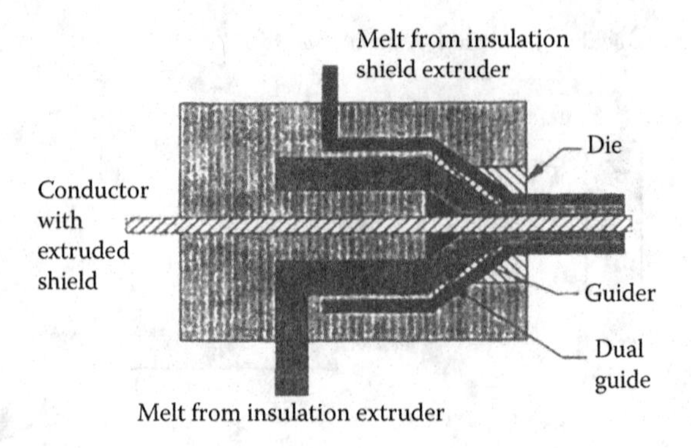

FIGURE 11.10 Dual style crosshead.

All three extruders feed into a single head for "true triple" extrusion. There are numerous lines now in service in North America, and the world in general, that make use of these triple heads.

11.4.5 Extrusion Heads

Extrusion heads are designed to "channel" the extrudate onto the wire or cable passing through the head. Dual and triple heads channel the flow of two or three extruders in a single head. Figure 11.10 shows a cut away view of a dual crosshead and Figure 11.11 shows a triple head.

11.4.6 Finishing

Finishing operations generally include the addition of wires, tapes, or braids (in medium-voltage cables, the addition of metallic shields) prior to jacketing.

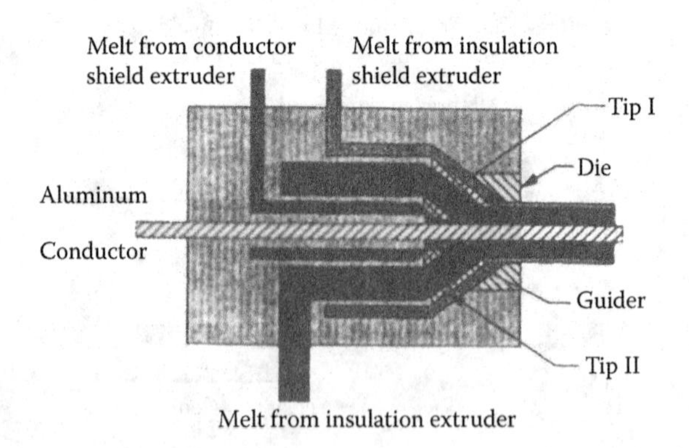

FIGURE 11.11 Triple style crosshead.

11.4.7 Assembly

In cases of covered overhead service cables and similar constructions, a number of single cables may be assembled as a group. This is done on cablers or twisters. The equipment has some similarity to the equipment discussed under stranding. For assemblies to be jacketed later and serve as multiconductor cables, it is common to add fillers to "round out" the assembly as well as use taping heads to apply binder and jacketing tapes in the same operation.

11.5 PAPER INSULATED CABLES

11.5.1 Paper Insulation

It has been found that up to a certain point, the mechanical strength of paper increases with its moisture content. Accordingly, prior to their use in the taping machine, pads (rolls) of paper are conditioned for a definite period in a room in which the temperature and humidity are controlled. This procedure ensures that all of the paper is in the same condition as it is being applied over the conductor and results in more uniformity in the taped insulation. When the paper is dried prior to impregnation, it shrinks uniformly. This allows for cables with sector conductors to be cabled without wrinkling.

11.5.2 Paper Taping

The importance of controlled tension in the taping process is realized when one is reminded that to have an optimum of electrical strength, paper insulation must be tightly applied, free from wrinkles, and other mechanical defects that non-uniformly applied layers of tapes would have. Close, automatic control must be accomplished.

In one method, the tape from the pad passes around a pulley that is geared to a small motor armature whose direction or rotation is opposite to the direction of tape feed. As the tape feeds along, the armature is revolved opposite to the direction it would take if turning freely and against the motor field torque. The pulley, therefore, exerts a back pull on the tape at all times and with a constant value. Torque must be regulated to the tension that is required. See Figure 11.12 for an example of a paper taping machine.

11.5.3 Cabling

A large cabling machine is used for assembling individually insulated conductors into two-, three-, or four-conductor cables. The cradles may be operated rigidly or in a planetary motion to accommodate the large diameter cabling bobbins. This reduces the bending stresses to which the paper is subjected. Facilities are provided for mounting smaller bobbins between cradles that may be used for fillers, smaller cables such as fiber optic, or tubes. Small packages of fillers may also be carried on the spindles. Guides and bushings are used for placing

FIGURE 11.12 Paper taping machine.

sector-shaped conductors in their proper position without undue strain on the insulation. Behind the assembly bushings, heads are mounted for applying paper tapes on nonshielded type cables, or in the case of shielded cables, intercalated binder tapes.

Metal binder tapes are spot-welded when a new pad of tape is inserted in a taping head. The cable is drawn through the machine by a large capstan to a take-up reel. The large diameters of the capstan and impregnating reels reduce the bending stresses in the insulation.

11.5.4 Impregnating Compounds

Paper cables have been impregnated with numerous compounds over the years. A few that have been used include:

- Type A: Unblended naphthenic-base mineral oil
- Type B: Naphthenic-base mineral oil blended with purified rosin
- Type C: Naphthenic-base mineral oil blended with a high molecular weight polymer
- Type D: Petrolatum blended with purified rosin
- Type E: Polybutene

When paper-insulated cable is impregnated with a dielectric fluid, the combination is better than either individual components alone and results in valuable characteristics:

1. High initial electrical resistivity
2. Low rate of deterioration from high temperatures
3. Low power factor
4. Very flat power factor vs. temperature curve
5. Low ionization factor

Investigation had shown that the unblended mineral oil was the most stable oil from a chemical and electrical standpoint. Natural inhibitors in the petroleum afforded high oxidation stability.

These inhibitors are complex resins occurring naturally in crude petroleum. For the most part, they are eliminated in the refining process and necessarily so because they represent impurities. If the petroleum is overrefined, all these inhibitors are removed, resulting in a clear oil of high electrical characteristics but having unstable qualities. These resins act as antioxidants by taking up oxygen themselves from the oil and thus inhibiting oxidation deterioration. In the refining process used for this oil, a good balance is obtained between electrical characteristics and high thermal stability. Most of the oil impregnated, medium voltage cables were made with Type A compound. Types B, C, and D were more viscous than Type A and were suitable for long vertical runs or slopes with Type C being the most frequently used compound.

The predominant compound used since 1980 has been the synthetic material polybutene. Since this is not oil, it is proper to refer to these as fluid-filled cables.

11.5.5 Drying and Impregnating

Assuming that the proper materials have been selected and good mechanical construction employed, the electrical characteristics of the completed cable depend primarily upon the drying and impregnating process.

It has been established by many laboratory investigations that oil, under heat and exposure to air, rapidly loses it desirable insulating properties. Also, the presence of residual air and moisture is harmful to impregnated paper insulation. Thus, paper-insulated cables are dried and impregnated in a closed system.

The functional principles of this closed system are:

1. Transfer of hot impregnating fluid from a vacuum tank to another tank under vacuum without exposure to air
2. Use of relatively high fluid pressure (85 to 200 pounds per square inch) during impregnation
3. Complete degassing and dehydration of the fluid
4. Use of extremely high vacuum (1 mm or less)

If there is more than one impregnating compound used in a plant, it is desirable to have separately assigned tanks for each material.

Prior to transfer to the impregnating tank, the fluid to be used is heated in its steam-jacketed storage tank where it is kept under vacuum. During this heating period, the fluid is agitated in order to maintain a uniform temperature.

In the center of each of the steam-jacketed vacuum and pressure impregnating tanks is a steam-jacketed cylinder of slightly smaller diameter than the hollow drum of the impregnating tanks. This reduces the amount of fluid subjected to heat during each impregnating cycle. Over the top of this cylinder, a large, circular baffle plate is mounted. When the fluid is admitted into the tank, it strikes this baffle where it forms a thin film. This affords an opportunity to subject the fluid to a second degassing treatment.

The impregnating of the paper can be considered to take place in two steps. First, the fluid flows back and forth between the tapes from the outside of the insulation toward the conductor. This is best accomplished by applying vacuum. Second, the fluid penetrates the fibers of the individual paper tapes. This is accomplished by using pressure.

11.5.6 Conditioning the Impregnating Fluid

When the fluid is subjected repeatedly to the temperatures used in impregnating, it gradually loses many of its desirable properties. A plant to condition the fluid is frequently installed at the factory to ensure uniform quality fluid. Each of the vacuum operations is performed under a vacuum of a precise absolute pressure. Automatically controlled electric heaters maintain the proper temperatures for each operation. By the use of this equipment, the fluid is maintained at all times at a quality equal to that of the new fluid. This, together with the impregnating control, results in cables of uniform quality.

11.5.7 Control of Impregnation

In the manufacturing of solid-type paper insulated cables, the general practice is to regulate the drying and impregnating process by setting up standard periods of time for each operation. Slight modifications are made for the particular size and type of cable to be treated.

Due to the inherent variations in the materials and manufacturing, dielectric measurements are made on the cable undergoing drying and impregnation. This control consists of periodic readings, giving an accurate measure of temperature and degree of impregnation. This established definite control throughout every step of the impregnating process.

Flexible electrical connections are made between the ends of the cable and the permanent terminals on the tank as the cable is placed in the tank.

The conductor resistance is converted to conductor temperature. Knowing the viscosity–temperature characteristics of the fluid, these combined data effect a control of the impregnating ability of the fluid. Dryness is determined by the constancy of the AC capacitance values. Thorough saturation is determined by the change in AC capacitance. Complete cooling is determined from the conductor temperature measurements.

11.5.8 Control of the Cooling Cycle

Uniform cooling at a defined rate produces high quality paper insulated cable. To accomplish this, a large refrigeration plant can be installed as part of the impregnating system. This enables the mill to cool the impregnated cables at a prescribed rate independent of the water supply temperature.

The cooling is brought into use after the impregnation cycle is completed. The temperature of the entire tank is reduced under this controlled cycle to room temperature. This is a gradual reduction that is made while the cable is still in the sealed tank. The seal is then broken and the cables, coated with a thick layer of fluid, and transferred to the lead press, or other sheathing process.

The lead presses that were used for most medium voltage cables in the past, could extrude lead under pressures of 3,000 tons. A lead charge of up to 2,000 pounds was placed in a melting kettle located adjacent to each press. The advantage of a large charge was that fewer stops had to be made. This stop could last for several minutes and for all of that time the cable in the die-block was subjected to high temperatures. Continuous extrusion techniques were also developed.

The melting kettle had automatic temperature control to keep the lead at the proper molten state. An agitator was used to keep the metal stirred so there was no separation in the mix. Means were provided to removing dross (oxidized lead) from the top of the molten mass. Over the lead, an atmosphere of inert gas, such as carbon dioxide or nitrogen, could be used to reduce the formation of oxides.

The lead sheathing process takes the impregnated cables through a steel die-block attached to the cylinder of the press. The die-block has a core tube having a diameter just large enough to receive the cable core and a die having an outer diameter of the sheath diameter. Pressure is exerted on the lead in its plastic state by a hydraulic ram or piston. The lead is extruded at temperatures of about 375°F to 400°F from the cylinder into the die-block and around the outer portion of the core tube. The lead is squeezed down over the cable to form a thick, continuous, homogeneous sheath. Pressure behind the lead tube forces the cable through the die-block.

When the cylinder is charged, it is overflowed and the exposed lead is allowed to congeal. The exposed lead is then skimmed off to level with the cylinder by means of a mechanical guillotine or cropping device that removes all traces of oxide. The hydraulic ram, having a rounded nose, is then immediately brought down onto the surface of the skimmed surface of the lead.

The press is started and some amount of lead pipe is extruded and checked to make sure the crystalline structure, welds, and ductility are satisfactory. After the cable is started through the press, a sample of lead sheathed cable is cut off and concentricity checked. As the cable leaves the die-block, it is water-cooled and either given a jacket or a coating of grease, as required.

The discontinuous type of extrusion presses with vertical rams and containers that had to be filled with liquid lead were largely replaced by continuous extrusion machines. The screw of these machines is vertical and lead is fed from the bottom end from a reservoir of lead. The extrusion temperature is maintained at about 300°C.

11.6 FACTORY TESTING

11.6.1 Electrical Tests for 100% Inspection of Low Voltage Cables

The Insulated Cable Engineers Association (ICEA) recognizes alternative test methods for electrical testing of single conductor or assemblies of single conductor secondary cables (up to 2,000 volts phase-to-phase) as called for in the appropriate standard. Jacketed assemblies of single conductor cables are tested between phases and ground in addition to a test on the single conductors. Sheathed or armored cables are also tested from each phase to the sheath or armor.

11.6.1.1 Alternating Current Spark Test

The cable conductor is grounded. The covered/insulated cable surface is passed through a close network of metallic bead chains or similar contact electrode. The electrodes are at an AC voltage potential selected on the basis of the type and thickness of the covering. In the event of a pinhole, skip, or other sufficiently weak spot (electrically speaking), a fault to the grounded conductor occurs. The fault triggers an alarm such that the operator can mark the fault for removal or repair.

11.6.1.2 Direct Current Spark Test

This is similar to the AC spark test except that direct current, higher potential values, and continuous circular electrodes are used.

11.6.1.3 Alternating Current Water Tank Test

The entire reel of finished cable is immersed in a water tank with only the cable ends protruding above the water. After a soak period to ensure that water has permeated the entire reel of cable, the cable conductors are energized at an AC voltage level that is dependent on material type and thickness. The test voltage is applied for 5 minutes. The water acts as a ground and during the soak period it is hoped that water infiltrates into any damage, pinhole, or electrically weak areas.

11.6.1.4 Direct Current Water Tank Test

Similar to the AC water tank test except at specified DC test voltages.

11.6.1.5 Insulation Resistance Test

When tested, the insulation resistance is to meet the specified standard. However, for many modern insulations, the test may not be meaningful. When conducted, in connection with the water tank test above and while still immersed, a bridge is used to read the insulation resistance. For modern insulations, the readings are so high that "apparent" differences, even though possibly huge, are meaningless and dependent on numerous factors unrelated to the insulation resistance.

11.6.2 Electrical Testing of Medium Voltage Cables

Except for the insulation resistance test described under Section 11.6.1.4, tests described in Section 11.6.1 are not generally applicable to medium voltage cables. The electrical 100% inspection tests are generally conducted in a dry environment on finished cables.

11.6.2.1 Alternating Current Voltage Withstand Test

This test is conducted from the conductor to the insulation shield, which acts as the outer electrode precluding the need for water immersion.

11.6.2.2 Partial Discharge Test

Unique to medium and higher voltage cables is the partial discharge test. ICEA standards require that such cables be subjected to a partial discharge while on the shipping reel.

The cable must be allowed to "rest" after manufacturing to allow any pressures that were developed during manufacturing to escape. It must be performed prior to the AC voltage test. Alternating current voltage is raised to an established level that is approximately four times the operating voltage. The voltage is reduced while the partial discharge level in picocoulombs is recorded. In the past, the voltage at which the measurable discharge extinguished (extinction voltage) was of particular interest. In modern cables, no measurable discharge is allowed throughout the test voltage range and time of application.

Partial discharge testing is extremely sensitive to defects in the cable capable of discharging as well as external electrical interference. Shielded rooms are provided to minimize this external noise.

11.6.3 Other Factory Tests

Factory tests involving dimensions, material properties before and after aging, chemical, environmental, qualification, and type tests are beyond the scope of this chapter and the standard(s) covering the cable type involved should be consulted (see Chapter 10 [Standards]).

REFERENCES

1. Landinger, C. C., 2001, adapted from class notes for "Understanding Power Cable Characteristics and Applications," University of Wisconsin–Madison.
2. Kelly, L., 1995, adapted from class notes for "Power Cable Engineering Clinic," University of Wisconsin–Madison.

12 Cable Installation

William A. Thue

CONTENTS

12.1 INTRODUCTION

Thomas A. Edison installed his earliest cables in New York City in 1882. The cables were placed in iron pipes in the factory and then were spliced together in the field every 20 feet in an egg-shaped splice casing. Other systems, such as by Brooks, Callender, and Crompton, were installed by 1885 where they also used short sections of iron conduit. American Bell Telephone Company installed the first flexible communication cables in 1882 and 1883 when cables were pulled into the conduit in the field. "Pumplogs" were first used for water supply lines, but were used in 1883 in Washington, DC, for telegraph cables. Tree logs were hollowed out, the exterior was trimmed to make them square, and the entire log was treated with creosote. These became the conduits of choice! So began the duct and manhole systems with the need to pull cables.

The *Underground Systems Reference Book* [1] of 1931 stated "It is necessary to inquire into the harmful effects of the pulling stress on the cable insulation. The conclusion that has been reached, based on tests and experience, is that satisfactory operation of the cable is assured, provided that it has suffered no mechanical injury." It was recommended that a coefficient of friction between 0.40 and 0.75 should be used and that the total tension should be limited to 10,000 pounds. Little other advice was offered.

Significant advances were made in the understanding of cable pulling calculations with the 1949 paper by Buller [2] and the 1953 work of Rifenberg [3]. These papers provide the basic data for making cable pulls in all situations encountered in the field. They provide excellent quantitative data when used to calculate pulling tensions.

Even in the 1957 version of the *Underground Systems Reference Book* [4], there was little additional guidance given for such an important consideration as sidewall bearing pressure (SWBP) for distribution type cables. It was generally felt that 100 pounds per foot was acceptable. Later this was increased to 300 pounds per foot with no laboratory test work to support that value. Experience was still relied upon and 300 pounds met those criteria. Since the runs in city streets were relatively straight and manholes were located at every street intersection, these conservative values did not pose a problem.

Pipe-type cable systems pointed the way to the importance of accurate tension calculations. Here the avoidance of a splice could impact the cost of the system very significantly. Rifenburg's work [3] included all the necessary options, but the allowable SWBP limits needed to be evaluated since the somewhat arbitrary value of 400 pounds had not actually been tested in a laboratory environment. The understanding of the need for such information led to the project for "Increasing Pipe Cable Section Lengths," Electric Power Research Institute (EPRI) Final Report EL-2847, March 1983 [5].

A significant increase in the understanding of cable pulls was reached with the completion of EPRI Final Report, EL-3333, "Maximum Safe Pulling Lengths for Solid Dielectric Insulated Cables," February 1984 [6]. A discussion of the results of these and other test work will be described in the following section.

12.2 DISCUSSION OF CABLE PULLS

Cable manufacturers have handbooks in print that describe methods for making safe cable pulls and for making the necessary calculations of pulling tensions. Computerized pull programs are available from suppliers of cable pulling compounds. Cable pull programs are available from EPRI [6]. There are many cable manufacturers, utilities, architect-engineering firms, and pulling equipment companies that also have programs.

An entirely new group of cable pulling compounds have become available since the EPRI project. They are able to achieve the very low coefficients of friction that their literature suggests—generally these lower values are for the higher SWBPs that are found in the field and in the newer test procedures.

12.2.1 Maximum Allowable Tension on Conductors

The maximum allowable tension on cable conductors that should be used during pulls must be based on experience as well as good engineering. Factors that have an impact on the value include type of metal, temper, and factors of safety. The limits have been set based on only the central conductor of the cable or cables. This quickly establishes one of the safety factors because all of the components of a cable provide some mechanical strength.

One obvious limit is to consider the mechanical stress level at which the conductor will permanently deform/stretch. Upper limits have generally been set well below the elongation value of the conductor metal. The classic approach has been to use the values shown in Table 12.1, but the spread in values shown represent present-day data from suppliers. Even higher values have been recommended and published by the Association of Edison Illuminating Companies (AEIC) [7].

12.2.2 Pulling Tension Calculations

The concept of the significant factors in a cable pull can be appreciated by looking at the equation for pulling a single cable in a straight, horizontal duct. The basic equation is:

$$T = W \times L \times f \tag{12.1}$$

where

T = tension in pounds
W = weight of one foot of cable in pounds
L = length of pull in feet
f = coefficient of friction for the particular duct material and outer layer of the cable.

It is obvious that the weight of the cable and the length of the pull can be determined with great accuracy. The one thing that varies tremendously is the value of the coefficient of friction—it can vary from 0.05 to 1.0. In test conditions, values as high as 1.2 have been recorded! Even when the materials used in the duct and jacket are known, the type and amount of lubricant can be an important factor in this variation.

TABLE 12.1
Maximum Allowable Pulling Tension on Conductors

Metal	Temper	Pounds per cmil
Copper	Soft (annealed)	0.008
Aluminum	Hard	0.008
Aluminum	3/4 Hard	0.006 to 0.008
Aluminum	1/2 Hard	0.003 to 0.004
Aluminum	Soft	0.002 to 0.004

The significance of this is that the accuracy of the calculation cannot come out to six decimal places even if you have a calculator or computer with that many places! It is also not wise to argue whether one method of tension calculation can attain an accuracy of 1% better than another when one considers the probable inaccuracy of the coefficient of friction.

12.2.3 Coefficient of Friction

Since this is a significant variable in all calculations, let us look at this early in the discussion of cable pulling. What do we mean by "coefficient of friction?" Historically, the test apparatus for friction determination consisted of a section of duct that was cut longitudinally in half. The open duct was mounted on an inclined plane. A short sample of cable was placed near the top end and the angle of inclination was increased until the cable started to move as the result of gravity. Using the angle at which movement began, the static coefficient of friction was calculated. Generally, the angle of inclination could be decreased and the cable would maintain its slide. Using this second angle, the dynamic coefficient of friction was obtained.

As described above, many of the earlier publications suggested that the coefficient of friction that should be used varied from 0.40 to 1.0. This was, of course, very safe for most situations.

The EPRI project [6] demonstrated that there were other important issues that needed to be established in making accurate determinations of the coefficient of friction such as the level of SWBP. This force is duplicated in present-day test methods by applying a force that pushes the cable down on the conduit. The interesting fact is that this actually reduces the coefficient of friction in most instances! The quantity and type of lubricant are important. Too much lubricant can increase the friction. A more viscous lubricant should be used with heavier cables, etc.

12.2.4 Sidewall Bearing Pressure

When one or more cables are pulled around a bend or sheaves, the tension on the cable produces a force that tends to flatten the cable against the surface. This force is expressed in terms of the tension from the bend in pounds divided by the bend radius in feet.

$$\mathrm{SWBP} = T_O / R \tag{12.2}$$

where

SWBP = force in pounds per foot
T_O = tension from the bend in pounds
R = inner radius of the bend in feet

SWBP is not truly a unit of pressure, but rather a unit of force for a unit of length. In the case of a smooth set of sheaves or bend, the unit is the entire length of contact. However, any irregularity, such as a bump on the surface, or a small radius sheave

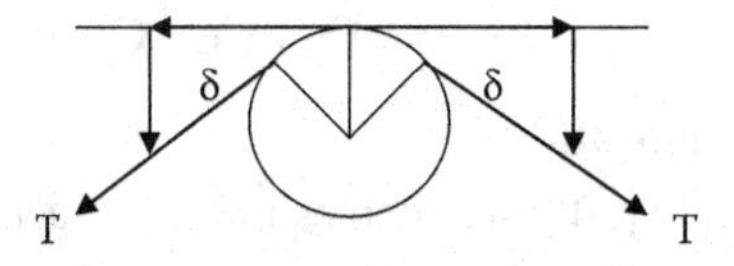

FIGURE 12.1 Pulling forces in horizontal bend.

with limited bearing surface (even though it may be part of a multisheave arrangement), reduces the effective bearing surface length. This must be taken into account in the calculation to prevent damage to the cable.

For multiple cables in a duct, the matter is complicated because of the fact that the SWBP is not equally divided among the cables. This situation is taken into account by using the weight correction factor which will be discussed later in this chapter.

Figure 12.1 shows the mathematical derivation for a horizontal bend of one cable—ignoring the weight of the cable. As the angle approaches zero, the force between the cable and the bend approaches unity.

$$\text{The force per unit length} = \frac{2T \sin \delta}{2R\delta} \tag{12.3}$$

where $\sin \delta = \delta$ for small angles and sidewall pressure,

$$T/R = \frac{2T \sin \delta}{2R\delta} \tag{12.4}$$

The *T*/*R* ratio is independent of the angular change of direction produced by the bend. It depends entirely on the tension from the bend and the bend radius with the effective bend radius taken as inner radius of the bend. Increasing the radius of the bend obviously decreases the SWBP.

SWBP limits that have been used historically are shown in Table 12.2. As with maximum pulling tension values, AEIC [7] has published limits that exceed the values shown in this table.

TABLE 12.2
Sidewall Bearing Pressure Limits

Cable Type	SWBP in Pounds per Foot
Instrumentation	100
600 V nonshielded control	300
600 V power	500
5 to 15 kV shielded power	500
25 to 46 kV power	300
Interlocked armored	300
Pipe-type	1,000

12.2.5 Pulling Multiple Cables in a Duct or Conduit

12.2.5.1 Cradled Configuration

A frequent requirement is to pull three cables into one duct. This brings the relative diameters of the cables into play with the inner diameter of the duct. If the cables are relatively small as compared with the duct diameter, the cables are said to be "cradled."

The outer two cables push in on the center cable, making it to seem heavier than it actually is.

The pulling calculation handles this by using a "weight correction factor" that increases the apparent weight of that center cable.

The SWBP on the center cable in Figure 12.2 is influenced by the other two cables. The effective SWBP for the cradled configuration may be calculated from:

$$\text{SWBP} = \left[(3W_C - 2)/3\right]T_O/R \tag{12.5}$$

See Equations 12.2 and 12.7 for definitions of the terms.

12.2.5.2 Triangular Configuration

When the diameter of each of the three cables is closer to one-third of the inner diameter of the duct, the situation is known as a "triangular" configuration.

In this situation, the top cable rides on the two lower cables without touching the duct wall.

The effect of this is that one cable effectively increases the weight of the two lower cables, but does not function as a longitudinal tension member. This means that one must use the cross-sectional area of only two of the cables in the maximum allowable tension determination in this example, not three cables.

The SWBP on the two lower cables in Figure 12.3 is influenced by the upper cable. The effective SWBP for the triangular configuration may be calculated from:

$$\text{SWBP} = W_C T_O/2R \tag{12.6}$$

The units are defined in Equations 12.2 and 12.7.

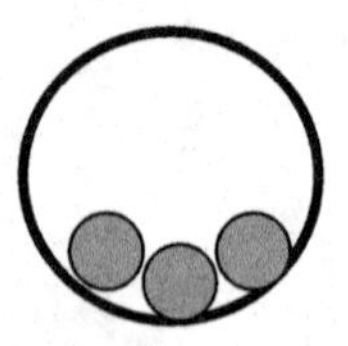

FIGURE 12.2 Cradled cables.

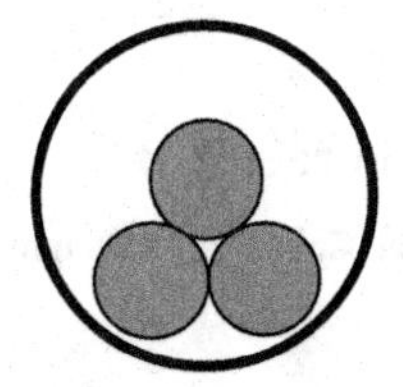

FIGURE 12.3 Triangular arrangement.

12.2.5.3 Weight Correction Factor

When two or more cables are installed in a duct or conduit, the sum of the forces developed between the cables and the conduit is greater than the sum of the cable weights.

Weight correction factor is therefore defined as:

$$W_C = \Sigma F / \Sigma W \tag{12.7}$$

where

W_C = weight correction factor (also merely "c")
ΣF = force between cable and conduit, usually in pounds
ΣW = weight of the cable with the same units as above

For the typical case of three cables of equal diameter and weights in a conduit of given size, the weight correction factor is higher for the cradled configuration than the triangular configuration (Table 12.3).

The mechanism for this relationship is shown in Figure 12.4.

It is always safer to anticipate the cradled configuration unless the cables are triplexed or if the clearance is near the 0.5-inch minimum.

The equations for calculating the weight correction factor are:

Cradled:

$$W_C = 1 + 4/3\left(d/D - d\right)^2 \tag{12.8}$$

Triangular:

$$W_C = \frac{1}{1 - \left[\left(d/D - d\right)^2\right]^{1/2}} \tag{12.9}$$

where D = diameter of inside of conduit and d = diameter of each cable.

12.2.5.4 Jamming of Cables

When the diameter of each of the three cables is about one-third of the inner diameter of the duct, a situation may occur where the cables may jam against the inside

TABLE 12.3
Weight Correction Factors

Configuration of Three Cables	Weight Correction Factor
Triangular	1.222
Cradled	1.441

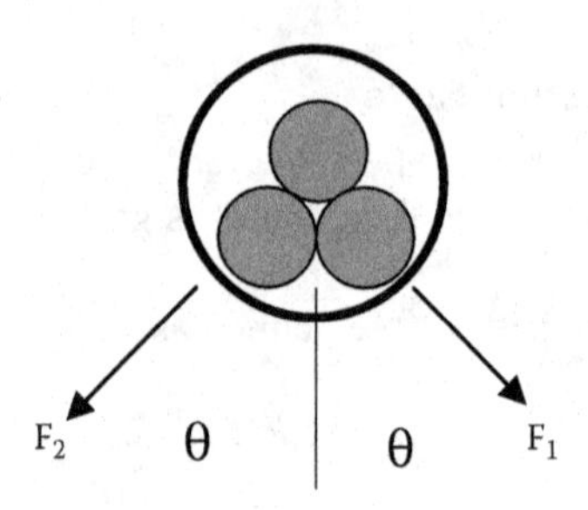

FIGURE 12.4 Weight correction.

of the conduit. This generally occurs when the cables go around a bend or a series of bends. The "center" cable may try to pass between the outer two cables. When the sum of the diameters of the three cables is just slightly larger than the inner diameter of the duct, jamming can occur. Jamming increases the pulling tension manyfold and can result in damaging the cable or even pulling the cables in two, breaking pull irons in manholes, etc. (Figure 12.5).

The "jam ratio" of the cables in this duct needs to be evaluated. The equation for finding the jam ratio of three cables in a duct may be determined by:

$$\text{Jam Ratio} = 1.05 \times D_d / D_c \tag{12.10}$$

where D_d = inside diameter of the duct or conduit and D_c = outer diameter of each of the three cables.

The factor of 1.05 has been used to account for the probable ovality of the conduit in a bend and to account for the cable having a slightly different diameter at any point. If precise dimensions are known, this 5% factor can be reduced.

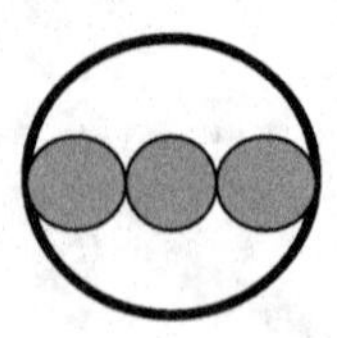

FIGURE 12.5 The condition that causes jamming for three cables in a conduit.

Where the jam ratio falls between 2.6 and 3.2, jamming is probable if there are bends in the run and unless other precautions are taken. To avoid any problems with jamming, it is wise to avoid pulls where the ratio is between 2.6 and 3.2.

How can jamming be avoided even though the calculation shows that the ratio indicates a significant probability of jamming? There are several solutions, some of which are obviously possible during the planning stages and some during the installation stage.

- Use a different size of cable or conduit to change the ratio.
- Have the cable triplexed (twisted together) at the factory.
- Tie the cables together in the field with straps.
- Use precautions at the feed point to keep a triangular configuration and allow no crossovers.

The National Electrical Code (NEC) ANSI C-1 requires that the total fill of a conduit be 40% or less for three cables in a conduit. This means that the cross-sectional area of all three of the cables cannot be more than 40% of the cross-sectional area of the conduit. Unfortunately, for 40% fill, the jam ratio is 2.74, which is in the lower danger ratio. An example of this situation is when three 1.095-inch diameter cables are installed in a conduit with an inner diameter of exactly 3.0 inches. (The actual inner diameter of a nominal 3-inch conduit is 3.068 inches!) If the designer tries to reduce the fill to say 38% to stay safely within the 40% limit, the jam ratio gets worse—2.81.

The NEC generally does not govern the utility practices; hence clearance limits are not based on percent fill. Utility practice considers that 0.5 inches of clearance is satisfactory for general pulls. Clearances as small as 0.25 inches have been successfully made when good engineering practices and careful field supervision are employed [7].

To complete this discussion of jam ratio, it is important to know that jamming *can* occur when more than three cables are installed in a conduit. A modification of the equation is necessary as shown:

$$\frac{3D}{n_1d_1 + n_2d_2 + n_3d_3 + n_4d_4 + \cdots} \tag{12.11}$$

where

D = conduit inner diameter (ID)
n_1, n_2, n_3 = number of cables of diameter 1, 2, 3, etc.
d_1, d_2, d_3 = diameters of cables in groups 1, 2, 3, etc.

Theoretically any combination of cable diameters that fall in the critical zone can jam. Field experience has shown that the probability of jamming decreases as the number of cables increases.

12.2.5.5 Clearance

Another consideration before cables are placed in a conduit is the amount of clearance between the cable or cables and the inside of the conduit. This may be quickly seen in the example of the three cables in a triangular configuration in Figure 12.1. The distance from the top cable to the inside "top" of the conduit is defined as the clearance.

12.3 PULLING CALCULATIONS

The previous sections have presented some of the fundamentals of pulling cables. Now let us see how those factors, plus a few more, come together when we actually calculate the tension on a cable or cables that are to be installed.

12.3.1 Tension Out of a Horizontal Bend

Bends in cable runs are a fact of life. The important point is that the friction and SWBP around that bend increase the tension from the bend with respect to the tension on the cable into the bend.

$$T_O = T_{IN} e^{cfa} \tag{12.12}$$

where

T_O = tension from the bend
T_{IN} = tension into the bend
c = weight correction factor
f = coefficient of friction
a = angular change of direction in radians

This is a simplified equation that ignores the weight of the cable. It is sufficiently accurate where the incoming tension at the bend is equal to or greater than 10 times the product of the cable weight per foot times the bend radius expressed in feet. The practical situation where T_{IN} is less than 10 times the product of the weight and radius is where the cable is being fed at low tension into a large radius bend [8]. In such a case, the equation becomes:

$$T_O = T_{IN} \frac{\left[e^{cfa} + e^{-cfa}\right]}{2} + \left[(T_{IN})^2 + (wr)^2\right]^{1/2} \frac{\left[e^{cfa} - e^{-cfa}\right]}{2} \tag{12.13}$$

In order to fully appreciate the effect of the impact of the exponent of e^{cfa}, Table 12.4 shows the multiplier of the tension for various values of the exponent. This also shows the significance of keeping the coefficient of friction down to a low value—especially when you have multiple bends in a run.

It is essential to remember that the exponent of e has three terms: weight correction factor, coefficient of friction, and angle of deflection. Sometimes this exponent is shown only as two factors because of limitations of older computer programs. In order to make that workable when multiple cables are installed, it is necessary to multiply the selected coefficient of friction by the weight correction factor. If you

TABLE 12.4
Multiplier of Tension Values for Various Exponents

cfa	45° Bend	90° Bend
0.1	1.08	1.17
0.2	1.17	1.37
0.3	1.27	1.60
0.4	1.48	2.19
0.5	1.60	2.56

therefore have a situation where three cables will be cradled, for instance, the *cfa* value (from Table 12.4) is 1.442 times the coefficient of friction. Putting this another way, if you consider the proper coefficient of friction to be 0.2, and the cables will be in a cradled configuration, you must use a *cfa* value of 0.3. If there is only one cable, it means that you would use the 0.2 value of *cfa* for that same coefficient of friction since W_C for one cable is unity.

A large number of bends in a run can literally multiply the tension exponentially! This is one of the reasons that many installation practices keep the number of 90° bends to a maximum of three or four.

12.3.2 Which Direction to Pull?

There are always two possible directions that a cable can be pulled for any run—just as a cable always has two ends. Let us go through the calculations of an example that has one bend so that we can see how the pulling tension can vary (Figure 12.6).

Given: 1 × 1,000 kcmil copper cable
Weight of cable = 6 pounds per foot
Coefficient of friction = 0.5

Pulling from A to D:

Tension at A = 0
Tension at B = 300 × 6 × 0.5 = 900 pounds
$e^{cfa} = e^{1 \times 0.5 \times 1.5708} = 2.713^{0.7854} = 2.19$
Tension at C = 900 × 2.19 = 1,971 pounds
Tension at D = −10 × 6 × 0.5 + 1,971 = 30 + 1,971 = 2,001

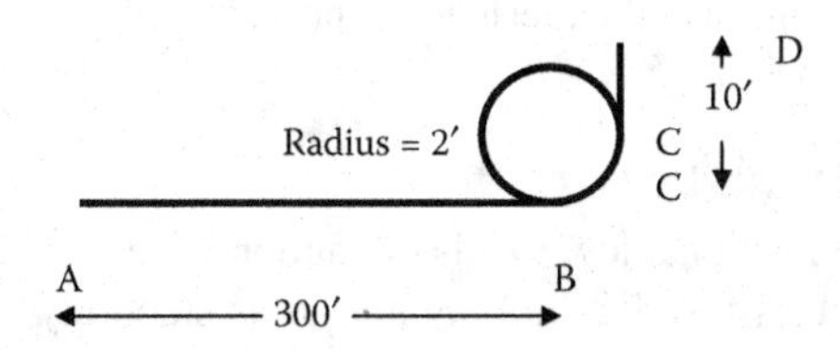

FIGURE 12.6 Pulling around a bend. Technical Conference, Portland, Oregon, June 1980.

This exceeds the long established 1,000-pound limit for a pulling grip, so this must be pulled with a pulling eye on the conductor. The established limit for a 1,000 kcmil copper conductor is based on the conductor kcmils multiplied by 0.008, or 8,000 pounds.

The maximum allowable tension from the bend at point C results in a SWBP of 1,971/2, or 985.5 pounds per foot, which is above the value generally agreed to by the manufacturers.

Pulling from D to A:

Tension at D = 0
Tension at C = $10 \times 6 \times 0.5 = 30$ pounds
$e^{cfa} = e^{1 \times 0.5 \times 1.5708} = 2.713^{0.7854} = 2.19$
Tension at B = $2.19 \times 30 = 65.7$ pounds
Tension at A = $300 \times 6 \times 0.5 + 65.7 = 965.7$ pounds

This is satisfactory in all respects. The total tension does not exceed 1,000 pounds and the tension from the bend is well below accepted levels. A cable grip may be used to pull the cable from D to A.

From this example, it can be seen that it is always preferable to set up the reel as close to the bend as possible.

12.4 CABLE INSTALLATION RESEARCH

Pipe-type cable pulling was addressed by the EPRI in Final Report EL-2847, March 1983, entitled "Increasing Pipe Cable Section Lengths" [5]. SWBPs of up to 1,000 pounds per foot were recommended.

The fact that the suggested values for pulling tension and related considerations of extruded distribution cables had developed only from an understanding of past successful pulls made it seem reasonable to look at extruded dielectric cables using laboratory and field generated data. EPRI undertook Research Project 1519 in the late 1970s. The work was published as Final Report EL 3333 [6] in March 1983. Additional insight into pulling considerations was accomplished through the publication of AEIC CG5-2005 [7], "Underground Extruded Power Cable Pulling Guide", in May 1990.

The values of SWBP, allowable maximum tension on the conductors, and maximum allowable tension on a pulling basket, are much less conservative than the level generally accepted by cable manufacturers.

Since there are obvious advantages for a utility to make longer pulls based on this data, a good deal of caution and experience is necessary.

12.4.1 Research Results

Before reviewing the results, a few words of caution is needed. Field conditions are seldom ideal. The actual installation may not go as smoothly as was planned. For instance, when one believes that the cable may be pulled in one continuous motion, the actual pull may be made in a series of starts and stops. This alters the coefficient of friction because of the unplanned start, with the cable probably already far into the duct. When one anticipates excellent lubrication application, the amount of lubrication may be more or even less than that planned. Therefore,

- Do not blindly use these values to their upper limits.
- Be conservative.
- Follow the cable manufacturer's recommendations.

The most dramatic findings of this EPRI research project over previously accepted values were:

- Pulling tension on both copper and aluminum conductors was about 50% greater than older limits.
- Allowable SWBP was as much as 200% greater than previously recommended.
- Coefficient of friction was about 50% lower—with values as low as 0.05.
- Basket grips could be used to pull 10 times (or more) the force previously recommended as long as tension limits on the conductor are not exceeded.

12.5 FIELD EXPERIENCE

While the calculations have been found to be very accurate, field experience has shown that conditions may exist in the field that may greatly affect the success of the cable pull. A new installation can be calculated and results of the cable pull will follow the calculations fairly well. However, older conduits may have been exposed to unknown conditions. The following work practices and field conditions should be considered to make a successful installation without damaging the cable.

- Since buried conduit is not visible, verification of the location route should be made before setting up the cable pulling equipment. Conduit maps are helpful and a brief survey of the area to see if there has been new construction, which may affect the pull, is often required. Checking the areas where the pulling equipment and reels of cable are to be set up for the pull will often times save a lot of field modifications. Choosing these locations is especially important in a congested area or narrow streets or alleys.
- Older conduits may be damaged, contain stones and dirt, or have sections that have shifted leaving ridges. It is therefore recommended that the older conduits be inspected for these conditions. This can be done by pulling a mandrel through the conduit and inspecting its condition upon leaving the conduit pull.
- Measurements should be made while the cable is being installed so as to monitor any conditions that could cause possible damage. Equipment that contains a dynamometer to indicate pulling tensions is available. Some equipment contains charting facilities, which will record the pulling tensions as the pull progresses. These measurements are very important when installing transmission cable in a pipe.

- Feeding tubes are flexible metal tubes, which guide the cable toward the conduit or pipe. Their purpose is to supply smooth arc so that the cable is not damaged when entering a manhole. They are often used when installing transmission cable. This flexible metal tube must be sized to match the diameter of the conduit or pipe into which the cable is being installed.
- Lubricating compounds or fluids are used to reduce the coefficient of friction as previously mentioned. These come in a variety of viscosities and contain lubricants that must be compatible with the cable being installed. As an example paper-insulated lead covered (PILC) cable with a bare lead sheath was installed with a lubricant consisting of a hydrocarbon based grease. This lubricant, while reducing the pulling tension, also provided corrosion protection to the lead. Conduits that contain this grease may require cleaning before installing a cable with an outer coating with semiconducting properties. Otherwise, the semiconducting properties of the new cable will deteriorate with time. Prelubrication of the conduit will ensure a better distribution of the lubricating material and ultimately reduce the pulling tension for the installation.
- Attachments to the cable for pulling operations vary with the amount of tension expected and the type of cable being installed. A woven basket grip is very useful for short pulls of limited tension. These grips are very popular when pulling three single conductor cables. The basket grip fits around all three conductors and as tension is applied during installation, the basket fits more tightly against the three conductors. It is important that each cable of the bundle has a sealed end so that water cannot enter the strands during installation. This precaution is needed as some conduits contain standing water, and that water can be forced into the stranding of the conductor during the cable installation. PILC cable requires a lead seal to be installed over single-conductor and three-conductor cables. It is necessary for the conductors to be wiped into the pulling eye for strength.
- Installing cables in a high-rise building offers some unique situations. Foremost is the place that you are going to set up the pulling equipment as well as the reel of cable being installed. Choice can be made to install the reel of cable in the upper floors or to feed the cable in from the lower floors. Installation of vertical pulls must take into account the type of grips that will be used as well as intermittent grips if required. Methods of holding cable when intermittent grips are used require special attention in order to generate enough slack to get grips to hold their respective weight. Clamping devices used to support vertical cables must be designed so that they do not damage the cable while providing the function of holding the cable.
- Lines used to pull cable vary mostly by the amount of tension they may be subjected to during installation. Hemp and braided plastic ropes may be used on short pulls. Longer pulls, such as those on high-pressure fluid-filled pipe cable installations will require steel rope. Steel rope is also available with a plastic coating.
- Installation of a cable in a conduit, which already contains cable, is not recommended. Called a pull-by, it is very difficult to prevent damage to the

existing cable or the cable being installed. Not only is it difficult to guard the cable against damage, it is also difficult to detect that damage incurred. Damage may occur to the jackets, sheaths, and neutral inside the conduit without detection.

REFERENCES

1. *Underground Systems Reference Book,* 1931, NELA.
2. Buller, F. H., August 1949, "Pulling Tension During Cable Installation in Ducts or Pipes," General Electric Review, Vol. 52 (No. 8), pp. 21–33, Schenectady, NY.
3. Rifenberg, R. C., December 1953, "Pipe-line Design for Pipe-type Feeders," AIEE Transactions on Power Apparatus and Systems, Vol. 72, Part III.
4. *Underground Systems Reference Book,* EEI, 1957.
5. "Increasing Pipe Cable Section Lengths," EPRI Final Report EL-2847, March 1983, EPRI, P.O. Box 10412, Palo Alto, CA 94303-0813.
6. "Maximum Safe Pulling Lengths for Solid Dielectric Insulated Cables," EPRI Final Report EL-3333, February 1984, EPRI, P.O. Box 10412, Palo Alto, CA 94303-0813.
7. *Underground Extruded Power Cable Pulling Guide*, AEIC CG5-2005, Association of Edison Illuminating Companies, P.O. Box 2641, Birmingham, AL 35291-0992.
8. Kommers, T. A., June 1980, "Electric Cable Installation in Raceways," Pulp and Paper Industry Technical Conference, Portland, OR.

13 Splicing, Terminating, and Accessories

James D. Medek

CONTENTS

13.1 INTRODUCTION

A fundamental concept that needs to be established early in this chapter is that when the terms "splice" and "joint" are used here, they are one and the same! "Cable splicers" have been around for over 100 years, but officially in the Institute of Electrical and Electronic Engineers (IEEE) Standards, when you join two cable ends together, you make a joint [1–3].

The basic dielectric theory that has been previously described for cables in Chapter 2 also applies to joints and terminations. Some repetition of those concepts may be presented so that this will be a stand-alone treatment and some repetition is constructive.

This chapter will address the design, application, and preparation of cables that are to be terminated or spliced together. The application of this material will cover medium voltage cable systems in particular with higher and lower voltage applications being mentioned in particular designs and applications. Joints and terminations are expected to perform all of the functions of the cable on which they are installed.

13.2 TERMINATION THEORY

13.2.1 INTRODUCTION TO FIELD PLOTS

Figures that show the electric fields in a cable, splice, or termination are depicted in two ways. Figure 13.1 shows the electric flux lines in a shielded single conductor cable. These lines radiate outward from the center of the conductor toward the grounded shield. The lines are closer together near the conductor, which demonstrates the fact that the electric stress is higher near the conductor. The lines get farther apart near the shield and this shows that the voltage stress is lower near that area.

A plot of equal potential lines (also called equipotential lines) is shown in Figure 13.2. These lines are at right angles to the flux lines described previously. The conductor is shown as having 100% of the impressed voltage and the equipotential lines appear as concentric circles around the conductor. The 75% voltage circle is rather close to the conductor and the subsequent 50% and 25% circles are each spaced a bit farther than the first one. This shows that the voltage difference for a given distance from the conductor is greater there than the same spacing near the shield.

13.2.2 PURPOSE OF A TERMINATION

A termination is a way of preparing the end of a cable to provide adequate electrical, mechanical, and environmental properties. A discussion of the voltage distribution at a cable termination serves as an excellent introduction to this subject.

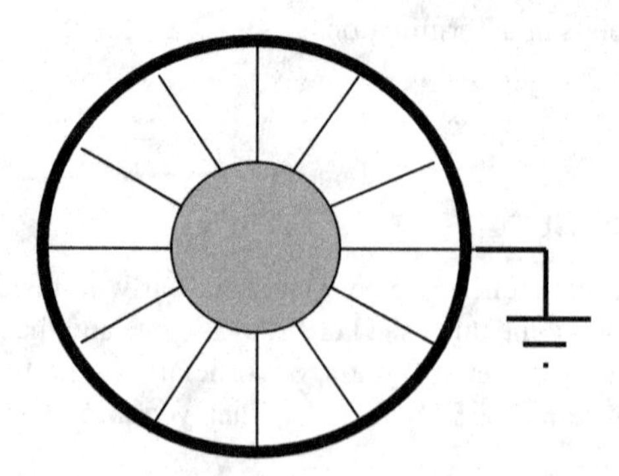

FIGURE 13.1 Electric flux lines in cable.

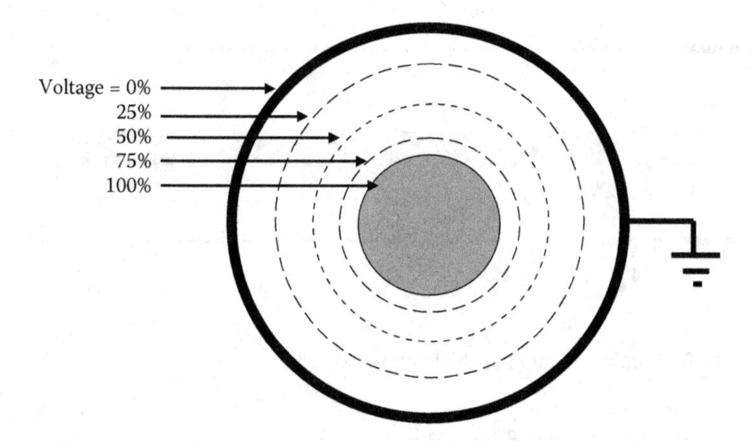

FIGURE 13.2 Equipotential lines in cable.

Whenever a medium or high voltage cable with an insulation shield is cut, the end of the cable must be terminated so as to withstand the electrical stress concentration that is developed when the geometry of the cable has changed. Previously, the electrical stress was described as lines of equal length and spacing between the conductor shield and the insulation shield. As long as the cable maintains the same physical dimensions, the electrical stress will remain consistent. When the cable is cut, the shield ends abruptly and the insulation changes from that in the cable to air. The concentration of electric stress is now at the end of the conductor and insulation shield.

In order to reduce the electrical stress at the end of the cable, the insulation shield is removed for a sufficient distance to provide the adequate leakage distance between the conductor and the shield. The distance is dependent on the voltage involved as well as the anticipated environmental conditions. The removal of the shield disrupts the coaxial electrode structure of the cable. In most cases, the resulting stresses are high enough to cause dielectric degradation of the materials at the edge of the shield unless steps are taken to reduce that stress.

In this operation, the stress at the conductor is reduced by spreading it over a distance. The stress at the insulation shield remains great since the electrical stress lines converge at the end of the shield. The equipotential lines are very closely spaced at the shield edge. If those stresses are not reduced, partial discharge may occur with even the possibility of visible corona. Obviously, some relief is required in most medium voltage applications.

13.2.3 Termination with Simple Stress Relief

To produce a termination of acceptable quality for long life, it is necessary to relieve voltage stresses at the edge of the cable insulation shield. The conventional method of doing this has been with a stress cone.

A stress cone increases the spacing from the conductor to the shield. This spreads out the electrical lines of stress as well as provides additional insulation at this high

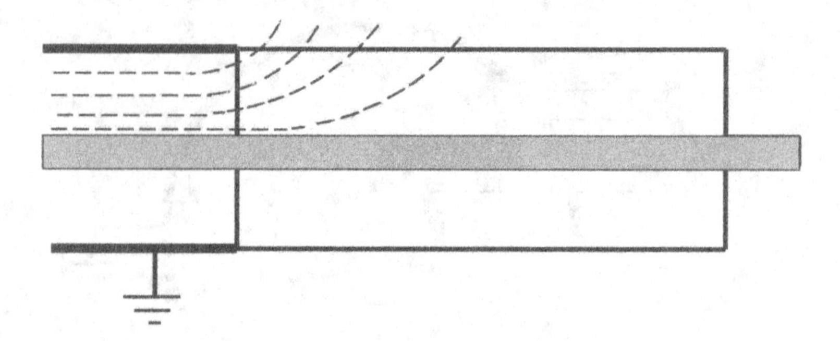

FIGURE 13.3 Equipotential plot, shield removed.

stress area. The ground plane gradually moves away from the conductor and spreads out the dielectric field—thus reducing the voltage stress per unit length. The stress relief cone is an extension of the cable insulation. Another way of saying this is that the equipotential lines are not concentrated at the shield edge as they were in Figure 13.3 and are spaced farther apart. Terminations that are taped achieve this increase in spacing by taping a conical configuration of insulating tape followed by a conducting layer that is connected electrically to the insulation shield. When stress cones are premolded at a factory, they achieve the same result with the concept built into the unit (see Figure 13.4).

Environmental conditions play a significant role in the length of a termination. Figure 13.5 shows the leakage or creepage distance of a simple stress cone termination. The total distance across any termination defines its leakage distance. A termination with skirts has a creepage distance that includes the whole surface from ground to the energized portion. Figure 13.5 shows the creepage length for a termination.

Experience in a particular area is needed to make good judgments as to this length. When additional leakage distance over the insulation is required, skirts can be placed over the conductor and insulation shield. These skirts can be built into the termination as shown in Figure 13.6 or added in a separate field assembly operation.

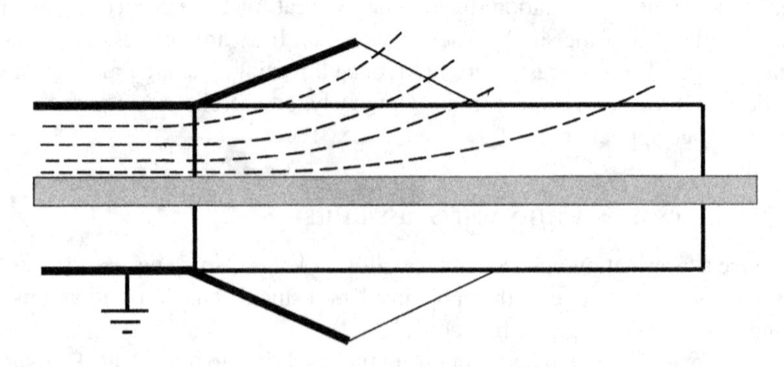

FIGURE 13.4 Stress cone.

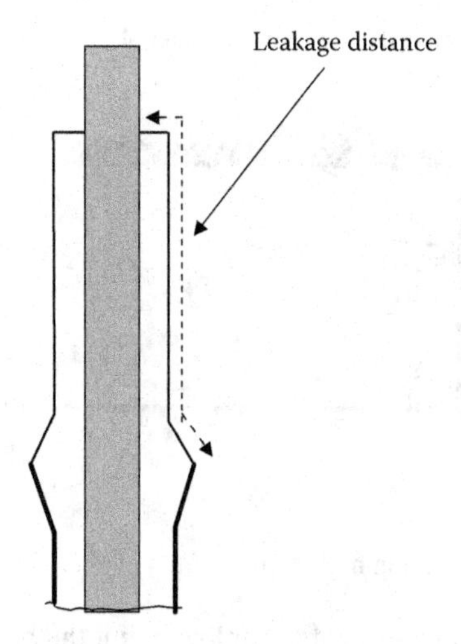

FIGURE 13.5 Creepage/leakage distance.

13.2.4 Voltage Gradient Terminations

Electrical stress relief may come in different forms. A high permittivity material may be applied over the cable end as shown in Figure 13.7. This material may be represented as a long resistor connected electrically to the insulation shield of the cable. By having this long resistor in cylindrical form extending past the shield system of the cable, the electrical stress is distributed along the length of the tube. Stress relief is thus accomplished by utilizing a material having a controlled resistance [4] or

FIGURE 13.6 Cold shrink termination with skirts.

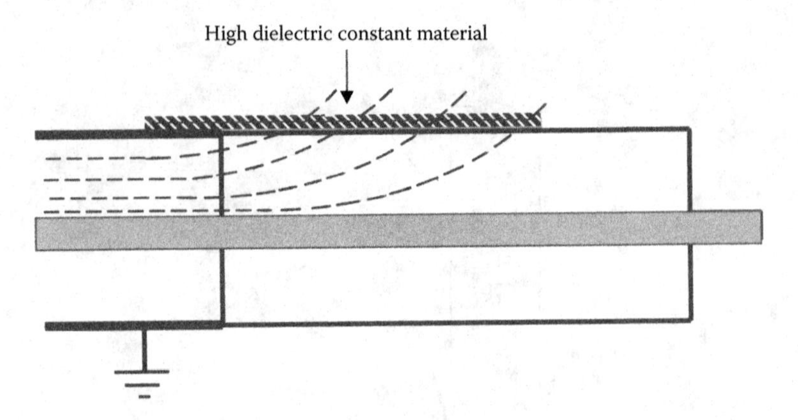

FIGURE 13.7 Stress cones using high K material.

capacitance. Other techniques may be employed, but the basic concept is to utilize a material with, say, a very high resistance or specific dielectric constant to extend the lines of stress away from the cable shield edge.

An application of a series of capacitors for stress control is frequently used on high and extra high voltage terminations (138 to 765 kV). These specially formed doughnut-shaped capacitors are used to provide the stress relief. The capacitors are connected in series, as shown in Figure 13.8, and distribute the voltage in a manner that is similar to the high permittivity material that was discussed previously. A stress cone may also be constructed in series and below the capacitors.

13.3 TERMINATION DESIGN

13.3.1 Stress Cone Design

The classic approach to the design of a stress relief cone is to have the initial angle of the cone to be nearly zero degrees and take a logarithmic curve throughout its length. This provides the ideal solution, but was not usually needed for the generous dimensions used in medium voltage cables. There is such a very small difference between a straight slope and a logarithmic curve for medium voltage cables that, for hand buildups, a straight slope is completely adequate.

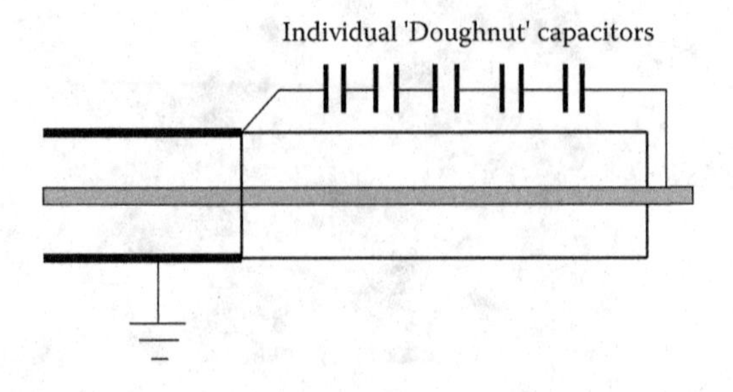

FIGURE 13.8 Capacitive graded termination.

In actual practice, the departure angle is in the range of 3 to 7 degrees. The diameter of the cone at its greatest dimension has generally been calculated by adding twice the insulation thickness to the diameter of the insulated cable at the edge of the shield.

13.3.2 Voltage Gradient Design

Capacitive graded materials usually contain particles of silicon carbide, aluminum oxide, or iron oxide. Although they are not truly conductive, they become electronic semiconductors when properly compounded. They do not have a linear E = IR relationship, but rather have the unique ability to produce a voltage gradient along their length when potential differences exist across their length. This voltage gradient does not depend on the IR drop, but on an exchange of electrons from one particle to another.

Resistive graded materials contain carbon black, but in proportions that are less than the semiconducting materials used for extruded shields for cable. They also provide a nonlinear voltage gradient along their length.

By proper selection of materials and proper compounding, these products can produce almost identical stress relief as a stress cone. One of their very useful features is that the diameter is not increased to that of a stress cone. This makes them a very valuable tool for use in confined spaces and inside devices such as porcelain housings.

13.3.3 Paper-Insulated Cable Terminations

Cables that are insulated with fluid impregnated paper insulation exhibit the same stress conditions as those with extruded insulations. In the buildup of the stress cone, insulating tapes are used to make the conical shape and a copper braid is used to extend the insulation shield over the cone, as shown in Figure 13.9. Similar construction is required on each phase of a three-conductor cable as it is terminated.

The field application of installing stress relief on individual phases can be seen in Figures 13.9 and 13.10. The type of termination is consistent on all types of paper-insulated lead cables (PILC) whether they are enclosed in a porcelain enclosure, a three-conductor terminating device, or inside a switch or transformer compartment (Figure 13.11).

A critical part of the design is the material used to fill the space inside the porcelain or other material that surrounds the paper cable. Since the cable is insulated with a dielectric fluid, it is imperative that the filling compound inside the termination be compatible with the cable's dielectric fluid. In gas-filled cable designs, the termination is usually filled with the same gas as the cable, but a dielectric fluid may be used in conjunction with a stop gland.

13.3.4 Lugs

The electrical connection that is used to connect the cable in a termination to another electrical device must be considered. Generally called a "lug," this connector must

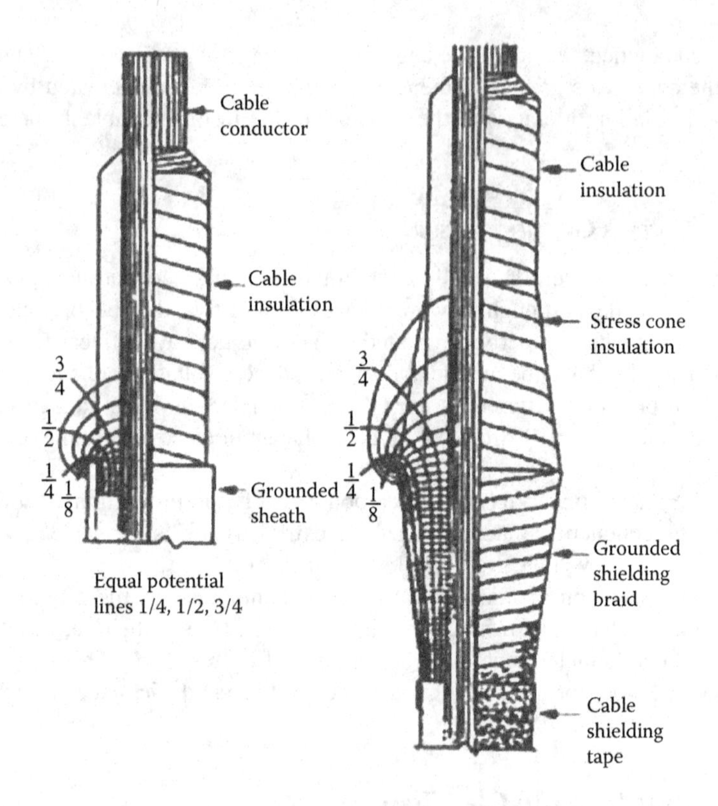

FIGURE 13.9 Equal potential lines, PILC cable with and without stress cones.

be able to carry the normal and emergency currents of the cable, it must provide good mechanical connection in order to prevent from becoming loose and creating a poor electrical connection and it must seal out water from the cable. The water seal is accomplished by two forms of seals. Common to all terminations is the need to keep water out of the strands. Many early connectors were made of a flattened section of tubing that had no actual sealing mechanism and water could enter along the pressed seam of the tubing. Sealing can be accomplished by filling the space between the insulation cutoff and the lug base with a compatible sealant or by purchasing a sealed lug. Figure 13.12 shows one type of terminal lug.

The other point that requires sealing is shown in Figure 13.12 that is common to most PILC cable terminations. Here the termination has a seal between the end of the termination and the porcelain body. Another seal that is required is at the end of the termination where the sheath or shield ends. Moisture entering this end could progress along inside the porcelain and result in a failure.

13.3.5 Separable Connectors (Elbows)

One of the most widely used terminations for cables is the "elbow," as it was originally called, but is more properly called a separable connector. It is unique in that it

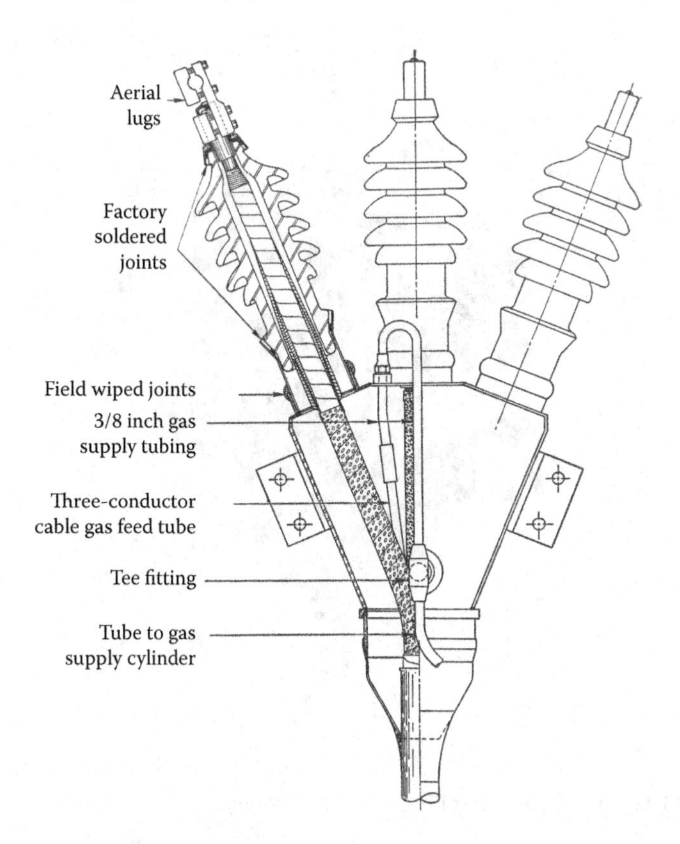

FIGURE 13.10 Gas-filled termination.

has a grounded surface covering the electrical connection to the device on which it is used. Used as an equipment termination, it provides the connection between the cable and the electrical compartment of a transformer, switch, or other device. Since the outer surface is at ground potential, this type of termination allows the personnel to work in close proximity to the termination. Another design feature is the ability to operate the termination as a switch. This may be done while the termination is energized and under electrical load. While elbows are available that cannot be operated electrically, this discussion will deal with the operable type that is shown in Figure 13.13. This figure shows a cutaway of a separable connector followed by a brief description of the parts.

The insulating portion of the elbow is made of ethylene–propylene diene monomer (EPDM) rubber with an outer covering of similar material that is loaded with carbon black to make it conductive. The inner semiconducting shields are made of the same material as the outer semiconducting layer.

Probe: The probe consists of a metallic rod with an arc quenching material at the end that enters the mating part, the bushing. The metallic rod makes the connection between the connector and the bushing receptor. Arc quenching material at the tip of the probe quenches the arc that may be encountered when operating the elbow under

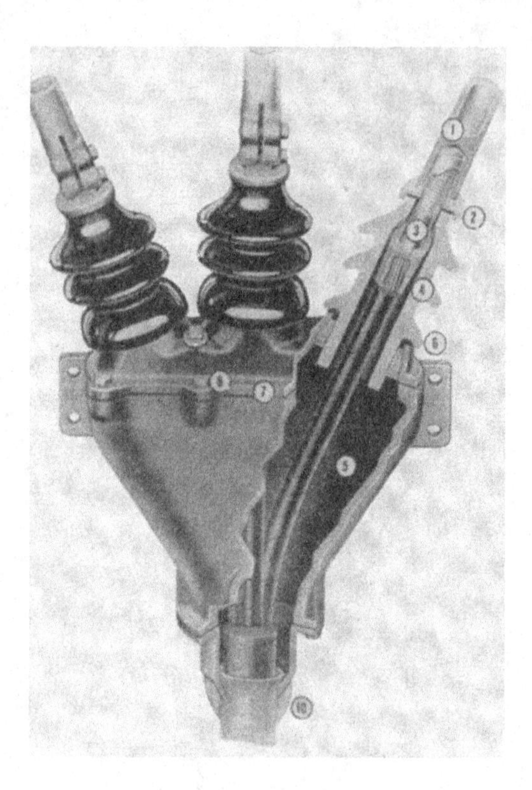

FIGURE 13.11 Three conductor PILC cable termination.

energized and loaded switching conditions. A hole in the metallic rod is used with a wire wrench to tighten the probe into the end of the cable connector.

Connector: The connector is attached to the conductor of the cable and provides the current path between the conductor and the metallic probe. It is compressed over the conductor to make a good electrical and mechanical connection. The other end has a threaded hole to accept the threaded end of the probe.

Operating Eye: This provides a place for an operating tool to be attached so that the elbow assembly can be placed into or removed from the bushing. It is made of metal and is molded into the conducting outer layer of the elbow.

Locking Ring: This is molded in the inner surface of the elbow and maintains the body of the elbow in the proper position on the bushing. There is a groove at the end of the bushing into which the locking ring of rubber must fit.

Test Point: Elbows may be manufactured with a test point that allows an approved testing device to determine if the circuit is energized. The test point is in the form of a metallic button that is molded into the elbow body and is simply one plate of a capacitor. It is supplied with a conductive rubber cap that serves to shunt the button to ground during normal service. The molded cap must be removed when the energization test is performed. A second use of the test point is a place to attach a faulted circuit indicator—a device made for test points that may be used to localize a faulted section of circuit for the purpose of reducing the

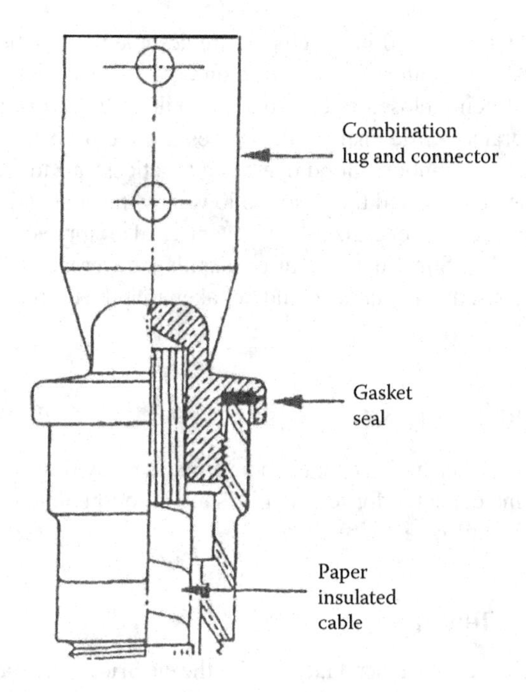

FIGURE 13.12 Terminal lug.

time of circuit outage. When in use, the indicator can remain on the elbow during normal service.

Test Point Cap: Covers and grounds the test point when a test point is specified.

Grounding Eye: This is provided on all molded rubber devices for the purpose of ensuring that the outer conductive material stays at ground potential.

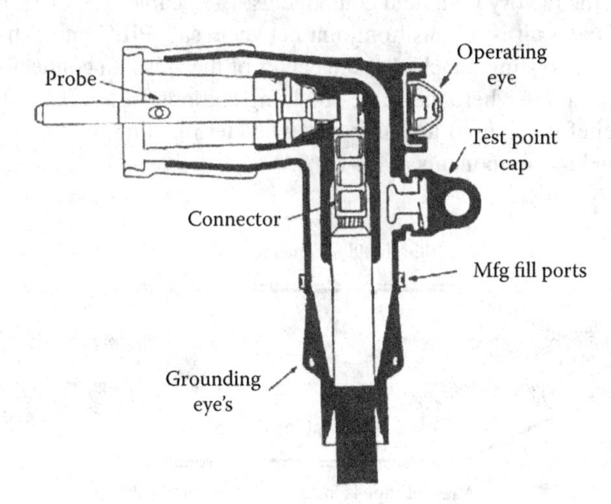

FIGURE 13.13 Load-break elbow. Courtesy Elastimold Div of Thomas and Betts Corp., Memphis, Tennessee.

Operating/Switching: Load-break elbows are designed to function as a switch on energized circuits. They can safely function on cables carrying up to 200 amperes and are capable of being closed into a possible fault of 10,000 amperes. Since this elbow can be operated while energized, devices are required to keep the internal surfaces free of contamination. Good operating practices call for cleaning the mating surfaces of the bushing and the elbow followed by the application of lubricant—while both devices are de-energized! Lubricant is also applied when assembling the elbow on the cable. Some manufacturers supply a different lubricant for the two applications and consequently care should be taken that the correct lubricant is used in each application.

13.4 SPLICING/JOINTING

As was mentioned earlier in this chapter, a termination may be considered to be half of a joint. The same concerns for terminations are therefore doubled when it comes to designing and installing a splice.

13.4.1 Jointing Theory

The ideal joint achieves a balanced match with the electrical, chemical, thermal, and mechanical characteristics of its associated cable. In actual practice, it is not always economically feasible to obtain a perfect match. A close match is certainly one of the objectives.

The splicing or joining together of two pieces of cable can best be visualized as two terminations connected together. The most important deviation, from a theoretical view, between joints and terminations is that joints are more nearly extensions of the cable. The splice simply replaces all of the various components that were made into a cable at the factory with field components. Both cable ends are prepared in the same manner unless it is a transition joint between, say, PILC and extruded cables. Instead of two lugs being attached at the center of the splice, a connector is used. At each end of the splice where the cable-shielding component has been stopped, electrical stress relief is required just as it was when terminating. Figure 13.14 shows a taped splice and its components.

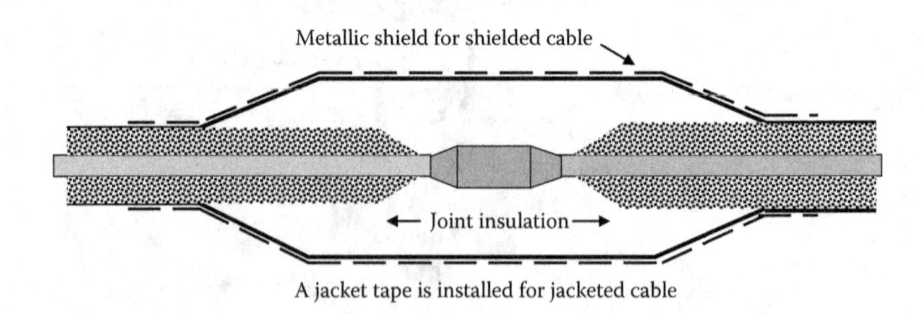

FIGURE 13.14 Taped splice.

Connector: Joins the two conductors together and must be mechanically strong and electrically equal to the cable conductor. In this application, the ends of the connector are tapered. This provides two functions:

1. It provides a sloping surface so that the tape can be properly applied and no voids are created
2. Sharp edges at the end of the connector are not present to cause electrical stress points

Penciling: On each cable being joined, you will notice that the cable insulation is "penciled" back. This provides a smooth incline for the tape to be applied evenly and without voids.

Insulation: In this application, rubber tape is used. Tape is applied to form the stress relief cone at each end of the splice. The overlapped tape continues across the connector to the other side. The thickness at the center of the splice is dictated by the voltage rating.

Conducting Layer: Covering the insulation is a layer of conducting rubber tape that is connected to the insulation shield of the cable at both ends of the splice.

Metallic Shield: A flexible braid is applied over the conducting rubber tape and connects to the factory metallic portion of the cable on each end. This provides a ground path for any leakage current that may develop in the conducting tape.

While not shown in Figure 13.3, there must be a metallic neutral conductor across the splice. This may be in the form of lead, copper concentric strands, copper tapes, or similar materials. It provides the fault current function of the cable's metallic neutral system.

13.4.2 Jointing Design and Installation

13.4.2.1 Cable Preparation

This is the most important step in the entire operation and is the foundation upon which reliable joints and terminations are built. Improperly prepared cable ends provide inherent initiation of failures.

The acceptable tolerances of cable end preparation are dependent upon the methods and materials used to construct the device. Common requirements include a cable insulation surface that is free of contamination, imperfections, and damage caused by such things as shield removal. A smooth surface for extruded dielectric insulations minimizes contamination and moisture adhering to the surface. If the insulation shield can be removed cleanly, there is no reason to use an abrasive on the surface. If a rough surface remains, it must be made smooth.

A cleaning solvent may be needed to obtain an uncontaminated surface. The proper choice of solvent must consider the effect on the person performing the application as well as the cable materials on which it is being applied.

One of the most critical areas is the edge where the insulation shield is removed. A cut into the insulation cannot be remedied and must be removed. For premolded devices, the edge of the shield must be "square"—perpendicular to the cable axis. Metallic connections to the metal shield of the cable must not damage the underlying cable components. Cutting into the strands of wire, or even of greater importance, into a solid conductor, cannot be tolerated.

The concern about the length of the termination is somewhat modified for a joint since the environmental concerns of a termination (external creepage path) are really not a concern for joints. The internal creepage path in a joint is certainly of concern. If you study the literature of the 1950s [5], one finds that the internal creepage path for a paper-insulated cable operating at 15 kV would be 1 inch for every 1 kV. When you look at the design of a premolded joint, you find that the same class of cable has a joint with about 1 inch of creepage—TOTAL. Both of those values are correct. Why is there such a difference? The paper cable was joined using hand applied tapes, either of paper or varnished cloth. There was an air path between the two insulating layers, so the 1 inch per kilovolt was correct. Premolded joints fit very solidly over the cable insulation and use its elastic pressure to maintain the seal. Experience has shown that the approximate 7 kV per inch is reliable for such joints.

13.4.2.2 Connecting the Conductors

Cable conductors are generally either copper or aluminum. Copper is a very forgiving metal in a splice and many methods of connecting two copper conductors together are possible: compression, welding, heat-fusion, soldering (even twisting for overhead conductors 75 years ago), etc. Aluminum connections are not as tolerant as copper. Great care must be taken to match the compression tool, die, and connector with each other for aluminum conductors. As conductor sizes approach 1,000 kcmil these concerns must be addressed more completely. One of the facts involved in the larger size conductors is that, on utility systems especially, the feeder cables are more prone to extended periods of high temperature operation as well as emergency overloads. The operation of the connector must be stable throughout load cycling and be capable of carrying the maximum amount of current without causing thermal degradation of the joint.

The connector metal should be the same as the cable where this is possible. There are situations where this cannot be done, such as the case where a copper conductor is to be connected to an aluminum conductor. It is acceptable to use an aluminum connector over a copper conductor, but a copper connector must not be used over an aluminum conductor because during load cycles, the relative rate of expansion of the two metals causes the aluminum to extrude out and results in a poor connection.

The shape of the connection is always of importance if the connection is not in a shielded area such as those that exist in all premolded splices. In order to minimize voltage stress at the connection for all of those other conditions, special connectors are required for medium and higher voltage cables. Tapered shoulders and filled indents are hallmarks of these connectors. Semiconducting layers are generally specified over these connectors.

13.4.2.3 Insulation for Joints

The material used as the primary insulation in a joint must be completely compatible with the materials in the cable. The wall thickness and its interfaces with the cable insulation must safely withstand the intended electrical stresses. The old rule-of-thumb for paper-insulated cables was that the insulation thickness of the cable was "doubled". In other words, the designs called for putting a layer equal to twice the factory thickness over the cable insulation. In premolded devices, the thickness is usually about 150% of the factory insulation. The joint insulations for hand taped joints (called self-amalgamating tapes) are predominately made from ethylene–propylene rubber but were originally made of a butyl or polyethylene base as long as the thermal properties matched the cable insulation. Premolded joints are almost always made of ethylene rubber compounds. Heat shrink joints are made of polyolefin compounds that have the property of being able to be expanded after being cross-linked using irradiation. The greater diameter remains until heat is applied to the product as it is in place over the connection.

It is not always a good idea to put on too much insulation. Besides being good electrical insulation, these materials are also good heat barriers. Too much insulation can reduce the ability of the joint to carry the same current as the cable without overheating the center of the joint.

13.4.2.4 Shield Materials

These materials must be compatible with the rest of the cable as well as have adequate conductance to drain off the electrostatically induced voltages, charging currents, and leakage currents. Electromagnetically induced currents and fault currents must also be safely handled across the joint area.

Joint semiconducting shields, like cable shields, achieve their ability to perform the task of electrical shielding by having a considerable amount of carbon black (about 30%) compounded into the material. These particles do not actually have to touch each other in order to be conductive.

13.4.2.5 Jackets for Joints

They must provide physical strength, seal against moisture entry into the splice, and resist chemical and other environmental attacks. It is important to use a jacket over the splice when jacketed cables are spliced together since corrosion of metallic neutrals or shields may concentrate at this point.

13.4.2.6 Premolded Splices

The manufactured splice shown in Figure 13.15 has essentially the same component requirements of the taped splice. These devices are designed to cover the range of medium voltage cable sizes. It is essential that the specified cable diameters of the splice be kept within the specified size range of each of the cables. The body of the splice must be slid over one of the cable ends prior to the connector being installed. It is finally repositioned over the center of the joint.

The components of this type of splice are listed as follows:

Connector: The connector shown indicates that it was mechanically pressed on the conductors. In addition, the ends of the connector are not tapered, nor is that a requirement when covered by a connector shield.

Cable End Preparation: The insulation of the cable at the connector is now cut at right angle to the conductor. In the taped splice, a penciled end was required for proper application of tapes. However, in this design, there is no taping required and

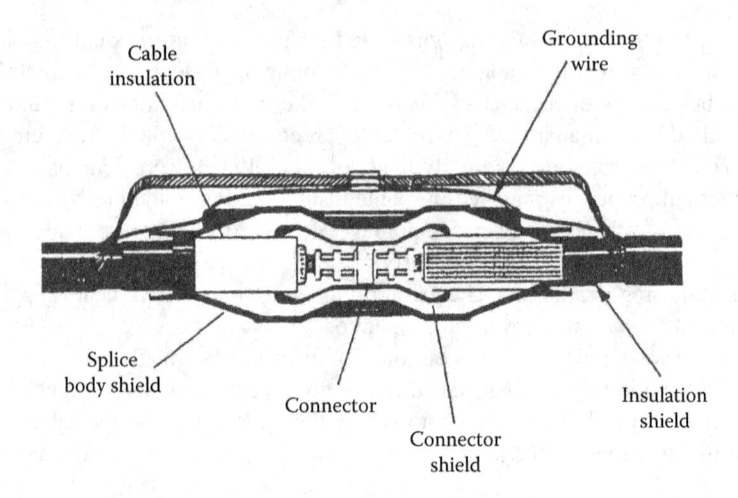

FIGURE 13.15 Premolded splice. Courtesy of Thomas and Betts Corp., Memphis, Tennesseee.

consequently a pencil is not required. A chamfer is required to remove any sharp edges of the cable to prevent the scratching of the inner surface of the splice housing and to make it easier to slide the splice body into position.

Connector Shield: This component is not found in this form in a taped splice, but is critical to the performance of a premolded splice. It is composed of conducting rubber material. In order to nullify the sharp edges of the connector and the air that is between the connector and the cable insulation, this connector shielding must make electrical contact with the cable conductor to eliminate any voltage difference that exists. When the connector shield makes contact with the connector and the cable is energized, both the connector and the shield material are at the same potential. With this design, no discharges can occur in the air or at the connector edges. Figure 13.16 gives an expanded view of the connector shield and its application over a connector.

Insulation: The EPDM insulation is injected between the connector shield and the outer conducting shield of the splice body.

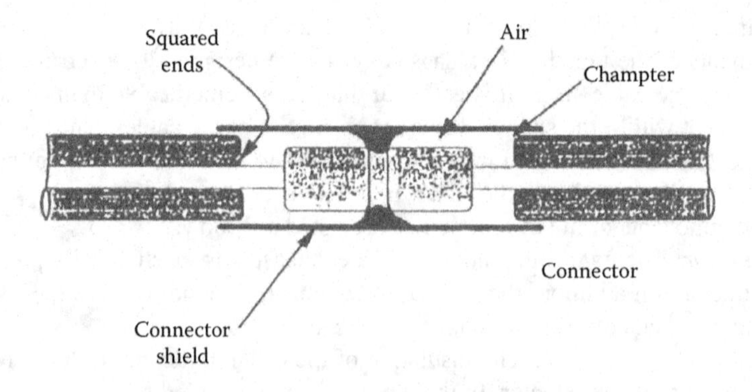

FIGURE 13.16 Detail of connector area of premolded splice.

Splice Body Shield: This is a thick layer of conducting rubber. It is designed to overlap the cable's conducting insulation shield on each end of the splice and to provide stress relief at both cable ends.

Grounding Eye: At each end of the splice, a grounding eye is required on all medium voltage premolded devices and they must be connected to the cable neutral. This provides a parallel path for the grounding of the splice body shield.

Neutral Across Splice: This generally consists of concentric strands from the cable that are twisted together and joined at the center of the splice. The wire used to make an electrical connection to the molded eye is shown connected to the neutral connector and to the concentric strands at each end of the splice.

In an actual field application, these strands should spiral around the splice and be in contact with the outer layer of the splice. This facilitates fault locating by providing a reliable metallic ground for the splice shield.

Adapters: Some designs of premolded splices incorporate adapters. They extend the range of cables that can be accommodated in a specific housing. They also permit the jointing of two widely different cable sizes and they also may enable the user to minimize the inventory of housings. Positioning of the adapter is important so that electrical stress points are not introduced. Adapters are applied before the connector is installed and as in the case of other premolded splices, the body must also be installed prior to the connection and the entire assembly is moved into place.

13.5 ALTERNATE DESIGNS

There are many ways of producing a joint or termination for medium voltage cables: hand applied tapes, combinations of premolded stress cones and hand applied tapes, premolded, cold shrink, resin filled, and heat shrink—just to mention some of them. The proper choice of which termination to select for a given application must consider factors such as cost of materials, time of installation, frequency of use, reliability, space requirements, and skill of the installer. Obviously, there is no universal solution to the wide variety of needs and conditions that are encountered in the field. The proper selection of a joint must consider all of these.

Field molded splices are constructed somewhat like a conventional taped splice. The insulating material is an uncured rubber material that may come in the form of tape or preformed sections. When the proper amount of material has been applied, the splice body is enclosed in a temperature controlled "oven" that confines the expanding material to the proper temperature and time required to cure the tapes. Stress relief is maintained by utilizing a conductive paint over the connector and the insulation after it is cured. The remainder of the shielding and neutral system is restored in the normal manner.

Heat shrink splices are available as a series of heat shrinkable tubes. Some may be preassembled by the manufacturer to reduce the number that must be handled in the field. The tubes must be slipped over the cable prior to connecting the conductors. After positioning each tube over the connected cables, heat is applied to shrink the tube snugly over the underlying surface and soften any mastic material used in the assembly. Stress relief is generally provided by stress control tubes that are also shrunk into place so that ends of the stress control tube overlap both cable insulation shields. The joint is finished in the normal manner.

Cold shrink splices are similar except that a removable liner is pulled out and the tube collapses over the underlying surface. As the name implies, no heat is required. Medium voltage joints contain all the necessary electrical components.

13.6 SELECTION OF JOINTS AND TERMINATIONS

When making a decision as to the best choice of devices to purchase, here are some of the questions and opportunities that should be considered:

- Are the components of the device compatible with the cable being spliced or terminated?
- Did the device pass all the tests that were specified so that it meets the requirements of the electrical system involved?
- Are codes applicable in the decision to use the chosen device?
- Review all safety requirements involved in the construction, application, and installation of the device.
- Will the device meet the mechanical requirements of the installation?
- Can the device be assembled with the tooling that is already available or are special tools required?
- Consider the positioning of the device. Splices are not recommended for installation at bends in the cable. Terminations are normally installed in an upright position. Other positions are possible but require special attention.
- Environmental conditions are of importance to attain expected life of any device. Heat may affect the ampacity of the device. Cold may have an effect on the assembly during installation. Contaminants are critical to the leakage path of a termination.
- Moisture is always the enemy of an underground system and must be controlled in construction and installation.
- Are there any existing work practices or procedures that will conflict with the application of this device?
- Investigate the economics of using different devices such as emergency reserve parts, training new personnel after original installation personnel have gone.
- The cost of the device and the cost of installation.
- Will the device do the job as well or better than what is presently used?

13.7 FAILURES ANALYSIS

Failures happen in spite of all the effort to properly design, make, and operate cable systems. A useful document to aid in the collection of necessary information after a failure is available from IEEE: IEEE 1511-2004, "Guide for Investigating and Analyzing Power Cable, Joint, and Termination Failures" [6].

REFERENCES

1. Medek, J. D., 2001, adapted from class notes for "Understanding Power Cable Characteristics and Applications," University of Wisconsin–Madison.
2. Balaska, T. A., January 1992, adapted from class notes for "Power Cable Engineering Clinic," University of Wisconsin–Madison.
3. Balaska, T. A., 1974, "Jointing of High Voltage, Extruded Dielectric Cables, Basics of Electrical Design and Installation," IEEE UT&D Conference Record, pp. 318–326.
4. Virsberg, L. G. and Ware, P. H., 1967, "A New Termination for Underground Distribution," IEEE Transactions, Vol. PAS-86, pp. 1129–1135.
5. *Underground Systems Reference Book*, EEI, 1957.
6. IEEE 1511-2004, "Guide for Investigating and Analyzing Power Cable, Joint, and Termination Failures."

14 Ampacity of Cables

Carl C. Landinger

CONTENTS

14.1 INTRODUCTION

Ampacity is the term conceived by William Del Mar in the early 1950s when he became weary of saying "current carrying capacity" too many times. The American Institute of Electrical Engineers (AIEE) and The Insulated Power Cable Engineers Association (IPCEA) published the term "ampacity" in 1962 in the "Black Books" of *Power Cable Ampacities* [1]. The term is defined as the maximum amount of current a cable can carry under the prevailing conditions of use without sustaining immediate or progressive deterioration. The prevailing conditions of use include environmental and time considerations.

Cables, whether only energized or carrying load current, are a source of heat. This heat energy causes a temperature rise in the cable which must be kept within limits that have been established through years of experience. The various components of a cable can endure some maximum temperature on a sustained basis with no undue level of deterioration.

There are several sources of heat in a cable, such as losses caused by current flow in the conductor, dielectric loss in the insulation, current in the shielding, sheaths, and armor. Sources external to the cable include induced current in a surrounding conduit, adjacent cables, steam mains, etc.

The heat sources result in a temperature rise in the cable that must flow outward through the various materials that have varying resistance to the flow of that heat. These resistances include the cable insulation, sheaths, jackets, air, conduits, concrete, surrounding soil, and finally to ambient earth.

In order to avoid damage, the temperature rise must not exceed those maximum temperatures that the cable components have demonstrated that they can endure. It is the careful balancing of temperature rise to the acceptable levels and the ability to dissipate the heat that determines the cable ampacity.

14.2 SOIL THERMAL RESISTIVITY

The thermal resistivity of the soil, rho, is the least known aspect of the thermal circuit. The distance for the heat to travel is much greater in the soil than the dimensions of the cable or duct bank, so thermal resistivity of the soil is a very significant factor in the calculation. Another aspect that must be considered is the stability of the soil during the long-term heating process. Heat tends to force moisture out of soils, increasing their resistivity substantially over the soil in its native, undisturbed environment. This means that measuring the soil resistivity prior to the cable being loaded can result in an optimistically lower value of rho than will be the situation in service.

The first practical calculation of the temperature rise in the earth portion of a cable circuit was presented by Dr. A. E. Kennelly in 1893 [2]. His work was not fully appreciated until Jack Neher and Frank Buller demonstrated the adaptability of Kennelly's method to the practical world.

As early as 1949, Jack Neher described the patterns of isotherms surrounding buried cables and showed that they were eccentric circles offset down from the axis of the cable [3]. This was later reported in detail by Balaska, McKean, and Merrell after they ran load tests on simulated pipe cables in a sandy area [4]. They reported very high resistivity sand next to the pipes. Schmill reported the same patterns [5].

Factors that affect the drying rate include type of soil, grain size and distribution, compaction, depth of burial, duration of heat flow, moisture availability, and the watts of heat that are being released. A lengthy debate regarding the main concern of this drying has been in progress for over 20 years: the temperature of the cable/earth interface or the watts of heat that is driven across that soil. An excellent set of six papers was presented at the Insulated Conductors Committee Meeting in November 1984 [6].

In situ tests of the native soil can be measured with thermal needles. Institute of Electrical and Electronics Engineers (IEEE) Guide 442 outlines this procedure [7]. Black and Martin have recorded many of the practical aspects of these measurements [8].

14.3 AMPACITY CALCULATIONS

Dr. D. H. Simmons published a series of papers in 1925 with revisions in 1932, "Calculation of the Electrical Problems in Underground Cables" [9]. The National Electric Light Association (NELA) published the first ampacity tables in the US that covered PILC cables in ducts or air in 1931. In 1933, the Edison Electric Institute (EEI) published tables that expanded the NELA work to include other load factor conditions.

The major contribution was made by Jack Neher and Martin McGrath in their June 1957 classic paper [10]. The AIEE-IPCEA "black books" [1] are tables of ampacities that were calculated using the methods that were described in their work. Those books have now been revised and were published in 1995 by IEEE [11]. IEEE also sells these tables in an electronic form [12].

The fundamental theory of heat transfer in the steady-state situation is the same as Ohm's law where the heat flow varies directly as temperature and inversely as thermal resistance:

$$I = \sqrt{\frac{T_C - T_A - \Delta T_d}{R_{el} \times R_{th}}} \times 10^{-3} \tag{14.1}$$

where

I = current in amperes that can be carried (ampacity)
T_C = maximum allowable conductor temperature in °C
T_A = ambient temperature of ambient earth in °C
ΔT_d = temperature rise due to dielectric loss in °C
R_{el} = electrical resistance of conductor in ohms/foot at T_C
R_{th} = thermal resistance from conductor to ambient in thermal ohm feet

14.3.1 Heat Transfer Model

Cable materials store as well as conduct heat. When operation begins, heat is generated, which is both stored in the cable components and conducted from the region of higher temperature to that of a lower temperature. A simplified thermal circuit for this situation is equivalent to an R–C electrical circuit (Figure 14.1).

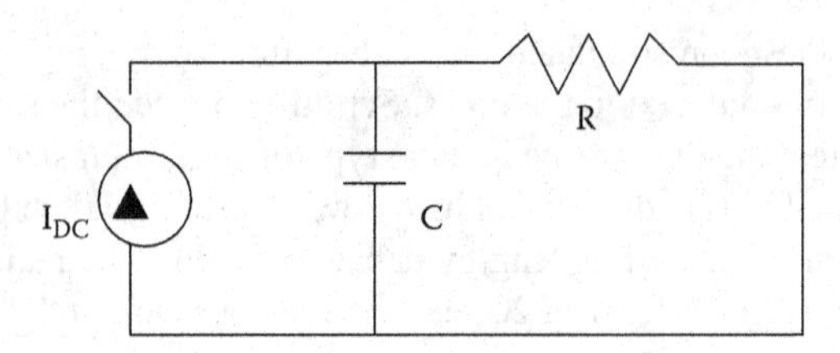

FIGURE 14.1 Simplified thermal circuit.

At time $t = 0$, the switch is closed and essentially all of the energy is absorbed by the capacitor. However, depending on the relative values of R and C, as time progresses, the capacitor is fully charged and essentially all of the current flows through the resistor. Thus, for cables subjected to large swings in loading for short periods of time, the thermal capacitance must be considered (see Section 14.4 of this chapter).

14.3.2 Load Factor

The ratio of average load to peak load is known as load factor. This is an important consideration since most loads on a utility system vary with the time of the day. The effect of this cyclic load on ampacity depends on the amount of thermal capacitance involved in the environment.

Cables in duct banks or directly buried in earth are surrounded by a substantial amount of thermal capacitance. The cable, surrounding ducts, concrete, and earth all take time to heat (and to cool). Thus, heat absorption takes place in those areas as the load increases and permits a higher ampacity than if the load had been continuous. Of course, cooling takes place during the dropping load portions of the load cycle.

For small cables in air or conduit in air, the thermal lag is small. The cables heat up relatively quickly, i.e., in 1 or 2 hours. For the usual load cycles, where the peak load exists for periods of 2 hours or more, load factor is not generally considered in determining the ampacity.

14.3.3 Loss Factor

The loss factor may be calculated from the following formula when the daily load factor is known:

$$LF = 0.3(lf) + 0.7(lf)^2 \tag{14.2}$$

where LF = loss factor, and lf = daily load factor per unit.

Loss factor becomes significant at a specified distance from the center of the cable. This fictitious distance, D_X, derived by Neher and McGrath, is 8.3 inches (211 mm^2). As the heat flows through the surrounding medium beyond this diameter, the effective rho becomes lower and hence the explanation of the role of the loss factor in that area.

14.3.4 Conductor Losses

When electric current flows through a material, there is a resistance to that flow. This is an inherent property of every material and the measure of this property is known as resistivity. The reciprocal of this property is conductivity. When selecting materials for use in an electrical conductor, it is desirable to use materials with as low a resistivity as is consistent with cost and ease of use.

Copper and aluminum are the ideal choices for use in power cables and are the dominant metals used throughout the world.

Regardless of the metal chosen for a cable, some resistance is encountered. It therefore becomes necessary to determine the electrical resistance of the conductor in order to calculate the ampacity of the cable. See Chapter 3 for details of the conductor loss calculation.

14.3.4.1 Direct Current Conductor Resistance

This subject has been introduced in Chapter 3. Some additional insight is presented here that applies directly to the determination of ampacity. The volume resistivity of annealed copper at 20°C is (Table 14.1):

$$\rho_{20} = 0.017241 \text{ ohm mm}^2/\text{meter} \tag{14.3}$$

In ohms-circular mil per foot units this becomes:

$$\rho_{20} = 10.371 \text{ ohm-circular mil/foot} \tag{14.4}$$

Conductivity of a conductor material is expressed as a relative quantity, i.e., as a percentage of a standard conductivity. The International Electrotechnical Commission (IEC) in 1913 adopted a resistivity value known as the International Annealed Copper Standard (IACS). The conductivity values for annealed copper were established as 100%.

TABLE 14.1
Direct Current Resistivity at 20°C

Metal	Volume Conductivity Percentage IACS	Volume Resistivity Ohms-cmil/ft	Volume Resistivity Ohms-mm²/m	Weight Resistivity Ohms-lb/mile²
Soft copper	100.00	10.371	0.01724	875.2
Hard copper	96.16	10.785	0.01793	910.15
Copperweld	39.21	26.45	0.043971	2046.3
1350-H19	61.2	16.946	0.02817	434.81
5005-H19	53.5	19.385	0.03223	497.36
6201 T81	52.5	19.755	0.03284	506.85
Alumoweld	20.33	51.01	0.08401	3191.0
Steel	5.0	129.64	0.21551	9574.0

An aluminum conductor is typically 61.2% as conductive as an annealed copper conductor. Thus, a #1/0 AWG (53.5 mm²) solid aluminum conductor of 61.2% conductivity has a volume resistivity of 16.946 ohms—circular mil per foot and a cross-sectional area of 105,600 circular mils. Thus, the direct current (DC) resistance per 1,000 feet at 20°C is:

$$R_{DC(20)} = 16,946 \times 1,000/105,60($$

$$R_{DC(20)} = 0.1605 \text{ ohms}$$

To adjust the tabulated values of conductor resistance to other temperatures that are commonly encountered, the following formula applies:

$$R_{T2} = R_{T1}\left[1+\alpha\left(T_2 - T_1\right)\right] \tag{14.5}$$

where

R_{T2} = DC resistance of conductor at new temperature
R_{T1} = DC resistance of conductor at "base" temperature
α = temperature coefficient of resistance

Temperature coefficients for various copper and aluminum conductors at several base temperatures are shown in Table 14.2.

14.3.4.2 Alternating Current Conductor Resistance

When the term "AC resistance of a conductor" is used, it means the DC resistance of that conductor plus an increment that reflects the increased apparent resistance in the conductor caused by the skin-effect inequality of current density. Skin effect results in a decrease of current density toward the center of a conductor. A longitudinal element of the conductor near the center is surrounded by more magnetic lines of force than is an element near the rim. Thus, the counter electromotive force (emf) is greater in the center of the element. The net driving emf at the center element is thus reduced with consequent reduction of current density.

Methods for calculating this increased resistance have been extensively treated in technical papers and bulletins, Reference [10] for instance.

TABLE 14.2
Temperature Coefficients for Conductor Metals

Metal	0°C	20°C	25°C	30°C
61.2% Al	0.00440	0.00404	0.00396	0.00389
100.0% Cu	0.00427	0.00393	0.00385	0.00378
98.0% Cu	0.00417	0.00385	0.00378	0.00371
96.0% Cu	0.00408	0.00377	0.00370	0.00364

14.3.4.3 Proximity Effect

The flux linking a conductor due to nearby current flow distorts the cross-sectional current distribution in the conductor in the same way as the flux from the current in the conductor itself. This is called proximity effect. Skin effect and proximity effect are seldom separable and the combined effects are not directly cumulative. If the distance between the conductors exceeds 10 times the diameter of a conductor, the extra I^2R loss is negligible.

14.3.4.4 Hysteresis and Eddy Current Effects

Hysteresis and eddy current losses in conductors and adjacent metallic parts add to the effective alternating current (AC) resistance. To supply these losses, more power is required from the cable. They can be very significant in large ampacity conductors when the magnetic material is closely adjacent to the conductors. Currents greater than 200 amperes should be considered to be large for these effects.

14.3.5 Calculation of Dielectric Loss

As has been seen in Chapter 4, dielectric losses may have an important effect on ampacity. For a single-conductor, shielded and for a multiconductor cable having shields over the individual conductors, the following formula applies:

$$W_d = 2\pi fCnE^2F_p \times 10^{-6} \tag{14.6}$$

and

$$C = 7.354\varepsilon / \text{Log}_{10}\left(D_O/D_I\right) \tag{14.7}$$

where

f = operating frequency in hertz
n = number of shielded conductors in cable
C = capacitance of individual shielded conductors in micro-micro farads/ft
E = operating voltage to ground in kilovolts
F_p = power factor of insulation
ε = dielectric constant of the insulation
D_O = diameter over the insulation
D_I = diameter under the insulation

14.3.6 Metallic Shield Losses

When current flows in a conductor, there is a magnetic field associated with that current flow. If the current varies in magnitude with time, such as with 60 hertz alternating current, the field expands and contracts with the current magnitude. In the event that a second conductor is within the magnetic field of the current carrying conductor, a voltage, which varies with the field, will be introduced in that conductor.

If that conductor is part of a circuit, the induced voltage will result in current flow. This situation occurs during operation of metallic shielded conductors. Current flow in the phase conductors induces a voltage in the metallic shields of all cables within the magnetic field. If the shields have two or more points that are grounded or otherwise complete a circuit, current will flow in the metallic shield conductor.

The current flowing in the metallic shields generates losses. The magnitude of the losses depends on the shield resistance and the current magnitude. This loss appears as heat. These losses not only represent an economic loss, but they also have a negative effect on ampacity and voltage drop. The heat generated in the shields must be dissipated along with the phase conductor losses and any dielectric loss. Recognizing that the amount of heat that can be dissipated is fixed for a given set of thermal conditions, the heat generated by the shields reduces the amount of heat that can be assigned to the phase conductor. This has the effect of reducing the permissible phase conductor current. In other words, shield losses reduce the allowable phase conductor ampacity.

In multiphase circuits, the voltage induced in any shield is the result of the vectoral addition and subtraction of all fluxes linking the shield. Since the net current in a balanced multiphase circuit is equal to zero when the shield wires are equidistant from all three phases, the net voltage is zero. This is usually not the case, so in the practical world there is some "net" flux that will induce a shield voltage/current flow. In a multiphase installation of shielded, single-conductor cables, as the spacing between conductors increases, the cancellation of flux from the other phases is reduced. The shield on each cable approaches the total flux linkage created by the phase conductor of that cable (Figure 14.2).

As the spacing, S, increases, the effect of phases B and C is reduced and the metallic shield losses in the A phase are almost entirely dependent on the A phase magnetic flux.

There are two general ways that the amount of shield losses can be minimized:

- Single point grounding (open circuit shield)
- Reduction of the quantity of metal in the shield

The open circuit shield presents other problems. The voltage continues to be induced and hence the voltage increases from zero at the point of grounding to a maximum at the open end that is remote from the ground. The magnitude of voltage is primarily dependent on the amount of current in the phase conductor. It follows that there are two current levels that must be considered: maximum normal current and maximum fault current in designing such a system. The amount of voltage that can be tolerated depends on safety concerns and jacket designs.

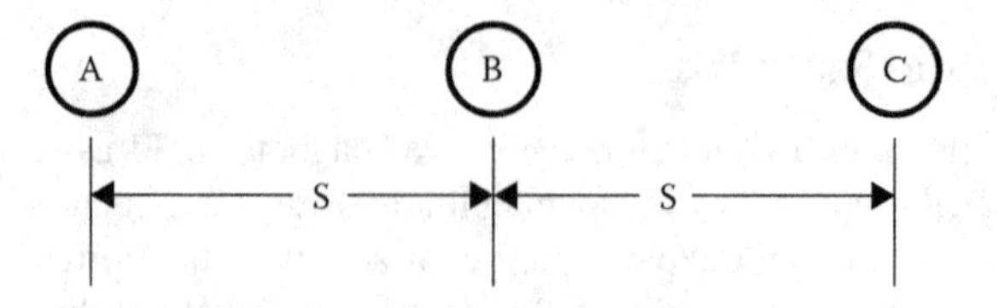

FIGURE 14.2 Effect of spacing.

Another approach is to reduce the amount of metal in the shield. Since the circuit is basically a one-to-one transformer, an increase in the resistance of the shield gives a reduction in the amount of current that will be generated in the shield. As an example, a 1,000 kcmil (507 mm^2) aluminum conductor, three 15 kV cables with multiground neutrals that are installed in a flat configuration with 7.5 inch (190.5 mm^2) spacing. A cable with one-third conductivity neutral will have four times as much current in the shields as a one-twelfth neutral cable. If the phase conductors are carrying a balanced 600 amperes, the outside, lagging phase cable will have 400 amperes in the shield. A similar cable configuration with one-twelfth neutral will have only 100 amperes. The total current is reduced from 1,000 amperes to 700 amperes. This translates to an increase of ampacity of roughly 25% for the reduced neutral cables.

In order to take shield losses into account when calculating ampacity, it is necessary to multiply all thermal resistances in the thermal circuit beyond the shield by one plus the ratio of the shield loss to the conductor loss. This incremental thermal resistance reflects the effect of the shield losses.

The shield loss calculations for cables in other configurations are rather complex, but very important. Halperin and Miller developed a method for closely approximating the losses and voltages for single-conductor cables in several common configurations.

14.4 TYPICAL THERMAL CIRCUITS

14.4.1 Internal Thermal Circuit for a Shielded Cable with Jacket

Thermal circuits will be shown in increasing complexity of the number of components. The symbols used throughout will be:

$\bar{R}$ = thermal resistance (pronounced R bar) in thermal ohm-feet
Q = heat source in watts per foot
$\bar{C}$ = thermal capacitance

The subscripts throughout are:

C = conductor
I = insulation
S = shield
J = jacket
D = duct
SD = distance between cable and duct
E = earth

14.4.2 Single Layer of Insulation, Continuous Load

The internal thermal circuit is shown in Figure 14.3 for a cable with continuous load. The conductor heat source passes through only one thermal resistance. This may be an insulation, a covering, or a combination as long as they have similar thermal resistances. Note that these circuits stop at the surface of the cable. The remainder of the thermal circuit will be added in examples that follow.

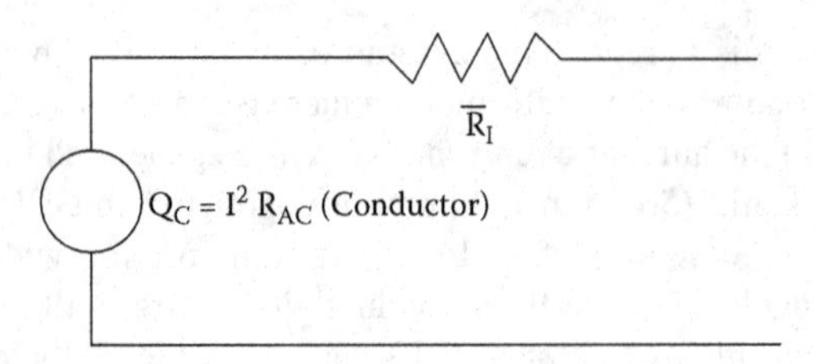

FIGURE 14.3 Continuous load flow.

Figure 14.3 shows a continuous load flowing through one layer of insulation. The heat does not travel beyond the surface of the cable in this example.

14.4.3 Cable Internal Thermal Circuit Covered by Two Dissimilar Materials, Continuous Load

In this example, the continuous load flows through two dissimilar materials, but the heat still stays at the surface of the last layer of insulation (Figure 14.4).

14.4.4 Cable Thermal Circuit for Primary Cable with Metallic Shield and Jacket, Continuous Load

This thermal diagram (Figure 14.5) shows a primary cable with its several heat sources and thermal resistances still with a constant load where p and (1−p) divide the thermal resistance to reflect Q_i.

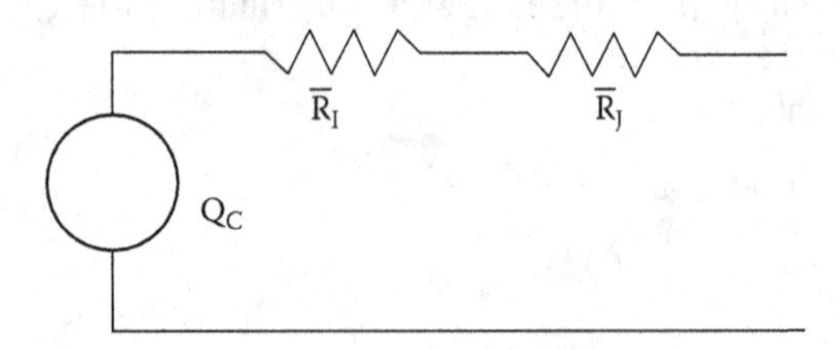

FIGURE 14.4 Load flow through two materials.

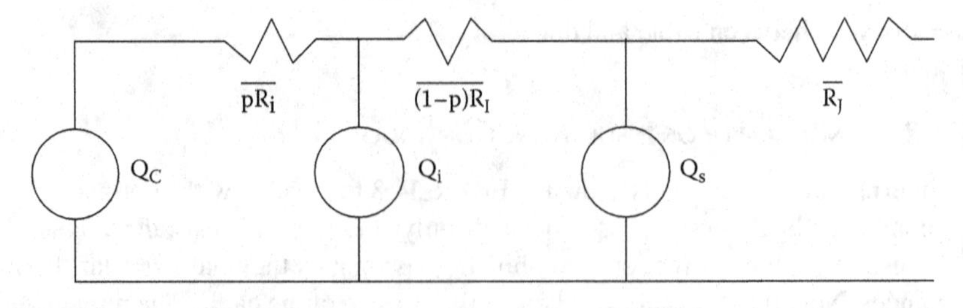

FIGURE 14.5 Shielded cable with several heat sources.

14.4.5 Same Cable as Example 3, but with Cyclic Load

Figure 14.6 shows the same cable as in Figure 14.5, but the cyclic load is accounted for with the capacitors that are parallel to the three heat sources.

14.4.6 External Thermal Circuit, Cable in Duct, Continuous Load

Figure 14.7 shows the resistances that are external to the cable.

14.4.7 External Thermal Circuit, Cable in Duct, Time Varying Load, External Heat Source

See Figure 14.8, where H_X = external heat Source.

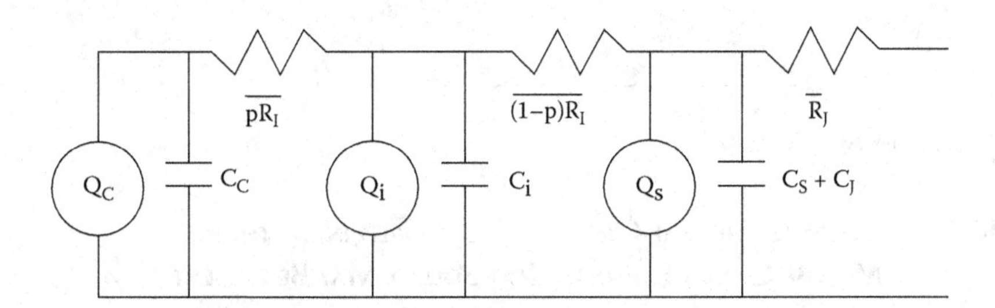

FIGURE 14.6 Cyclic load has been added.

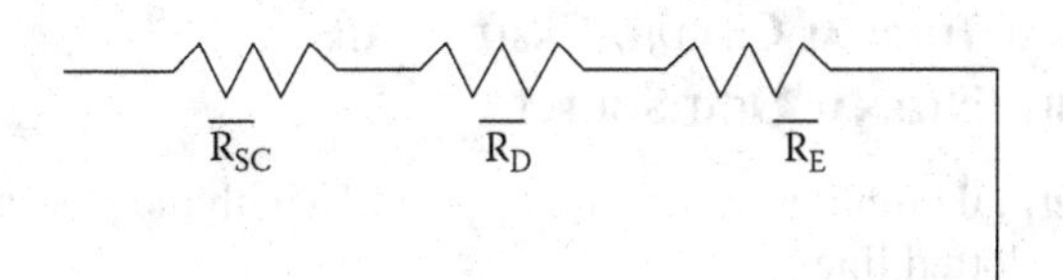

FIGURE 14.7 External circuit.

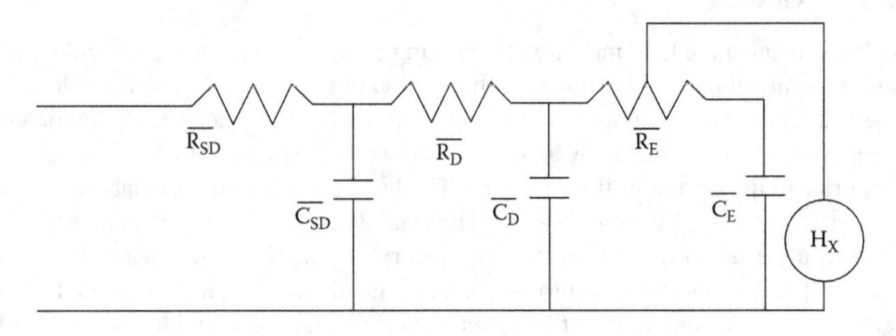

FIGURE 14.8 External heat source added.

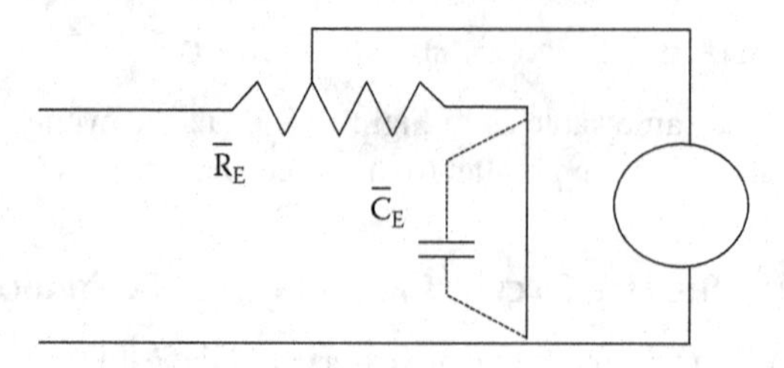

FIGURE 14.9 External thermal circuit.

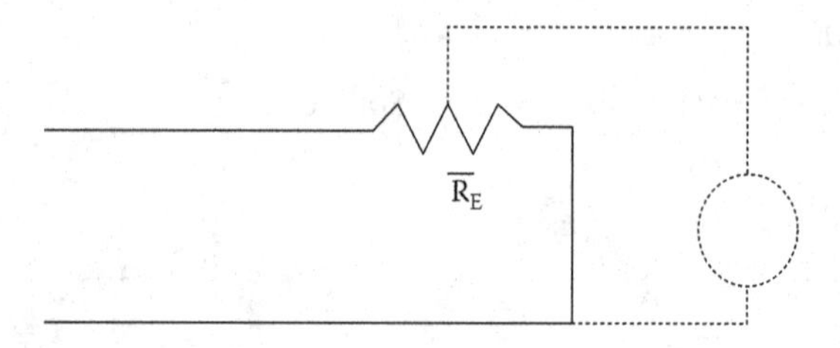

FIGURE 14.10 Cable in air.

14.4.8 External Thermal Circuit, Cable Buried in Earth, Load May Be Cyclic, External Heat Source May Be Present

The depiction of possible cyclic load and external heat source is shown in Figure 14.9 by dotted lines.

14.4.9 External Thermal Circuit, Cable in Air, Possible External Heat Source

The external thermal circuit is shown in Figure 14.10 with the possible external heat source shown by dotted lines.

14.5 SAMPLE AMPACITY CALCULATION

14.5.1 General

Methods to calculate the ampacity of operating cables continue to be a popular subject for technical papers. Fortunately, the portion of the work that had been done by slide-rule and copious quantities of notepaper has been replaced with computers. Manipulations were handled by assuming intermediate values of the various parameters prior to the advent of the computer. The hand calculations were laborious, but the user did get a feel for the concept. The availability of tables and computer programs could lead to quick, but possibly incorrect, answers. The Neher–McGrath paper [10] is the best reference to use before a hand calculation is attempted. As a matter of fact, you should read that paper even if you have decided to use any available tables or programs!

The following simple example of a calculation is presented with the intent of giving insight into the process.

The general equation that has been previously given in Equation 14.1.

Another form of this equation recognizes the other possible heat sources that have been indicated in the thermal circuit diagrams. Equation 14.1 expands to:

$$I = \sqrt{\frac{T_C - \left(T_A + \Delta T_d\right)}{R_{AC}\left(\overline{R_i} + \overline{R_{TH}}\right)}} \tag{14.8}$$

where

ΔT_d = temperature rise due to dielectric loss in °C

$\overline{R_i}$ = to account for thermal resistance of insulation and/or coverings between the conductor and the first heat source beyond the insulation.

$\overline{R_{TH}}$ = is the thermal resistance to ambient adjusted to account for additional heat sources such as shield loss, armor loss, steam lines, etc.

14.6 AMPACITY TABLES AND COMPUTER PROGRAMS

14.6.1 TABLES

The *IEEE Standard Power Cable Ampacity Tables*, [11,12], IEEE Standard 835-1994, is a book (or electronic version) that contains over 3,000 tables in 3,086 pages. Voltages range from 5 kV to 138 kV. Although there are situations that are not covered by these tables, this is an excellent beginning point for anyone interested in cable ampacities.

ICEA P-54-440, *Ampacities of Cables in Open Top Trays* covers ampacities in trays that is not covered in IEEE Standard 835.

The early ampacity tables *AIEE S-135, Volumes 1 & 2*, IPCEA Pub. P-46-426, 1962 continue to be used for appropriate situations.

It is important to point out that situations in the US falling under the jurisdiction of the National Electric Code are covered by the ampacity tables found in NFPA 70, National Electrical Code. Similar codes may be expected to exist in every country outside the US and the reader is advised to consult them as appropriate.

Manufacturers have also published catalogs that cover the more common situations [13,14].

14.6.2 COMPUTER PROGRAMS

Most of the large cable manufacturers and architect/engineering firms have their own computer programs for ampacity determination. This is an excellent source of information when you are engineering a new cable system. These programs generally are not for sale.

There are commercially available programs throughout North America. These are especially useful when you need to determine the precise ampacity of a cable, for instance, in a duct bank with other cables that are not fully loaded. The general cost of one of these programs is about $5,000 in US dollars.

14.7 "AMPACITIES" UNDER SHORT CIRCUIT CONDITIONS

Short circuit conditions commonly involve large inrushes of current. The heating associated with these can be extremely destructive and hazardous. Calculations to determine the capability of cables to withstand these currents (and durations) are generally handled by other calculation methods and assumptions.

In North America, some common methods are found in:

ICEA P-32-382 "Short Circuit Characteristics of Insulated Cables" latest edition

ICEA P-45-482 "Short Circuit Performance of Metallic Shields and Sheaths of Insulated Cable" latest edition

EPRI EL-3014, Project 1286-2, final report 4/83, "Optimization of Design of Metallic Shield–Concentric Conductors of Extruded Dielectric Cables Under Fault Conditions"

Similar methods are published in IEC 60949-9, "Calculations of Thermally Permissible Short Currents Taking Into Account Non-Adiabatic Heating Effects" and other national standards.

14.8 RELATIONSHIPS BETWEEN AMPACITY AND VOLTAGE DROP CALCULATIONS

Factors that impact ampacity calculations also impact voltage drop calculations. The AC and DC resistances of conductors increase with increasing temperature, thereby increasing voltage drop. Similarly, shield, sheath, and pipe losses have the effect of increasing the apparent conductor resistance and impacting inductive reactance. An excellent reference that discusses these relationships is reference [9].

REFERENCES

1. *Power Cable Ampacities,* 1962, AIEE Pub. No. S-135-1 and IPCEA Pub. No. P-46-426.
2. Kennelly, A. E., 1893, "On the Carrying Capacity of Electrical Cables...", *Minutes,* Ninth Annual Meeting, Association of Edison Illuminating Companies, New York, NY.
3. Neher, J. H., 31 January–4 February 1949, "The Temperature Rise of Buried Cables and Pipes," AIEE Paper No. 49-2, Winter General Meeting, New York, NY.
4. Balaska, T. A., McKean, A. L. and Merrell, E. J., 19–24 June 1960, "Long Time Heat Runs on Underground Cables in a Sand Hill," AIEE Paper No. 60-809, Summer General Meeting.
5. Schmill, J. V., 30 January–4 February 1966, "Variable Soil Thermal Resistivity – Steady State Analysis," IEEE Paper. No. 31 TP 66-14, Winter Power Meeting, New York, NY.
6. Insulated Conductors Committee Minutes, Appendices F-3, F-4, F-5, F-6, F-7, and F-8, November 1984.
7. *IEEE Guide for Soil Thermal Resistivity Measurements*, IEEE Std. 442-1979.

8. Black, W. Z. and Martin, M. A. Jr., 1–8 February 1981, “Practical Aspects of Applying Thermal Stability Measurements to the Rating of Underground Power Cables,” IEEE Paper No. 81 WM 050-4, Atlanta, GA.
9. Simmons, D. M., May–November 1932, “Calculation of the Electrical Problems of Underground Cables,” *The Electrical Journal,* East Pittsburgh, PA.
10. Neher, J. H. and McGrath, M. H., October 1957, “The Calculation of the Temperature Rise and Load Capability of Cable Systems,” AIEE Transactions, Vol. 76, Pt. III, pp. 752–772.
11. *IEEE Standard Power Cable Ampacity Tables,* IEEE Std. 835-1994 (hard copy version).
12. *IEEE Standard Power Cable Ampacity Tables,* IEEE Std. 835-1994 (electronic version).
13. *Engineering Data for Copper and Aluminum Conductor Electrical Cables,* 1990, EHB-90, The Okonite Company.
14. *Power Cable Manual,* 1997, Second Edition, The Southwire Company.

15 Thermal Resistivity of Soil

Deepak Parmar

CONTENTS

15.1 INTRODUCTION

The determination of the ampacity of electric power cables has been discussed in Chapter 14. In that chapter, considerable emphasize has been placed on the "internal" aspects of the cable and the generation of heat inside the cable. This chapter will emphasize the external heat flow considerations that are required to carry the internal heat to the ultimate heat sink, the ambient earth's surface.

The first practical calculation of the temperature rise in the earth portion of a cable circuit was presented by Dr. A. E. Kennelly in 1893 [2]. His work was not fully appreciated until Neher and Mr. Grath [23] demonstrated the adaptability of Kennelly's method to cable engineering. As early as 1949, Jack Neher described the patterns of isotherms surrounding buried cables and showed that they were eccentric circles offset down from the axis of the cable [3]. This was later reported in detail by Balaska, McKean, and Merrell after they ran load tests on simulated pipe cables in a sandy area in New York [9]. They reported very high resistivity sand adjacent to the pipes. Analyses of the thermal resistivity of soils by Sinclair [3], Adams and Baljet [4], Black and Martin [5], and others have clearly shown that migration of moisture from the backfill soil is a critical element in the ampacity of cable systems [8]. Although a moisture content value per se is not a part of the ampacity calculation, it has the highest impact on the thermal resistivity of soil.

In-situ tests to determine the thermal resistivity of the native soil can be conducted by thermal needle method. IEEE Guide 442 outlines this procedure. Black and Martin have recorded many of the practical aspects of these measurements in reference [5].

15.2 MECHANISM OF HEAT TRANSFER IN SOIL

Heat flows through a soil primarily by conduction through mineral particles, and secondarily by conduction and convection through the moisture or air that occupies the pore space between solid particles. Thermal resistivity depends on soil composition and texture, water content, density, and various other factors to a lesser degree. This complex interrelationship does not lend itself to a simple formula; rather testing must be carried out on any given soil to determine its resistivity. Note that once a cable is installed, the soil moisture is the only parameter that changes significantly with time.

There are many factors that affect the thermal resistivity of soils:

1. Soil Composition
   a. Mineral type and content
   b. Organic content
   c. Chemical bonding between particles
2. Texture
   a. Grain size distribution
   b. Grain shape
3. Water/Moisture Content
   a. Degree of saturation
   b. Porosity

4. Dry Density
   a. Porosity
   b. Solid content
   c. Inter particle contacts
   d. Pore size distribution
5. Ambient Temperature
6. Other Factors
   a. Dissolved salts and minerals
   b. Changes in water levels

Soil is a composite consisting of solid mineral grains, typically making point-to-point contact, and pore space filled with water and air. The thermal resistivity of a given soil mass is a function of the intrinsic resistivities of its components. These may range from 12°C-cm/W for quartz mineral, to 40°C-cm/W for limestone, to 165°C-cm/W for water, to about 500°C-cm/W for organics (Table 15.1). Even certain highly compacted soils can have up to 30% voids between solid particles, which in a dry state are filled with very high resistivity air (about 4,500°C-cm/W).

Quartz is three to five times more conductive than other minerals. Reactive clay minerals enhance particle bonding; flaky mica particles can be indicative of a loose microstructure. In addition, grain shape (round, elongated, platy) and angularity influence soil density, grain contacts, and microstructure. A significant organic content (say >4%) can substantially increase resistivity.

Soil texture refers to the soil grain size, shape, and particle size gradation. Since most of the heat conduction is through the solid particles and their contacts, the resistivity is minimized for soils that maximize these contacts—Figure 15.2. Well-graded materials and those consisting of crushed particles (i.e., angular as opposed to rounded or uniform grains—Figure 15.1) generally have more particle contacts and compact to higher densities. This has the added benefit of retarding moisture migration because

TABLE 15.1
Soil Components

Material	Thermal Resistivity
Quartz, Silica	12
Granite	30
Limestone	40
Sandstone	50
Shale	60
Shale (friable)	200
Mica	170
Ice	45
Water	165
Organics	~500
Air	~4,500

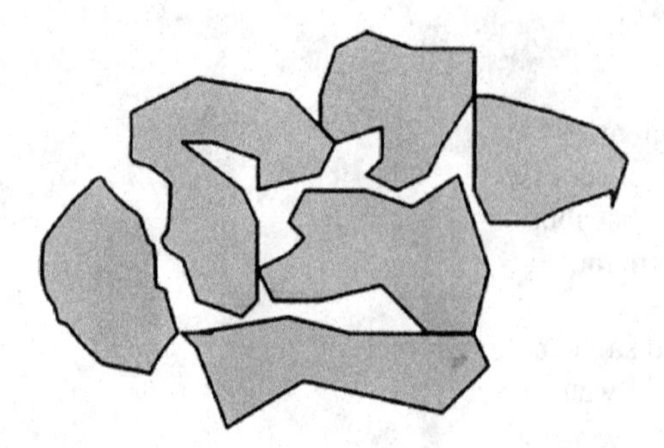

FIGURE 15.1 High thermal resistivity soil (low density and high porosity).

of the smaller pore spaces. To ensure backfill thermal performance, particle gradation limits are specified based on a sieve analysis.

In a wet soil, water provides an easy path for heat conduction over these thermal bridges. As the soil dries, discontinuities develop in the heat conduction path and the thermal resistivity increases.

Densification increases mineral grain contacts and displaces air (i.e., lowers the porosity), thereby reducing the soil resistivity, most notably at low moisture contents. Well-graded soils are potentially denser because smaller grains can efficiently fill the spaces between the larger particles (Figure 15.2). The specification of a minimum dry density for a corrective thermal backfill is one factor in obtaining the required thermal performance.

For a given soil, the major influence on the thermal resistivity is the moisture content. In a dry state, the pore spaces are filled with air (~4,500°C-cm/W); thus extremely restricted heat conduction paths exist only along the mineral contacts. As water (~165°C-cm/W) replaces air, the soil resistivity is substantially lowered (as much as three to eight times) as the relatively good heat conduction paths are expanded (thermal bridges, Figure 15.3). This is illustrated by the *thermal dryout curve* (thermal resistivity vs. soil moisture content, Figure 15.4). A soil that is better

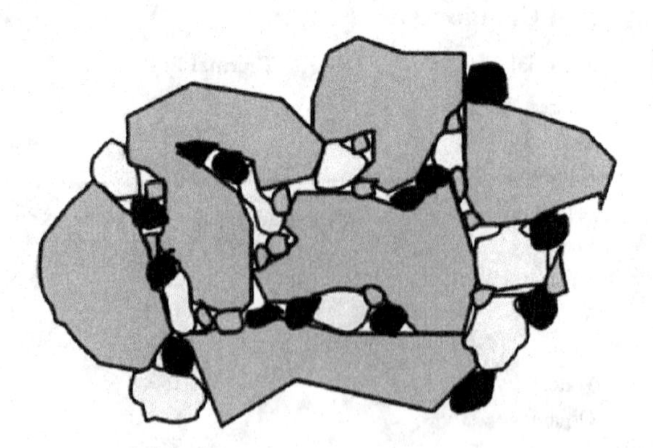

FIGURE 15.2 Low thermal resistivity soil (higher density and lower porosity).

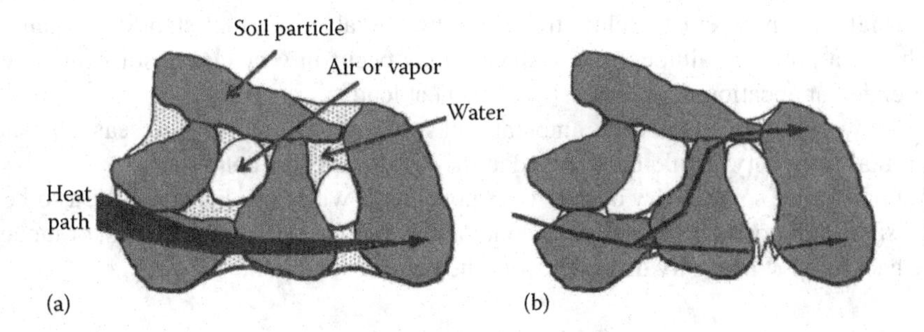

FIGURE 15.3 Influence of water (moisture content) on thermal resistivity. (a) *Wet soil*: high water content provides an easy path for heat conduction ("thermal bridges"), therefore the soil thermal resistivity is low. (b) *Damp soil*: as soil dries, discontinuities develop in the heat conduction path due to low water content, therefore themal resistivity increases.

able to retain its moisture, as well as able to efficiently re-wet when dried, will have better thermal performance characteristics.

Cables that are installed under the groundwater table will experience a constant and low soil thermal resistivity; but, if the water table should ever drop below the cable level, then the potential for thermal drying of the soil and attendant high

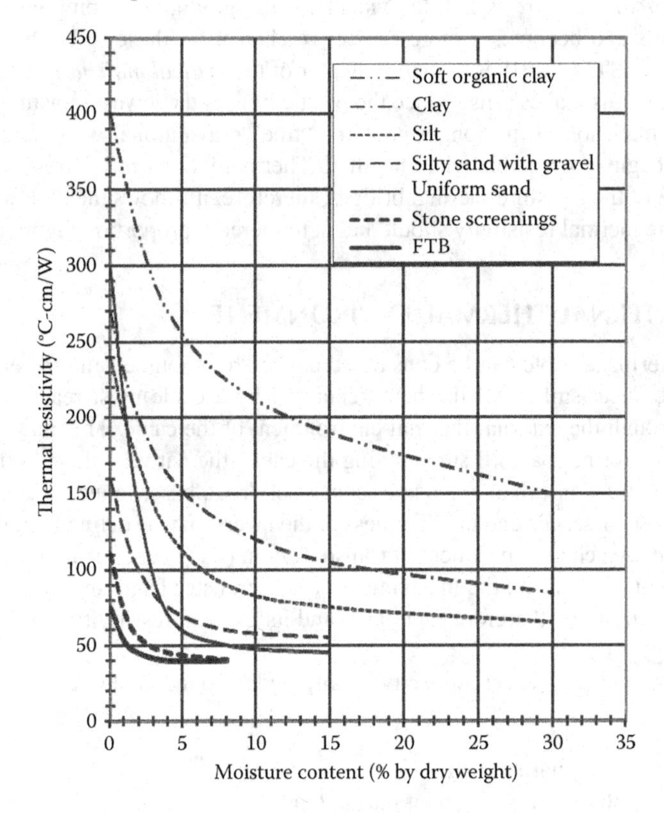

FIGURE 15.4 Impact of moisture on ampacity.

thermal resistivity exist. Below the groundwater table, thermal stability is generally not a problem, although some drying may occur in very clayey soils under an extended application of an excessive cable heat load.

Above the water table, the ambient soil moisture varies with the seasons. The thermal resistivity should be assessed at the driest expected conditions.

For a native soil, the dry density does not change with time, but spatially, a soil of the same composition and moisture content may exist at different densities; or it may be backfilled at a density different from the natural condition.

15.2.1 Critical Soil Moisture

At higher soil moistures, the *thermal dryout curve* is relatively flat. At lower moistures, the curve slopes upward more steeply, such that a small amount of soil drying gives rise to a larger increase in thermal resistivity. The *critical moisture* is defined as the point below which the thermal resistivity begins to increase disproportionately. Well-graded and granular soils have a sharp knee in their *thermal dryout curves* and the *critical moisture* is clearly defined (Figure 15.4). For these soils, as a rule, the *critical moisture* may be taken at the point where the wet (flat line) thermal resistivity has increased by 10%.

The *critical moisture* is not so evident for the gradually sloping curve of fine grained soils and becomes a more subjective choice; for these soils, the *Atterberg* plastic limit/ASTM D4318 is a good indicator of the *critical moisture.*

Intuitively this makes sense, since the plastic limit is the lowest moisture at which the soil is malleable. It no longer contains "free" gravitational water and at lower moistures begins to crumble when handled. Thermally this limit corresponds to the breakdown of the moisture thermal bridges; therefore, it follows that below the plastic limit, the thermal resistivity should begin to increase proportionally more.

15.3 EXTERNAL THERMAL ENVIRONMENT

In simple terms, a cable can be considered as the "heat source" and the earth's surface as the "heat sink." All the heat generated by a cable must reach the earth's surface through the external thermal environment of the cable. This may consist of a corrective thermal backfill surrounding the cable, the native soil, and other materials. Installations in urban areas may consist of "road base," concrete and asphalt. Some of these materials and the thickness of the layers may be defined (regulated) by other agencies such as department of transportation (DOT) and local municipalities. For all practical purpose, the mechanism of heat transfer from underground cables is by *conduction* and therefore each layer and its thermal resistivity will impact the cable rating.

In order to calculate the ampacity of any underground cable, it is necessary to know the

1. Soil thermal characteristics
2. Earth ambient temperature at burial depth
3. Surface or air temperature

15.4 NATIVE SOIL THERMAL RESISTIVITY

An aspect that must be considered is the long-term stability of the soil during the heating process. Heat tends to force moisture out of soils, increasing their resistivity substantially over the soil in its native, undisturbed environment. This means that measuring the soil resistivity prior to the cable carrying load current can result in an optimistically low value. Therefore, it is essential to conduct "thermal dryout characterization" of the soil and that of the backfill. An appropriate value of the thermal resistivity is determined from this dryout curve. Factors that affect the drying rate and thermal stability will be discussed.

Common components of soils with their thermal resistance (rho) in °C-cm/watts are shown in Table 15.1. The values for water and air are based on conduction—not for moving air or water.

The earth that surrounds a cable system can have great variations in thermal resistance along its route. This can be the result of natural soil variations as well as moisture migration caused by heat produced current flow. The backfill material that surrounds a cable system is an extremely critical element in achieving optimum heat dissipation of the system.

Along a cable route, the native soil thermal resistivity can vary significantly as a function of the distance and also of depth. In addition, these values can change from, say, 50°C-cm/W–60°C-cm/W at a moisture content of 10% to, say, about 150°C-cm/W at 10% and as high as 350°C-cm/W in totally dry condition (0% moisture content) (Figure 15.4).

Organic soils such as "peat" and "top soil" or soils with vegetation or root matter exhibit high thermal resistivity, both in wet and especially in dry condition. These type of soils have relatively high moisture content but fairly low density. They are very porous (permeable) and therefore will dry relatively easily. The high porosity results in high voids (air content) and thus very high resistivity. If native soil is considered to be used as a nonclassified backfill over a good quality "corrective thermal backfill envelope" or over a "concrete duct-bank," it must be free of all organic matter.

The impact of soil thermal resistivity on ampacity of cables directly buried in a native soil is shown in Figure 15.5. While actual values for rho and amperes vary widely, this relationship holds for any cable system, regardless of the voltage. The change in the soil thermal characteristic (primarily the thermal resistivity) is purely a function of the heat output of an energized cable system and not the voltage. A low voltage cable system may have numerous cables in a single trench or only a few cables carrying very high current. The *resultant net heat output per unit length of the trench is the prime factor that induces the soil moisture migration (drying) that results in higher thermal resistivity.*

On wind and solar farm projects, the cables are operated at relatively low voltages (34 kV). However, the risk of cable failure is quite high if the native soil that is used to backfill the trenches is not installed at the specified density in order to meet the design thermal resistivity.

The effective thermal resistivity of a combination of imported backfill and the native soil surrounding a trench is shown in Figure 15.6. Optimized thermal characteristics and dimensions of the envelope of the controlled backfill around the cable

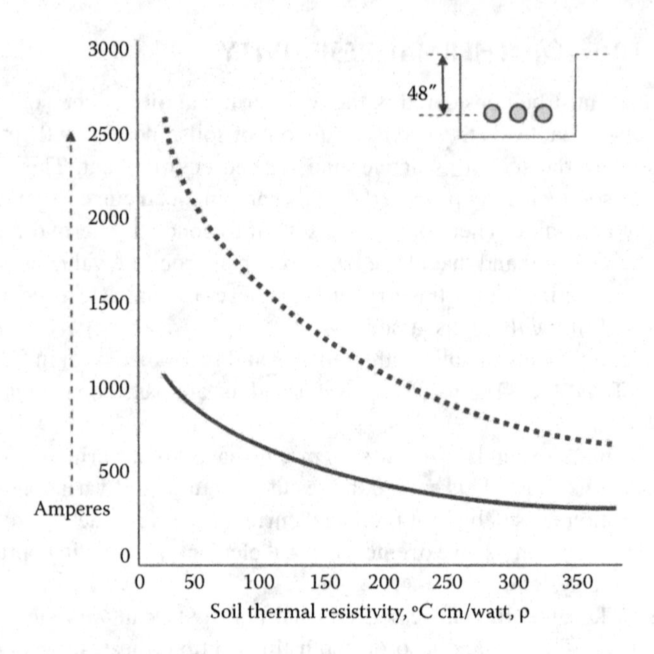

FIGURE 15.5 Effect of soil thermal resistivity on ampacity.

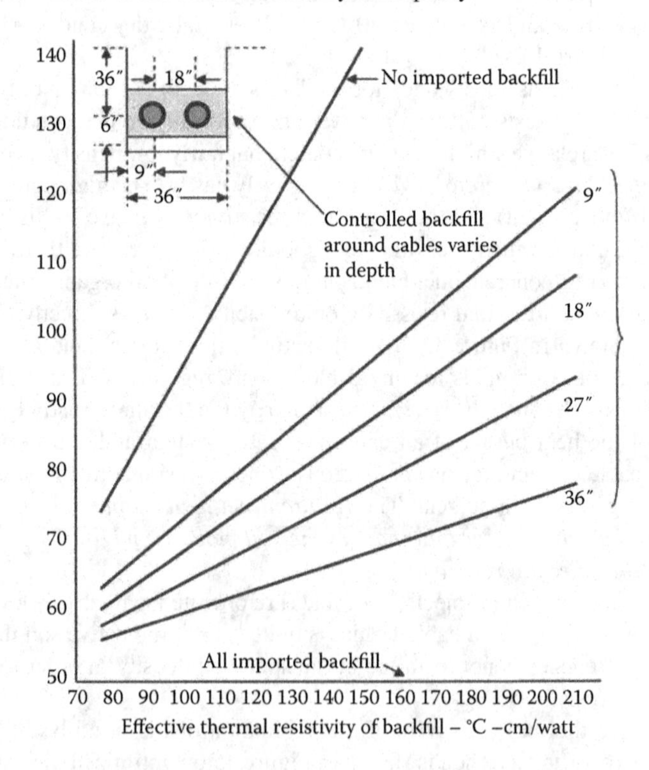

FIGURE 15.6 Effective thermal resistivity.

will produce a lower effective thermal resistivity. Most computer-based ampacity programs are capable of handling the numerous variables and parameters for such situations.

15.5 SEASONAL TEMPERATURE CHANGES AT OPERATING DEPTH

Another important factor that influences the ampacity of a cable is that earth temperatures change with depth. Ground temperature changes on a seasonal basis, following the air temperature but with a much reduced amplitude. The soil temperature varies significantly in the top 3 ft (1 m) and the variation decreases rapidly with the depth as indicated by the typical curves in Figure 15.7. Temperatures are quite constant below 4 m (13 ft) depth. Also there is a time lag between the maximum air temperature and the maximum soil temperature at a given depth. Local data is available from government environment, weather, and agriculture agencies.

15.6 SOIL THERMAL STABILITY

Thermal stability is a concept used to describe the ability of a moist soil to maintain a relatively constant thermal resistivity when subjected to an imposed heat load, thus allowing a power cable not to exceed a safe operating temperature. Thermal instability (or thermal runaway) occurs when a soil is unable to sustain the heat from a cable and the soil progressively dries resulting in a substantial increase in the soil thermal resistivity and attendant increase in the cable operating temperature. If moisture is not replenished or current reduced, the ultimate result may be cable failure due to overheating.

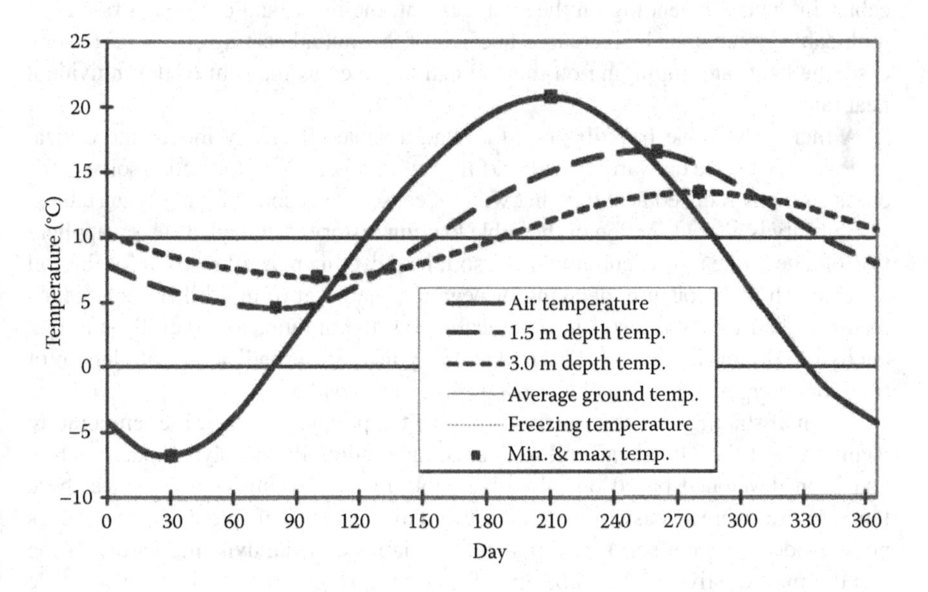

FIGURE 15.7 Seasonal temperature change with depth.

Heat flow through a moist soil takes place predominantly by conduction. At high moisture levels, liquid fills the gaps between soil particles and provides a continuous medium, which makes an efficient thermal conductor (Figure 15.3). When an appreciable heat rate is emitted by a heat source, there is a tendency for the liquid to vaporize near the heat source and to condense in the cooler regions away from it (moisture migration). On the other hand, there is significant liquid return flow by capillary suction, which tends to maintain a balanced liquid moisture distribution throughout the soil (thermal stability—region A in Figure 15.3).

If the amount of liquid drops below the *critical moisture content*, then the imposed heat causes the liquid thermal bridges between the soil particles to break down more rapidly than the capillary suction can replace them. This in turn increases the thermal gradient in the soil, forcing more moisture migration, resulting in a significant increase in the thermal resistivity around the heat source (thermal instability—region B in Figure 15.3).

15.6.1 Heat Rate

The heat transfer rate at the cable–soil interface is the driving force tending to cause moisture migration in the soil. Depending on the soil type and moisture conditions, a minimum heat rate must be exceeded before thermal instability conditions can arise. Larger diameter cables may not promote moisture migration as they dissipate heat over a larger surface area. That is, although a thermal probe, single-conductor cable, or pipe-type cable, may generate the same heat rate (watts per centimeter), the heat flux or watts density (watts per centimeter squared) at the soil interface is less for the larger diameter. The effective diameter is the diameter of an individual cable for directly buried cables and the pipe or conduit diameter for pipe-type cables or cables in ducts. Depending on the arrangement, the individual cable heat rate for a multicable system may be increased because of the mutual heating effect. As a worst case, the heat rate impinging on the soil can be taken as the sum of the individual heat rates.

A thermally stable backfill placed around a cable effectively moves the critical cable–soil interface outward to the backfill–soil interface. Thus, the native soil experiences a greatly reduced heat flux that will not cause it to become thermally unstable.

Stability is also a function of the cable loading history. Although a steady application of a heat rate may eventually dry a soil, load cycling may allow time for the soil to replenish lost moisture, delaying or negating the onset of instability. Emergency loadings tend to dry the soil near the cable, but the duration is generally not long enough to do much damage. Thermal stability testing can indicate a safe length of time for emergency loadings for a given cable–soil system.

Thermal stability is not a function of cable temperature. Rather, the temperature is an effect of the cable heat–soil resistance interaction. Previously, ampacity tables had been developed based on allowable cable interface temperatures, since these temperatures were an easy thing to measure. This concept is incomplete since it does not consider different soil types, moisture variations, or the dynamic nature of the soil thermal resistivity. The ampacity tables may have generally provided safe cable designs and operations only because of a high degree of conservatism.

15.6.2 Soil Type and Density

Certain soils and soil conditions enhance the thermal stability of a cable system. A soil must be an efficient heat conductor and have the ability to resist moisture migration when subjected to a heat load, as well as re-wet quickly if dried. This is best accomplished with a well-graded sand to fine gravel (sound mineral aggregate), with a small percentage of fines (silt and clay), that can be easily compacted to a high density. For maximum density, the smaller grains efficiently fill the spaces between the larger particles, and the fines enhance the moisture retention. A sound mineral aggregate, without organics, and without porous or friable particles, ensures efficient thermal conduction. Porous soils will dry quickly and therefore they are unstable; although a clayey soil dries very slowly, it cannot be easily re-wetted and therefore it is also a poor soil.

15.6.3 Soil Moisture

Above the water table, the natural soil moisture content is not constant but varies in response to climatic conditions. In the operation of a cable, the soil has the greatest potential for induced instability when it is in its driest state. At this point, the added heat load from a cable may create an unstable situation. The stability of a soil must be assessed at its lowest expected ambient moisture content; but this is frequently not an easy parameter to determine.

15.7 SOIL DRYING TIME CONCEPT

Figure 15.8 indicates a typical temperature–log time response of a thermal probe inserted in a moist soil and emitting an appreciable heat rate. The slope of the initial straight line is proportional to the thermal resistivity of the moist, thermally stable

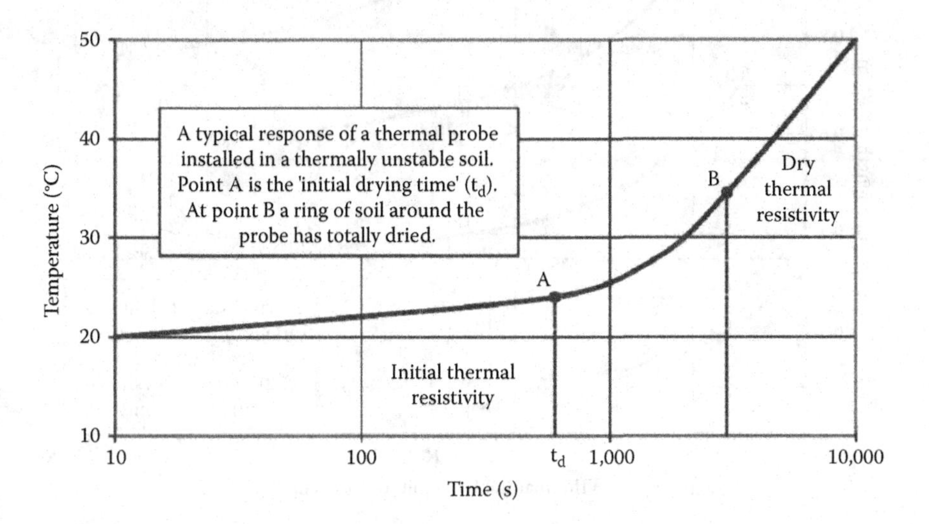

FIGURE 15.8 Thermal stability limit.

soil. When sufficient moisture migration has taken place to cause a substantial drying of the soil (*critical moisture content*), the line begins to curve upward. This point, called the *initial drying time*, is the onset of thermal instability. Further heating will result in rapidly increasing resistivity and the complete drying of the soil around the probe, as indicated by the final straight line portion of the curve, whose slope is proportional to the resistivity of the dry soil.

Analytically the initial drying times for different diameter heat sources dissipating the same constant heat rate (watts per centimeter), in identical soils, are related by the ratio of the diameters squared (the Fourier number). Thus, the initial drying time determined in the laboratory using a small diameter thermal probe can be extrapolated to the initial drying time for a large diameter, operating underground power cable:

$$t_1/t_2 = d_1^2/d_2^2 \quad \text{or} \quad t_1/d_1^2 = t_2/d_2^2$$

The parameter t/d^2, called the *thermal stability limit*, is a constant and depends only on the heat rate and initial moisture content for a given soil. Test results are presented as a plot of heat rate vs. *thermal stability limit* for a range of initial moistures (Figure 15.9).

The initial drying time is a conservative estimate of the time to instability for an installed cable. First, it assumes the constant application of the full heat load, when in fact an operating system is subjected to load cycles. Second, it only indicates when an unstable condition first appears, but a cable would not experience significant heating for some time. Third, it assumes no moisture replenishment, which could occur

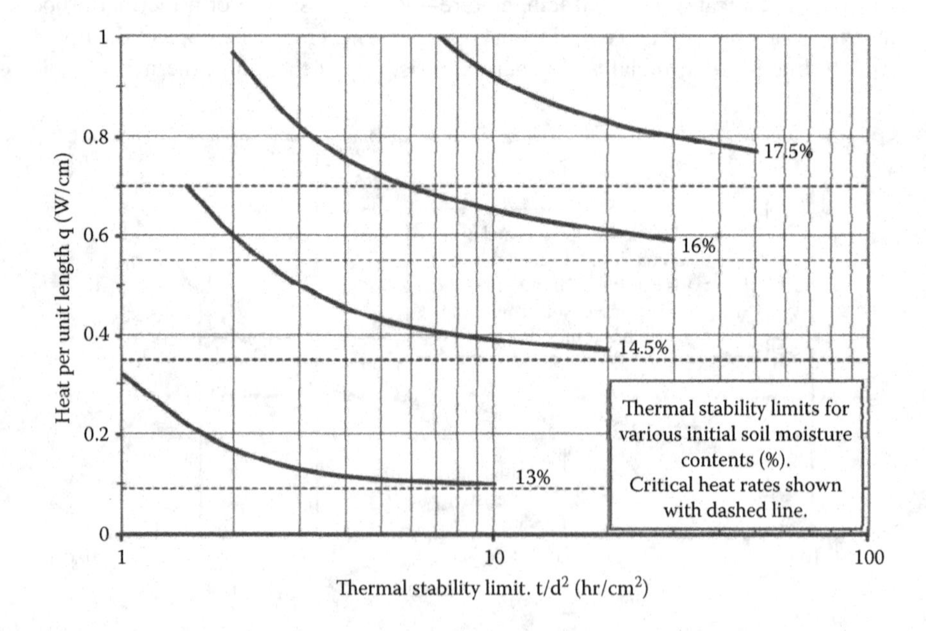

FIGURE 15.9 Thermal stability limits for various initial moisture contents.

in the field. Overall, though, a conservative estimate of the drying time may be a reasonable, safe value to use to calculate the *thermal stability limit*, since the lowest expected ambient soil moisture content is not usually well known and may be less than the initial moisture contents chosen for the tests.

15.8 CRITICAL HEAT RATE

The analytical model also predicts the existence of a limiting heat rate. This is indicated by the asymptotes of the empirical curves in Figure 15.9. Below the *critical heat rate,* a poor thermal soil will remain stable for an essentially indefinite period of time.

Once a drying time has been determined in the laboratory and extrapolated for the cable, a soil can be considered stable if:

1. The drying time exceeds the longest expected drought.
2. The cable loading is below the *critical heat rate* for a substantial time, particularly when the soil is driest.

A search for a local source of an ideally graded, natural soil often proves difficult. Native soils frequently have a rather uniform size and hence dry out very quickly when a heat source is applied. Typical silica sand has 88% of the sand passing through a #40 sieve (0.42 mm or smaller) and with 0.6% silt. Dry densities of only about 100 pounds per cubic foot, mostly made of quartz crystals that have a thermal resistivity of about 12, are attainable with normal construction techniques. The thermal resistivity of these sands in their native state is about 60 to 80 rho as long as they have about 3% by weight of moisture. Typically, a load of about 30 watts per foot from a cable or duct bank is sufficient to dry such sand to less than 1% moisture where the rho can rise to 350–400.

Thermal runaway conditions of native soils have been experienced by many utilities involving transmission cables as well as single-circuit, direct buried distribution feeder cables. In one situation, the sand had less than half percent moisture adjacent to a direct buried feeder cable exiting a substation even though lawn irrigation sprinklers were only 2 ft above. With 4 hours of watering every night, the sand could not retain enough moisture to prevent thermal runaway of a buried three-conductor 500 kcmil copper feeder cable. The pattern of eccentric circles of dry sand was found to be in agreement with the paper of Balaska, Merrell, and McKean [9], which described a simulated transmission cable in a sand hill in New York.

In another example, the native soil around a pipe cable was found to be baked completely dry and adhering to the pipe's outer surface even though the cable was 30 ft under the surface of a bay. The material in this situation had a clay-like composition (marl) with a high amount of organic fillers.

Similar reports have been given regarding high thermal resistivity of soils around cables under waterways in Denmark, England [10], and the deepest water in Lake Champlain [11].

In the situation described in Denmark, it was stated that "Cooling conditions for a submarine cable are normally assumed to be very good and the ampacity is based on a low value of thermal resistivity of the seabed." Since the land section had a thermal resistivity of 43°C-cm/watt to 54°C-cm/watt, it was *assumed* that the value in the seabed was equally as low. After two joints failed in service, it was discovered in laboratory investigations that the seabed material contained high organic levels and that the thermal resistivity was 105°C-cm/watt. Needle probes into the seabed discovered a rho of 94°C-cm/watt.

In the London investigation [10], they found the silt in the bottom of canals to be as high as 118°C-cm/watt and that even higher values could be reached in the presence of heated cables.

The Lake Champlain 115 kV cables [11] were installed in 1958 and failed in 1969 at a depth of about 300 ft. A sample of the soil near the site of failure was sent to a laboratory for analysis. They found the silt to have an average value of rho of 90 to 100 even though the silt "was not tested in the condition that it was in the lake bottom…." The new cable rating was based on the lakebed silt to have a rho of 140°C-cm/watt.

The lesson to be learned here is that moisture can migrate from soils even in deep waterways. To maintain a low thermal resistivity soil in a seabed, water must be free to move through a porous or granular environment and have a limited level of organic material. It should also be pointed out that readings taken with thermal probes along a proposed route may give optimistically low values of thermal resistivity if the heat source is not left on long enough to detect moisture migration in that soil.

15.9 GOOD THERMAL SOILS

The *thermal dryout curve* of a good thermal soil has a sharp knee at low moisture content and thus the *critical moisture* can be clearly defined; also, the totally dry thermal resistivity is quite low. Furthermore, above the *critical moisture* the *thermal dryout curve* is quite flat, that is, the resistivity is fairly constant and relatively independent of the soil moisture (i.e., crushed stone screenings in Figure 15.4). For these soils the thermal stability may be treated as a simple "binary" concept that is, above the *critical moisture content* the soil is stable (fairly constant resistivity), whereas below this moisture it is unstable (rapidly increasing resistivity). In the thermally stable range, an imposed heat rate causes a local redistribution of the moisture but an equilibrium between vapor outflow and liquid return is established leaving the resistivity virtually unchanged because of the flat nature of the *thermal dryout curve*.

In a cable installation, good thermal soil should be placed next to the cable, therefore, under normal heat loads, the simple concept of *critical moisture* can be used as the indication of the thermal stability. For these backfills, above the *critical moisture*, realistic cable heat loads do not exceed the *critical heat rate* as this would result in very long drying times (in the order of months). An exceptional heat load, though, may cause a stable situation to become unstable. In this case the drying time test, as previously discussed, may also be carried out for the good thermal backfill.

Thus, if the lowest expected ambient soil moisture is above the *critical moisture*, then the good thermal soil may be considered to be stable, and the lowest expected ambient soil moisture is used to determine the design thermal resistivity from the *thermal dryout curve*. If the lowest expected moisture content is below the *critical moisture*, then an unstable condition exists and the totally dry thermal resistivity must be used for the design. For a good thermal soil, though, the dry resistivity is still quite reasonable for cable operation.

15.10 POOR THERMAL SOILS

For clayey soils, the thermal resistivity increases fairly steadily as the moisture decreases without a definite *critical moisture* (Figure 15.4). For these fine grained soils, moisture movement is a slow, continuous process because of the extremely low permeability of the liquid phase. Therefore, a liquid–vapor equilibrium is not attained in a short-term thermal loading, and for these poor thermal soils the question is one of the time required for a substantial increase in thermal resistivity for a given heat rate.

Even though a uniform sandy soil (Ottawa sand in Figure 15.4) has a fairly constant and low thermal resistivity above the *critical moisture*, it is a poor thermal soil because of the high dry resistivity. For these porous soils, moisture movement and drying take place very easily, and large moisture redistribution above the *critical moisture* is likely.

The drying time concept must be used if one wishes to analyze the thermal stability of poor thermal soils.

15.11 USE OF CORRECTIVE THERMAL BACKFILLS

Invariably, native soils do not make good thermal backfills because their thermal performance is poor, or they are difficult to backfill properly. In its natural state a native soil may exhibit the *thermal dryout curve* of a good thermal soil. But, if the native soil does not have the correct texture or moisture content, once it is excavated it is difficult to handle and re-compact efficiently, and a poor *thermal dryout curve* may result. Also contamination with topsoil and other organic soils is difficult to avoid. Importing a corrective thermal backfill, with its assured excellent thermal performance, may not be much more expensive than attempting to use a marginal native soil. In the long run, the operational reliability gained by placing a classified thermal backfill around the cable has advantages over the possible variability and uncertainty of re-compacted native soil.

The common practice has been to use granular bedding material around a cable. In the past, fairly uniform sands that may compact well but have a poor thermal performance have been used. Only a good thermal backfill is to be placed around the cable (Section 15.10). The thickness of the backfill envelope should be such that the heat flux impinging on the native soil is so low as to be inconsequential for instability considerations. Now only the thermal stability of the backfill must be considered, which can be established by applying the binary concept.

15.12 CORRECTIVE THERMAL BACKFILLS

The use of a well-designed corrective thermal backfill can significantly enhance the heat dissipation and increase the ampacity of an underground power cable, as well as alleviate thermal instability concerns. The corrective backfill will reduce the heat flux experienced by the native soil so that it will not dry out. A good backfill should be better able to resist total drying and also have a low dry resistivity if it is completely dried. Typically, a trench is filled with thermal backfill to a minimum height of 300 mm above the cable. If poor quality native soils are encountered along a route, the thickness of the backfill can be increased to maintain a low composite resistivity.

Generally, utilities have relied upon the suppliers of such backfill materials to meet acceptable thermal and mechanical performance characteristics. Very few utilities have set stringent specifications or have a quality assurance program for installed backfills, especially with respect to thermal performance. The extra cost involved in the use of a properly designed and installed thermal backfill is somewhat higher than that involved in the use of unclassified material (i.e., native soils, bedding sands), yet cable rating increases are significant and can amount to substantial revenue gains for a utility over the life of a cable. It is usually not cost-effective to modify the native soil by using additives to improve its thermal properties.

Placing corrective thermal backfill in a cable trench will reduce the effects of high native soil resistivities. A constant composite resistivity can be maintained along the whole cable route by adjusting the thickness of Fluidized Thermal Backfills™ (FTB) to balance the variations in the native soil resistivity. Other factors, such as trench dimensions, cable spacing, burial depth, ambient conditions, etc., should also be taken into account for optimum design. This can easily be done using computerized cable ampacity programs.

It is important to note that the adverse effects of poorly constituted or installed thermal backfills are not reflected in the performance of the cable system in the early stages when the loads may be low. However, temperatures may rise beyond allowable levels as the loadings increase. The remedial cost of removal and replacement of such backfills is very high, especially if the area is paved. On the other hand, the loss of revenues from de-rating a system may be even higher.

Generally, native soils do not make good thermal backfills because their thermal performance is poor, or they are difficult to properly re-compact in a cable trench. In its natural state, a native soil may exhibit the *thermal dryout curve* of a good thermal soil. But, if the soil does not have the correct texture or moisture content and once it is excavated, it is difficult to handle and re-compact efficiently (especially clayey soils), and a poor *thermal dryout curve* may result. Also, contamination with topsoil and other organic soils is difficult to avoid. Importing a corrective thermal backfill, with its assured excellent thermal performance, may not be much more expensive than attempting to use a marginal native soil. In the long run, the operational reliability gained by placing a classified thermal backfill around the cable has advantages over the possible variability and uncertainty of re-compacted native soil.

15.12.1 Compacted Granular Backfills

Over the years, many unsuitable sands have been used as thermal backfills (so called "thermal sands") because of the ease of installation. Almost any sand, when moist, will give a reasonably low resistivity. The crucial aspect is its thermal performance when subjected to cable heat loads and when partially or totally dried. Uniform bedding sands dry easily and can have dry thermal resistivities exceeding 250°C-cm/W.

Since most of the heat conduction is through the soil mineral particles and their contacts, one must ensure a soil mixture that maximizes these contacts, that is, a high density soil. Well-graded soils are denser because smaller grains can efficiently fill the spaces between the larger particles. A good thermal backfill is specified by a narrow range of gradings based on a sieve analysis, and should have a wet thermal resistivity of 35°C-cm/W to 50°C-cm/W and a dry resistivity of 90°C-cm/W to 120°C-cm/W.

For a given soil, the major influence on the resistivity is the moisture content. A soil that is better able to retain its moisture, as well as able to efficiently re-wet when dried, will have a better thermal performance. A small amount of very fine particles (i.e., silt and clay size) or cement enhances this property.

It is easy to conclude that, in general, a well-graded sand to fine gravel (sound mineral aggregate), free of organics, and with a small percentage of fines, would make an ideal thermal backfill when compacted to its maximum density. Some natural sands may be suitable, although they must often be blended, which can be costly. Most crushed stone screenings meet these gradation characteristics, but the sharpness of the stone chips may cause concerns about damage to cable insulation. For a specific backfill, the laboratory determination of an optimum standard *Proctor* density and moisture and *thermal dryout curve* are essential to confirm its thermal performance.

The addition of a binder can improve the thermal performance of a poorly graded sand. Thorough mixing and proper compaction are crucial to the performance. A small quantity of cement (~2%) enhances the resistivity and stability of granular backfills. The addition of chemical, wax, or latex binders has also been suggested. Although the thermal properties are improved, field application is not practical. They are expensive and difficult to handle and transport, as well as being environmentally incompatible. Basically, cement is the only practical binder that should be considered.

One often neglected factor about compacted backfills is the need for quality assurance during installation. If the gradation of the backfill is not correct, or it is not at the optimum moisture content, or not enough compaction effort is applied, or the backfill lifts are too thick, then the maximum density will not be achieved and the thermal performance is degraded. This is especially significant in the restricted areas around cable groups where proper compaction is very difficult. Yet it is precisely in these zones, adjacent to the cables, that proper compaction is most important to ensure maximum heat dissipation from the cables.

A classified granular soil (i.e., specified grading and mineral quality) is more expensive than unclassified pit-run material. The transportation cost, though, is the prime factor in the site delivered cost and therefore suppliers should be as close as possible to the project site. The installation and quality assurance costs must be added to the material and transportation costs to arrive at the final cost. When compared on a final cost basis, it may be less expensive to use an FTB.

15.12.2 Field Installation

Corrective backfills must meet several criteria, including thermal efficiency and engineering performance compatible with road base material (mechanical support, no settlement, resistance to erosion, and frost heave). The backfill is specified by a gradation of grain sizes and a dry density. Proper installation of a granular backfill is crucial in meeting the above criteria.

A granular backfill reaches a maximum density at a specific *optimum moisture content*; at higher or lower moistures a lower density is attained. This behavior is indicated by the standard *Proctor* laboratory test (Figure 15.10). Installation specifications for a particular material usually require that 95% of the maximum *Proctor* dry density be attained. In order to meet the required density, the contractor must use suitable compaction equipment, ensure optimum moisture, compact in thin lifts, and use several passes over the same area.

Compacting in thick lifts may be faster but it leaves low density fill material beneath a dense crust. This is why rigorous inspection is required. For smaller, trench-type compactors the lifts should not exceed 150 to 200 mm (6–8 inches), especially close to the cables where more care and effort are needed to ensure maximum density. Thicker lifts may be permitted as the backfill reaches the ground surface (Figure 15.10).

Usually, stock-piled backfill is not at the optimum moisture content. Generally, it is too dry and moisture must be mixed in or sprayed on each lift before compacting. For granular, noncohesive soils the optimum moisture is about 8% to 12% and the soil appears quite wet.

The commonly used trench compactors are plate vibrators, vibrating rollers, and dynamic impactors (jumping jacks). The vibratory types are better for granular soils while the impact types are more effective with clayey soils. The smaller hand operated compactors are preferred because they are easy to maneuver in narrow areas.

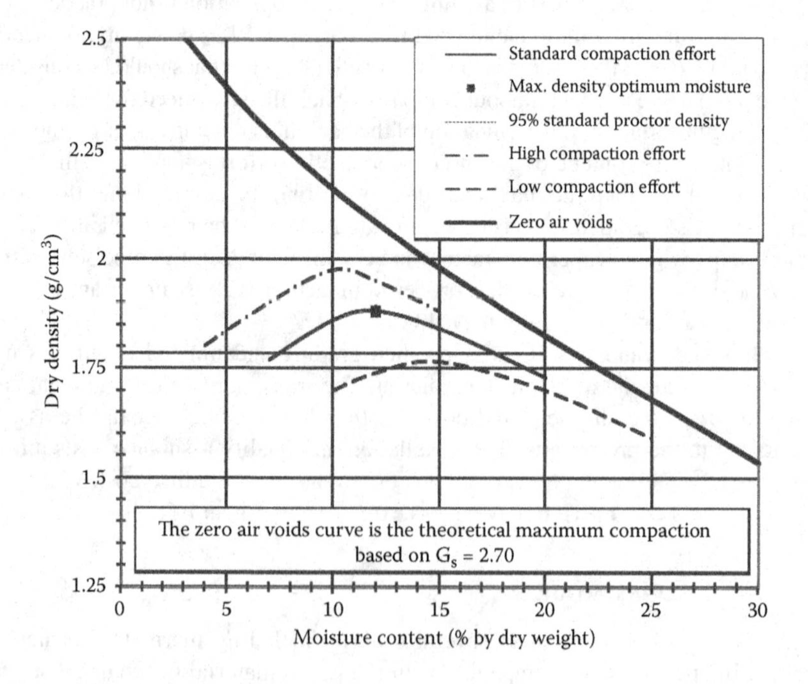

FIGURE 15.10 Soil compaction.

Flooding and hydraulic filling will not yield the same result as backfill may segregate. Ramming or drop weights are not acceptable because of poor control and the possibility of damage to cables.

15.12.3 Quality Control

The corrective backfill may be sourced from several suppliers. From each supplier, a specific material is chosen based on mineral quality, sieve analysis, standard *Proctor* density, and *thermal dryout curve*. For field installation, a minimum density (or maximum thermal resistivity) is specified for the specific backfill. A field quality assurance program is a must; otherwise, there is no way of ascertaining that the performance specifications are met.

An experienced on-site inspector should be able to visually check the consistency and quality of the material supplied. Occasional sieve analyses (two or three times) will confirm the grading.

The inspector should monitor the backfill moisture, lift thickness, and thoroughness of compaction. Regular in-situ moisture and density tests should be performed (say, every 20 m along the trench), according to ASTM procedures, by an independent firm. Troxler nuclear density devices are commonly used. The level of backscatter from a nuclear source can be related to density. Only certified personnel may use these devices. A sand cone method is still used. Standard sand is poured from a calibrated cone into a small hole dug in the backfill. The weight of the sand used can be related to a volume. The weight of the backfill removed divided by the volume is the density. If measured densities are significantly deficient, the backfill must be removed and re-compacted.

The thermal resistivity of the installed backfill may be tested periodically (say, every 100 m). If the backfill is compacted properly, then there should be no problem with the thermal properties. Preferably, an undisturbed *Shelby* tube sample should be obtained, which can be dried out in the laboratory to check the dry resistivity.

15.12.4 Fluidized Thermal Backfills (FTB)

FTB is a material engineered to meet specific thermal resistivity, thermal stability, strength, and flow criteria, as well as to offer construction advantages. Such a free flowing, controlled density fill is ideally suited for hard-to-access areas such as narrow trenches, small diameter tunnels, or areas congested with many underground services—basically where mechanical compaction is neither feasible nor practical. In addition, although a compacted backfill may provide a good thermal envelope, it is a labor-intensive process that requires strict quality control. While the material cost of FTB may be high, it should even be considered for general usage because of its assured quality and performance standards and because it can be installed very quickly, thereby speeding up construction and decreasing overall costs.

Commonly available "controlled density fills," "flowable fills," or "slurry backfills," which use large volumes of fly ash or sand, may meet the mechanical and flow requirements for trench backfilling, but they are completely unsuitable with respect to thermal performance. FTB should be designed and formulated by soil thermal specialists.

FTB is a slurry backfill consisting of a medium stone aggregate, sand, a small amount of cement for strength and particle bonding, water, and a fluidizing agent such as fly ash to impart a homogeneous fluid consistency. The component proportions are chosen, by laboratory testing of trial mixes, to minimize thermal resistivity and maximize flow without segregation of the components.

FTB will flow readily to fill all the voids, without vibration, yet harden quickly to a uniform density. Future settlements, if any, are negligible. It also affords mechanical protection for the cables and provides support for underground and surface facilities. FTB has good heat dissipation properties even when totally dry. Depending on the mix design, typical thermal resistivity values are 35°C-cm/W to 45°C-cm/W wet and 65°C-cm/W to 100°C-cm/W dry and thermal stability is excellent with a low *critical moisture* (less than 3%).

FTB is supplied as a ready-mix in concrete trucks and may be installed by pouring or pumping and usually does not require specific shoring or bulkheading. It solidifies by consolidation, with excess water seeping to the top. Pipes or ducts must be anchored or weighted down because of the buoyancy effect of the slurry. Regular FTB can be pumped up to 100 m, and greater distances with special modifications, using conventional concrete pumping equipment. Direct buried cables will not be damaged during installation, which may be a concern when using mechanical compaction devices. FTB solidifies quickly so that the ground surface may be reinstated the next day; but the low strength ~150 psi (1 MPa) affords easy "diggability" if the cables must be accessed in the future. If a higher strength is required, the cement content can be increased, and water adjusted, without degrading the thermal performance.

Quality control, during installation, is a matter of a few simple tests, such as random grain size analyses of the aggregates at the batch plant; air content (less than 2%) and slump (200 to 250 mm) measurements on the FTB before pouring. If the mix is too wet it must be rejected, therefore it may be ordered drier than required and water added at the site. Sample cylinders should be cast for strength, density, and thermal resistivity determinations in the laboratory when the FTB has hardened. For low strength FTB, two test cylinders for strength and two for resistivity should be cast for every 50 m of trench length. (Cylinders for high strength FTB tests should be cast as specified by ASTM for normal concrete.)

15.13 CONCRETE AS BACKFILL

Normal strength ~3,000 psi (20 MPa) concrete (100 mm slump, nonair entrained) also has good thermal performance characteristics (less than 100°C-cm/W, totally dry). The addition of air, fly ash, or porous aggregate will increase the thermal resistivity. For general applications, though, it does not flow readily, is more expensive than FTB, and cannot be removed easily if access to the cables is required. It may be used in duct banks or where high strength is a structural requirement [1,7].

Lean mix concrete (i.e., less cement, about 0.5 MPa), with a slump of about 100 mm, does not have the superior flow characteristics of FTB, and has a higher thermal resistivity because the crushed stone and sand aggregate combination is not ideal. For installation around cables, it requires either vibration or light compaction. Adding water to enhance the flow will increase the voids content, thus raising the thermal resistivity, and in the worst case, may lead to segregation of the mix components. Chemical additives, to increase the flow, are only effective at higher cement contents. Additives such as fly ash or bentonite when used in small quantities in lean mix concretes will substantially increase the flow characteristics.

15.13.1 Concrete for Cable Backfill

Cement bound sand (weak concrete) has been used as a backfill material around direct buried cables in Europe for many years. A typical material in use with a 12:1 sand/cement mix. Although this provides for some cable movement and relative ease for removal, these mixes do attain a rather high thermal resistivity when a load is applied for many months. A resistivity of about 105°C-cm/watt is typical. Lower values of rho may be attained by decreasing the ratio of sand to cement; in other words, by making the concrete more structurally sound.

Greebler and Barnett [12] reported that concrete around a laboratory installed duct bank had a rho of 85. This paper was the source of the 85 rho that may be found in the original "black books" of AIEE-IPCEA Ampacity Tables [6]. This concrete was poured above the floor in a wooden trough and was in an air conditioned building. This resulted in a poor structural cure and poor thermal resistivity. Brookes and Starr [13] found that the thermal resistivity of concrete around a buried pipe in New Jersey stayed at about 50 rho until the test was stopped after 24 months.

Nagley and Neese [14] found that the thermal resistivity of concrete around duct banks in the Chicago area varied from 38 to 53. An interesting discovery of their study was that a thin envelope of concrete around heavily loaded duct banks (about 1.25 inches rather than the more frequently used 2.5 inches) allowed the soil around the duct bank to dry out. The effective thermal resistivity of the surrounding earth soon reached unacceptably high levels.

The publication by the US Bureau of Reclamation in 1940, "Thermal Properties of Concrete," [7] showed that the thermal resistivity of concrete used in the Boulder (now Hoover) Dam varied from 28.5 to 42.5.

15.13.2 Florida Test Installation

To determine the actual resistivity of locally available structural concrete, a test facility was constructed in Florida. Two 8 1/8 inch steel pipes, covered with 0.5 inches of Somastic, were placed one above the other in a trench with 12 inches of separation. This configuration was chosen since it would permit evaluation of the effects of burial depth as well as provide more information on a configuration that might be required in a restricted area.

Thermocouples were installed on the pipe in each section, on the Somastic, 3 and 6 inches from the Somastic, and under the earth 6 inches from the outer surface of the concrete envelope. Figure 15.11 shows the test arrangement.

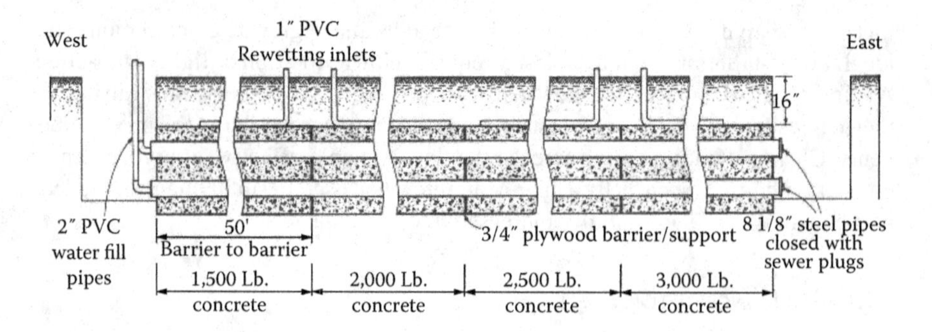

FIGURE 15.11 Arrangement of test envelope.

Four 50-foot sections were poured with various grades of commercially available structural grade concrete in a 6-inch envelope as shown in Figure 15.6. The compressive strengths of these four mixes were 1,500, 2,000, 2,500, and 3,000 pounds/in^2. All sections were allowed to cure for 40 days prior to any application of heat to simulate normal cure in the earth.

15.13.3 Test Results

The heat was increased from 15 to a high of 31 watts per foot in each pipe and continuing that load until the temperatures stabilized. Values of thermal resistivity were calculated for each pipe and its surrounding material separately by using Bauer's version [15] of the Kennelly formula [2].

The thermal resistivity of the concrete varied throughout the test period, but the general trend was a slight decrease in resistivity for the three higher strength mixes. The section that had the 1,500 pound/in^2 concrete was eliminated from the test after 3 months because the heater wires had burned out and the rho was increasing in that section (Table 15.2).

The 31 watts per foot level was chosen as the maximum because this resulted in an interface temperature that was considered at the time to be the maximum permissible value of 50°C. See Section 4.3 and the Minutes of the Insulated Conductors Committee of November 1984 for a six paper review of this thermal resistivity and interface issue and the effects of large heat levels (watts per foot of flux density) [16].

TABLE 15.2
Thermal Resistivity of Concrete Envelope

Concrete Strength	Top Pipe °C-cm/watt	Bottom Pipe °C-cm/watt	Average °C-cm/watt
3,000 pounds/in^2	37.9	44.4	41.2
2,500 pounds/in^2	41.1	40.4	40.8
2,000 pounds/ in^2	36.2	38.7	37.5
1,500 pounds/in^2	48.9	51.2	50.1

15.14 EARTH INTERFACE TEMPERATURE

Over the past 50 years, one of the dominant questions for thermal stability of the surrounding soil has been the earth interface temperature. The original Ampacity Tables were printed with both 50°C and 60°C columns so that the user could decide which value to use. The theory is that many soils will attain a balance between heat flow, moisture migration, and capillary action that will allow moisture to flow back into the soil near the heat source. Caution is advised on using an interface temperature in this manner without tests to confirm the stability of the soil involved.

Heat flux is considered to be one reason why moisture migrates away from a heat source. Heat flux is a function of the heat flowing out from the source (in watts/foot) and the cable surface diameter (in inches). The method usually used to determine stability is to place a thermal probe using the maximum heat anticipated from the circuit (e.g., 30 watts/foot) and measure the thermal resistance of the soil as a function; the time at which the slope changes give a good estimate of the time to dry out as well as the thermal resistivity of the dry soil.

$$t_c = t_p\left(D_c/D_p\right)^2 \tag{15.1}$$

where

t_c = time for soil near cable to dry, in minutes
t_p = time for soil near probe to dry, in minutes
D_c = diameter of cable or earth interface, in inches
D_p = diameter of probe, in inches

At the time of the described test, a maximum interface temperature of 50°C was chosen to be on the safe side. No adverse effects were noted on the surrounding soil for the loading simulated in the test.

15.15 ELEMENTS OF A ROUTE THERMAL SURVEY

15.15.1 Review and Planning

Review all available soil data along the proposed route (from government agencies, other utilities, borehole logs for existing tower lines, etc.) to select test locations (about 2 per kilometer; more frequent if soil conditions are variable).

Select a geotechnical company familiar with thermal testing, or a soil thermal specialist who can work with a local soil boring company. Once the borehole locations are identified and marked, underground services will have to be located and cleared, and appropriate permits obtained. (If additional geotechnical boreholes are being done, these can be used to define boundaries for soil types. This approach can reduce the number of thermal tests while still adequately characterizing the route.)

15.15.2 Field Testing

The field investigation, using a soil drill rig, can proceed in the standard geotechnical fashion, including continuous *Standard Penetration Tests* (SPT) and soil descriptions (USCS) and relative densities; only, at the appropriate depths, in-situ thermal testing and *Shelby* tube sampling are carried out by the soil thermal specialist. Keep a complete field log, also noting the depth of the groundwater table.

At each test location, obtain field soil descriptions, in-situ measurements of native soil thermal resistivities, and ambient temperatures (using the *THERMAL PROPERTY ANALYZER*™) at several levels up to the cable burial depth (say, the cable is to be buried at 1.5 m, then test at 0.5, 1.0 and 1.5 m).

Retrieve undisturbed, minimum 75 mm diameter thinwall, *Shelby* tube soil samples (say, from a depth of 1.2 to 1.6 m for the above example). Failing this, a bulk soil sample should be obtained from the augers or SPT and sealed to prevent moisture loss.

15.15.3 Laboratory Testing

In the laboratory, test the soil samples to determine: soil descriptions, moisture contents, dry densities, organic contents, *thermal dryout curves,* and *critical moistures* (optional: sieve gradations, *Atterberg* limits, standard *Proctor* density curves).

15.15.4 Analysis

Compare field and laboratory test results and evaluate any inconsistencies. Assuming that the same soil was sampled and field tested, problems are usually due to differences in density caused by sampling disturbance. Correlate the laboratory *thermal dryout curves* to the field results. If discrepancies cannot be resolved, then further sampling may be required.

Determine lowest expected ambient soil moistures and choose a design resistivity for the native soils from the *thermal dryout curves.*

15.15.5 Corrective Thermal Backfill and Thermal Stability

Source and test locally available granular materials suitable for thermal backfill (sieve gradation, standard *Proctor* density and optimum moisture, *thermal dryout curve,* and *critical moisture*), or develop a suitable FTB.

Choose a design resistivity for the backfill based on the lowest expected soil moisture, and choose a large enough thermal backfill envelope for the proposed cable so that the heat load on the native soil is inconsequential (use a computerized cable ampacity program). Also, using a cable ampacity program, the cable size and thermal backfill envelope can be optimized to minimize cost.

Examine the thermal stability of the proposed design. If the lowest expected soil moisture is safely above (say 25% greater than) the *critical moisture* of the backfill, then the system will be stable for normal heat loads and the design resistivities of the backfill, and native soil may be the resistivities associated with the lowest expected moisture level (add a safety factor).

If the lowest expected soil moisture is less than the *critical moisture* of the backfill, then the system will be unstable and the dry resistivity must be used as the design resistivity for the backfill. The design resistivity of the native soil can be the resistivity associated with the lowest expected soil moisture (add a safety factor) if the backfill envelope is large enough to allow only a marginal heat load on the native soil. For substantial heat loads, the drying time of the native soil must be investigated.

If the lowest expected soil moisture is above but near the *critical moisture* of the backfill, then the backfill may be considered unstable as a safe estimate; unless a drying time test indicates otherwise.

15.15.6 Quality Assurance

Since a thermal route survey only samples the soil in a few locations, during trench excavation, one may wish to make frequent on-the-spot thermal resistivity measurements of the native soil. If exceptionally poor soil conditions (potential "hot spots") are encountered, then the thickness of the thermal backfill envelope can be increased to maintain a uniformly low composite resistivity for the entire route.

Since the corrective thermal backfill is the most important component for dissipating the heat from the cable, provide specifications for composition and installation and make sure that rigid quality assurance is carried out during installation.

15.15.7 General

Numerous documents are available that provide useful information on this subject such as [8,17–29].

REFERENCES

1. Thue, W. A., 18–23 July 1971, "Thermal Resistivity of Concrete," IEEE CP 562 – WR, Summer Power Meeting, Portland, OR.
2. Kennelly, A. E., 1893, "On the Carrying Capacity of Electrical Cables…," *Minutes,* Ninth Annual Meeting, The Association of Edison Illuminating Companies, New York, NY.
3. Sinclair, W. A., et al, "Soil Thermal Characteristics in Relation to Underground Power Cables, Part IV," IEEE PA&S, Vol. 79, Part III, pp. 820–832.
4. Adams, J. I. and Baljet, A. F., "The Thermal Behavior of Cable Backfill Materials," IEEE 31 TP 67-9.
5. Black, W. Z. and Martin, Jr., M. A., 1–8 February 1981, "Practical Aspects of Applying Thermal Stability Measurements to the Rating of Underground Power Cables," IEEE Paper No. 81 WM 050-4, Atlanta, GA.
6. *Power Cable Ampacities*, AIEE Pub. No. S-135-1 and IPCEA Pub. No. P-46-426, 1962.
7. "Thermal Properties of Concrete," Bureau of Reclamation, Boulder Canyon Project, 1940.
8. *IEEE Standard Power Cable Ampacities,* IEEE 835-1994.
9. Balaska, T. A., McKean, A. L. and Merrell, E. J., 19–24 June 1960, "Long Time Heat Runs on Underground Cables in a Sand Hill," AIEE Paper No. 60-809, Summer General Meeting.
10. "Problems Connected with the Thermal Characteristics of Soils," Appendix XIII-c, Minutes of the Insulated Conductors Committee, 20–21 April 1964.

11. Cronin, L. D. and Tulloch, D. F., 15–17 November 1971, "Lake Champlain Submarine Cables," Appendix F-3, Minutes of the Insulated Conductors Committee.
12. Greebler, P. and Barnett, G. F., 1950, "Heat Transfer Study on Power Cable Duct and Duct Assemblies," AIEE Transactions, Vol. 69, Part I, pp. 857–867.
13. Brookes, A. S. and Starrs, T. E., October 1957, "Thermal and Mechanical Problems on 138 kV Pipe Cable in New Jersey," AIEE Transactions, PA&S, pp. 773–784.
14. Nagley, D. C. and Nease, R. J., September 1967, "Thermal Characteristics of Two Types of Concrete Conduit Installations," IEEE T-PAS 67, pp. 1117–1123.
15. Bauer, C. A. and Nease, R. J., February 1958, "A Study of the Superposition of Heat Fields and the Kennelly Formula as Applied to Underground Cable Systems," AIEE Transactions, Vol. 76, pp. 1330–1337.
16. Insulated Conductors Committee Minutes, Appendices F-3, F-4, F-5, F-6, F-7, and F-8, St. Petersburg, FL, November 1984.
17. Black, W. Z., Hartley, J. G., Bush, R. A. and Martin, M. A., 1982 and 1987, "Thermal Stability of Soils Adjacent to Underground Transmission Power Cables," EPRI EL-2595 & EL-5090, RP 7883.
18. Blackwell, J. H., 1954, "A Transient Flow Method for Determination of Thermal Constants of Insulating Materials in Bulk," *Journal of Applied Physics*, Vol. 25 (No. 2), pp. 137–144.
19. Boggs, S. A., Radhakrishna, H. S., Chu, F. Y., Ford, G. L., Griffin, J. D. and Steinmanis, J. E., 1981, "Soil Thermal Resistivity and Thermal Stability Measuring Instrument," EPRI EL-2128, RP 7861.
20. Carslaw, H. S. and Jaegar, J. C., 1959, "Conduction of Heat in Solids," Second Edition, Oxford University Press, London.
21. Farouki, O. T., 1981, "Thermal Properties of Soils," U. S. Army Corps of Eng., CRREL monograph 82-1.
22. Ford, G. L. and Steinmanis, J. E., 1981, "The Importance of Weather Dependent Processes on Underground Cable Design," *Proceedings of Symposium on Underground Cable Thermal Backfill*, pp. 157–166, Pergamon Press, Toronto.
23. Neher, J. H. and McGrath, M. H., 1957, "The Calculation of the Temperature Rise and Load Capability of a Cable System," AIEE Trans, Part III, Vol. 76, pp. 752–772.
24. Radhakrishna, H., 1981, "Fluidized Cable Thermal Backfill," Proceedings of Symposium on Underground Cable Thermal Backfill, pp. 34–53, Pergamon Press, Toronto.
25. Salomone, L. A., Kovacs, W. D. and Wechsler, H., 1982, "Thermal Behavior of Fine-Grained Soils," National Bureau of Standards, NBS BSS 149.
26. Shannon, W. L. and Wells, W. A., 1947, "Tests for Thermal Diffusivity of Granular Materials," *Proceedings of ASTM*, Vol. 47, pp. 1044–1055.
27. Steinmanis, J. E., 1981, "Thermal Property Measurements Using a Thermal Probe," *Proceedings of Symposium on Underground Cable Thermal Backfill*, pp. 72–85, Pergamon Press, Toronto.
28. "Guide For Thermal Stability Measurements," ICC subcommittee 12-44, unpublished, 1996.
29. Malten, K. C., Kirby, M. J. and Williams, J. A., 1995, "Guidelines for the Design and Installation of Transmission and Distribution Cables Using Guided Drilling Systems," EPRI TR-105850, RP 7925-01.

16 Sheath Bonding and Grounding

William A. Thue

CONTENTS

16.1 INTRODUCTION

This discussion provides an overview of the reasons and methods for reducing sheath losses in large cables. While calculations are shown, *all* of the details are not covered as completely as in the Institute of Electrical and Electronics Engineers (IEEE) Guide 575 [1]. A complete set of references is included in that standard. The reader is urged to obtain a copy of the latest revision of that document before designing a "single-point" grounding scheme. Another excellent reference is the 1957 *Underground Systems Reference Book* [2].

The terms sheath and shield will be used interchangeably since they have the same function, problems, and solutions for the purpose of this chapter.

- Sheath refers to a water impervious, tubular metallic component of a cable that is applied over the insulation. Examples are a lead sheath and a corrugated copper or aluminum sheath. A semiconducting layer may be used under the metal to form a very smooth interface.
- Shield refers to the conducting component of a cable that must be grounded to confine the dielectric field to the inside of the cable. Shields are generally composed of a metallic portion and a conducting (or semiconducting) extruded layer. The metallic portion can be tape, wires, or a tube.

The cable systems that should be considered for single-point grounding are systems with cables of 1,000 kcmils and larger and with anticipated loads of over 500 amperes. Fifty or more years ago, those cables were the paper-insulated transmission circuits that always had lead sheaths. Technical papers of that era had titles such as "Reduction of Sheath Losses in Single-Conductor Cables" [3] and "Sheath Bonding Transformers" [4]; hence, the term "sheath" is the preferred word rather than "shield" for this discussion.

16.2 CABLE IS A TRANSFORMER

Chapter 4 described how a cable is a capacitor. That is true. Now you must think about the fact that a cable may also be a transformer.

When alternating current flows in the "central" conductor of a cable, that current produces electromagnetic flux in the metallic shield when present, or in any parallel conductor. This becomes a "one-turn" transformer when the shield is grounded two or more times since a circuit is formed and current flows.

We will first consider a single, shielded cable:

- If the shield is only grounded one time and a circuit is not completed, the magnetic flux produces a voltage in the shield. The amount of voltage is proportional to the current in the conductor and increases as the distance from the ground increases (Figure 16.1).

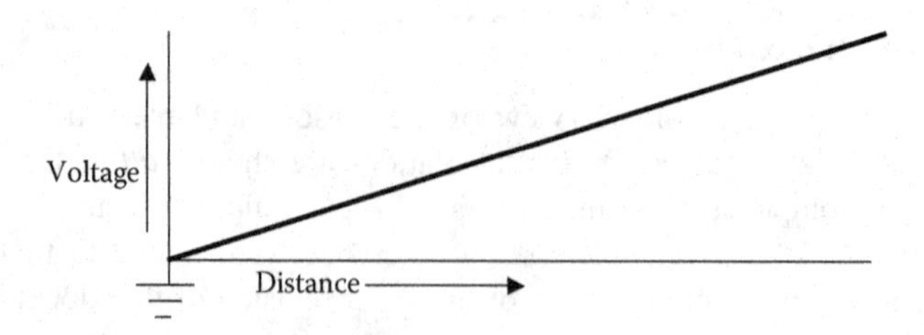

FIGURE 16.1 Single-point grounding.

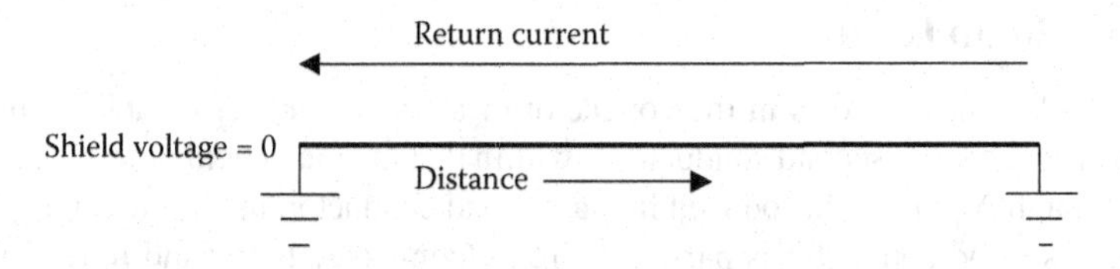

FIGURE 16.2 Two or more grounds.

- If the shield is grounded two or more times or otherwise completes a circuit, the magnetic flux produces a current flow in the shield. The amount of current in the shield is inversely proportional to the resistance of the shield. In other words, the current in the shield increases as the amount of metal in the shield increases. The voltage stays at zero (see Figure 16.2).

The voltage remains at zero, but the same current flows regardless of the distance between the grounds.

An important concept regarding multiple grounds is that the distance between the grounds has no effect on the magnitude of the current. If the grounds are 1 foot apart or 1,000 feet apart, the current is the same depending on the current in the central conductor and the impedance of the shield. In the case of multiple cables, the spatial relationship of the cables is also a factor.

One method of single-point grounding of a run of cable is to ground in the center of the run—leaving both terminals having energized shields. This reduces the standing voltage by half as compared with grounding one end of that circuit (Figure 16.3) [12].

16.3 AMPACITY

In Chapter 14, there is a complete description of ampacity and the many sources of heat in a cable such as conductor, insulation, shields, etc. See Chapter 15 to know how this heat must be carried through conduits, air, concrete, surrounding soil, and finally to ambient earth. If the heat generation in any segment is decreased, such as in the sheath, then the entire cable will have a greater ability to carry useful current. [5–7] (Note: The heat flow is always up toward the earth's surface.)

The heat source from the shield system is the one that we will concentrate on, in this discussions, as we try to reduce or eliminate it [13].

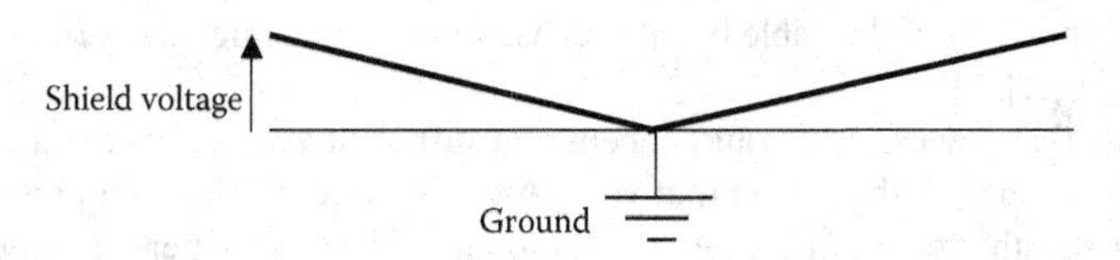

Note: Only half the voltage appears on the shield since the distance is halved

FIGURE 16.3 Center ground only.

16.3.1 Shield Losses

When an AC current flows in the conductor of a single-conductor cable, a magnetic field is produced. If a second conductor is within that magnetic field, a voltage that varies with the field will be introduced in that second conductor, in our case, the sheath.

If that second conductor is part of a circuit (connected to ground in two or more places), the induced voltage will cause a current to flow. That current generates losses that appear as heat. The heat must be dissipated the same way as the other losses. Only so much heat can be dissipated for a given set of conditions, so that these shield losses reduce the amount of heat that can be assigned to the phase conductor.

Let us assume that we are going to ground the shield at least two times in a run of cable. What is the effect of the amount of metal in the shield?

The following curves present an interesting picture of the shield losses for varying amounts of metal in the shield. These curves are taken from ICEA document P 53-426 [7]. As you can see, they were concerned about underground residential distribution (URD) cables where the ratio of conductivity of the shield was given as a ratio of the conductivity of the main conductor; hence, one-third neutral.

In the situation where 2,000 kcmil aluminum conductors are triangularly spaced 7.5 inches apart, the shield loss for a one-third neutral is 1.8 times the conductor loss!

For single-conductor transmission cables having robust shields, losses such as these are likely to be encountered in multipoint grounding situations and are generally not acceptable.

16.3.2 Shield Conductivity

The shield, or sheath, of a cable must have sufficient conductivity in metal to carry the available fault current that may be imposed on the cable. Single-conductor cables should have enough metal in their shields to clear a phase-to-ground fault and with the type of fusing or reclosing scheme that will be used. It is not wise to depend on the shields of the other two phases since they may be some inches away. You need to determine:

- What is the fault current that will flow along the shield?
- What is the time involved for the backup device to operate?
- Will the circuit be reclosed and how many times?

Too much metal in the shield of a cable section with two or more grounds is not a good idea. It involves additional cost to buy such a cable and the losses not only reduce the ampacity of the cable but also cause undue economic losses from the heat produced [10,11].

One way that you can test your concept of a sufficient amount of shield is to look at the performance of the cables that you have in service. Even if the present cable has a lead sheath, you can translate that amount of lead to copper equivalent. You will also need to consider what the fault current may be in the future. The Electric Power Research Institute (EPRI) has developed a program that does the laborious part of the calculations.

TABLE 16.1
Electrical Resistivity of Metals

Metal	Electrical Resistivity in Ohm-mm²/m × 10⁻⁸, 20°C
Copper, annealed	1.724
Aluminum	2.83
Bronze	4.66
Lead	22.0
Iron, hard steel	24.0

We can "convert" metals used in sheaths or shields to copper equivalent by measuring the area of the shield metal and then translate that area to copper equivalent using the ratio of their electrical resistivities (Table 16.1).

As an example, we have a 138 kV low pressure oil-filled (LPOF) cable that has a diameter of 3.00 inches over the lead and the lead is 100 mils thick.

The area of a 3.00 inch circle is = 7.0686 in^2

The area of a circle that is under the lead is:

Diameter = 3.00 – 0.100 – 0.100 = 2.80 in
Area = 1.4 × 1.4 × π = 6.1575 in^2
Area of the lead is 7.0686 – 6.1575 = 0.9111 in^2
The ratio of resistivities is 1.724/22.0 = 0.0784
The copper equivalent is 0.9111 in^2 × 0.0784 = 0.07139 in^2
To convert to circular mils, multiply inches squared by $4/\pi \times 10^6$ = 90,884 cmils
This lead sheath is between a #1/0 AWG (105,600 cmils) and a #1 AWG (83,690 cmils)

If the sheath increases to 140 mils and the core stays the same, we have:

The area of the sheath is = 7.4506 in^2
The area of lead is 7.4506 – 6.1575 = 1.2931 in^2
Multiply by the same ratio of 0.0784 = 0.1014
To convert to circular mils, multiply by $4/\pi \times 10^6$ = 129,106 cmils
This is almost a #2/0 AWG (133,100 cmils) copper conductor
Using the same concept, one can change from aluminum to copper, etc.

The allowable short-circuit currents for insulated copper conductors may be determined by the following formula:

$$\left[I/A\right]^2 t = 0.0297 \log_{10}\left[T_2 + 234/T_1 + 234\right] \tag{16.1}$$

where

I = short circuit current in amperes
A = conductor area in circular mils

t = time of short circuit in seconds
T_1 = operating temperature, 90°C
T_2 = maximum short-circuit temperature, 250°C

A well-established plot of current versus time is included in [9]. It is important to be aware that these results are somewhat negative since the heat sink of coverings is ignored and has not been addressed in Equation 16.1. On the other hand, the answers given are safer values.

16.3.3 Bonding Jumper Capability

A good connection must be made between the bonding jumper and the cable sheath to have enough capacity to take the fault current to ground or to the adjacent section—no matter how well-designed is the cable sheath. This is frequently a weak point in the total design. Another area of concern is the point of attachment of the bond wire to the sheath.

The bonding jumper should always be of the size greater than the equivalent sheath area and should be as short and straight as possible to reduce the impedance of that portion of the circuit. In all cases, the bonding jumper should be a covered wire, such as a 600 volt building wire, to improve its short-term ampacity.

16.4 MULTIPLE-POINT GROUNDING

16.4.1 Advantages

☺ No sheath isolation joints
☺ No voltage on the shield
☺ No periodic testing is needed
☺ No concerns when testing or looking for faults

16.4.2 Disadvantages

☹ Lower ampacity
☹ Higher losses

16.4.3 Discussion

Although you may have already decided to drop this concept, you should be aware of the consequences of a second ground or connection appearing on a run of cable that had not been planned. Such a second ground can complete a circuit and result in very high sheath currents that could lead to a failure of all of the cable that has been subjected to those currents. The higher the calculated voltage on the sheath, the greater the current flow may be in the event of the second ground. Periodic maintenance of single-point grounded circuits should be considered. If this is to be done, a graphite layer over the jacket will enable the electrical testing of the integrity of the jacket.

16.5 SINGLE-POINT AND CROSS-BONDING

To be precise, single-point grounding means only one ground per phase, as will be explained later. Cross-bonding also limits sheath voltages and demonstrates the same advantages and disadvantages as single-point grounding.

16.5.1 Advantages

☺ Higher ampacity
☺ Lower losses

16.5.2 Disadvantages

☹ Sheath isolation joints are required
☹ Voltage on sheath/safety concerns

16.5.3 Background

The term that was used to describe single-point grounding during the 1920s to 1950s was *open-circuit sheath*. The concern was to limit the *induced sheath voltage* on the cable shield. A 1950s handbook said that "The safe value of sheath voltage above ground is generally taken at 12 volts AC to eliminate or reduce electrolysis and corrosion troubles." The vast majority of the cables in those days did not have any jacket—just bare lead sheaths. Corrosion was obviously a valid concern. (Some US cable manufacturers still recommend 25 volts as the maximum for most situations.) The vastly superior jacketing materials that are available today have helped change the currently accepted value of "standing voltage" to 100 to 400 volts for normal load conditions. Since the fault currents are much higher than the load currents, it is usually considered that the shield voltage during fault conditions be kept to a few thousand volts. This is controlled by using sheath voltage limiters—a type of surge arrester.

16.5.4 Single-Point Bonding Methods

There are numerous methods of managing the voltage on the shields of cables with single-point grounding. All have one thing in common: the need for a sheath or shield isolation joint.

Five general methods will be explored:

- Single-Point Grounding
- Cross-Bonding
- Continuous Cross-Bonding
- Auxiliary Bonding
- Series Impedance or Transformer Bonding

Diagrams of each method of connection, with a profile of the voltages that would be encountered, are shown in Section 16.5.6.

There are other types of grounding schemes that are possible and are in service. Generally, they make use of special transformers or impedances in the ground leads that reduce the current because of the additional impedance in those leads. These were necessary years ago when the jackets of the cables did not have the high electrical resistance and stability that are available today.

16.5.5 Induced Sheath Voltage Levels

Formulas for calculating shield voltages and current and losses for single-conductor cables were originally developed by K. W. Miller in the 1920s [3]. The same general equations are also given in several handbooks. The table from Reference [7] is included as Figure 16.7. The difference in these equations is the use of the "j" term—to denote phase relationship—so only the magnitude of the voltage (or current) is determined. Each case that follows will include the formulas from reference [7,8].

The induced voltage in the sheath of one cable or for all cables in a circuit where the cables are installed as an equilateral triangle is given by:

$$V_{sh} = I \times X_m \tag{16.2}$$

where

V_{sh} = sheath voltage in microvolts per foot of cable
I = current in a phase conductor in amperes
X_m = mutual inductance between the conductor and the sheath

The mutual inductance for a 60 hertz circuit may be determined from the formula:

$$X_m = 52.92 \log_{10} S/r_m \tag{16.3}$$

where

X_m = mutual inductance in micro-ohms per foot
S = cable spacing in inches
r_m = mean radius of the shield in inches. This is the distance from the center of the conductor to the mid-point of the sheath or shield.

For the more commonly encountered cable arrangements such as a three-phase circuit, other factors must be brought into the equations. Also, A and C phases have one voltage while B phase has a different voltage. This assumes equal current in all phases and a phase rotation of A, B, and C.

Right-angle or "rectangular" spacing is a probable configuration for large, single-conductor cables in a duct bank. One arrangement is shown in Figure 16.4.

The induced shield voltages in A and C phases are:

$$V_{sh} = I/2\sqrt{3Y^2 + (X_m - A/2)^2} \times 10^{-6} \tag{16.4}$$

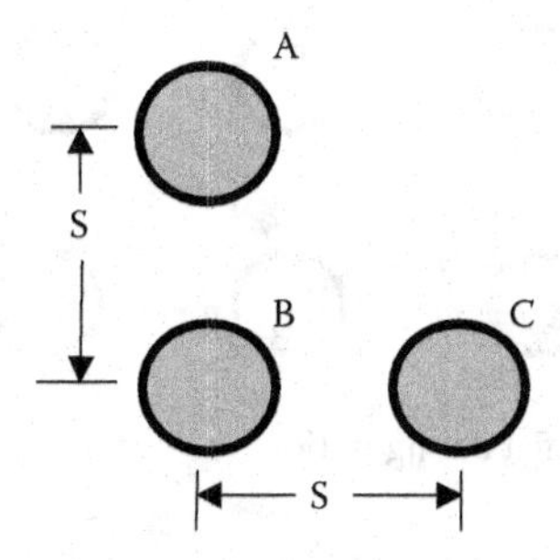

FIGURE 16.4 Right angle arrangement.

where

V_{sh} = sheath voltage on A and C phases in volts to neutral per foot
I = current in phases conductor in amperes
$Y = X_m + A/2$
$X_m = 52.92 \log_{10} S/r_m$ for 60 hertz operation
S = spacing in inches
A = 15.93 micro-ohms per foot for 60 hertz operation

The induced shield voltage in B phase is:

$$V_{sh} = I \times X_m \times 10^{-6} \tag{16.5}$$

A flat configuration is commonly used for cables in a trench, but this could be a duct bank arrangement as well (Figure 16.5).

The induced shield voltages in A and C phases are:

$$V_{sh} = I/2\sqrt{3Y^2 + (X_m - A)^2} \times 10^{-6} \tag{16.6}$$

where

V_{sh} = sheath voltage on A and C phases in volts to neutral per foot
I = current in phases conductor in amperes
$Y = X_m + A$ [This factor has changed from Equation 16.4)!!!]
$X_m = 52.92 \log_{10} S/r_m$ for 60 hertz operation

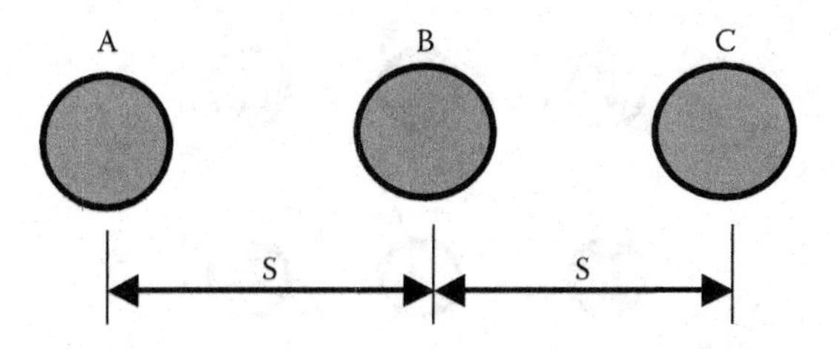

FIGURE 16.5 Flat arrangement.

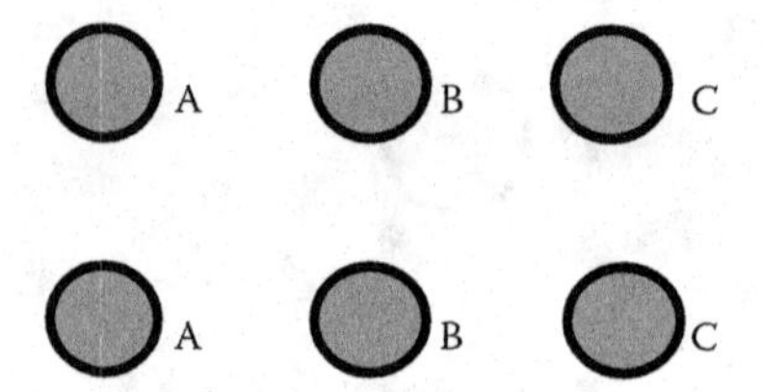

FIGURE 16.6 Two circuits, flat configuration.

S = spacing in inches
A = 15.93 micro-ohms per foot for 60 hertz operation

The induced shield voltage in B phase is the same as Equation 16.5 (Figure 16.6):

$$V_{sh} = I \times X_m \times 10^{-6} \tag{16.7}$$

The induced shield voltages in phases A and C are:

$$V_{sh} = I/2\sqrt{3Y^2 + \left(X_m - \mathrm{B}/2\right)^2} \times 10^{-6} \tag{16.8}$$

where

V_{sh} = sheath voltage on A and C phases in volts to neutral per foot
I = current in phase conductors in amperes
$Y = X_m + \mathrm{A} + \mathrm{B}/2$ [This factor has changed again!]
$X_m = 52.92 \log_{10} S/r_m$ for 60 hertz operation
S = spacing in inches
A = 15.93 micro-ohms per foot for 60 hertz operation
B = 36.99 micro-ohms per foot for 60 hertz operation

The induced shield voltage for B phase is (Figure 16.7):

$$V_{sh} = I\left(X_m + A/2\right) \times 10^{-6} \tag{16.9}$$

The induced shield voltages in phases A and C are similar to Equation 16.7, but Y changes:

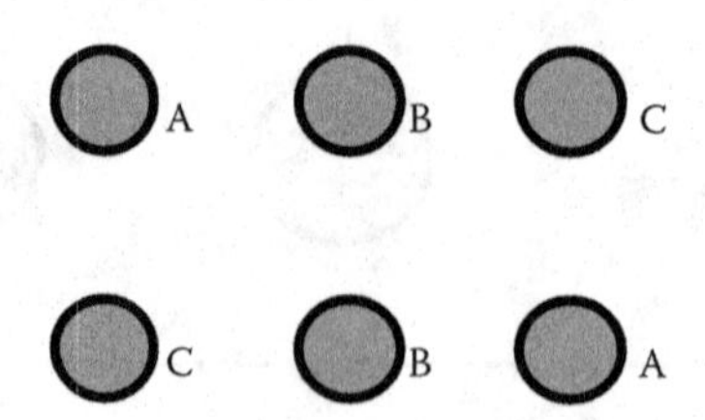

FIGURE 16.7 Two circuits, flat configuration, phases opposite.

$$V_{sh} = I/2\sqrt{3Y^2 + (X_m - B/2)} \times 10^{-6} \tag{16.10}$$

where

V_{sh} = sheath voltage on A and C phases in volts to neutral per foot
I = current in phases conductor in amperes
$Y = X_m + A - B/2$ [Now a minus, not +]
$X_m = 52.92 \log_{10} S/r_m$ for 60 hertz operation
S = spacing in inches
A = 15.93 micro-ohms per foot for 60 hertz operation
B = 36.99 micro-ohms per foot for 60 hertz operation

The induced voltage on B phase uses the same equation as Equation 16.9:

$$V_{sh} = I(X_m + A/2) \times 10^{-6} \tag{16.11}$$

16.5.6 Depiction of Bonding Methods

16.5.6.1 Multigrounded Cable Run—Two or More Grounds. Solid, Multigrounded Circuits

Three-conductor cables should always be solidly grounded in each manhole using a grounding conductor of sufficient size to carry the available fault current for the time dictated by the clearing relays and equipment. A similar grounding system should be utilized for most single-conductor cables, but ampacity concerns may dictate another grounding method.

A schematic diagram of a multigrounded system is shown in Figure 16.8.

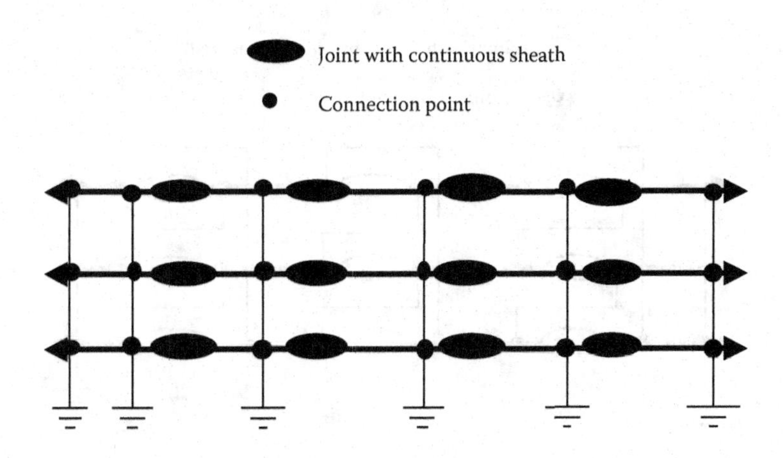

FIGURE 16.8 Multigrounded cable runs.

16.5.6.2 Cross-bonded Circuits

The goal of any shield isolation system is to reduce induced shield currents to the point that they will not seriously affect the ampacity of the circuit and to limit the sheath voltage to a safe value. For large single-conductor cables designed to carry 500 amperes or more, consideration should be given to the several alternative grounding systems that are available.

The most commonly used is cross-bonding where the cable circuit is divided into three equal sections (or six, or nine, etc.). The shield is solidly grounded at the beginning of the first section and at the end of the third section. The second section is isolated by means of shield "breaks" from the first and third sections and has its sheath *bonded* to other phases (see Figure 16.9).

The induced shield voltage from A phase is "cross-bonded" to B phase. This phase change reduces the resultant sheath voltage, and so on, for all sections of the circuit.

Bonding conductors must have sufficient capacity to carry the fault current that will be imposed and voltage resistance to keep the bonding jumper from being inadvertently grounded.

This method has the disadvantage of needing the joints evenly spaced through each triple section. When the joints are not evenly spaced, the voltage from one phase does not completely cancel out the other phase. This may not be critical, but it does mean that somewhat higher voltage levels will result.

16.5.6.3 Continuous Cross-Bonding

This is basically the same as the cross-bonding of Section 16.5.6.2 except that all of the joints have shield interruption provisions. Such a scheme is useful in situations where the triple sections are not practical from a field standpoint, such as for four sets of joints—where the matched sets of three are not attainable.

The sheath voltages are approximately the same as for cross-bonded circuits. A disadvantage of this system is that there are no solid grounds except at the terminations.

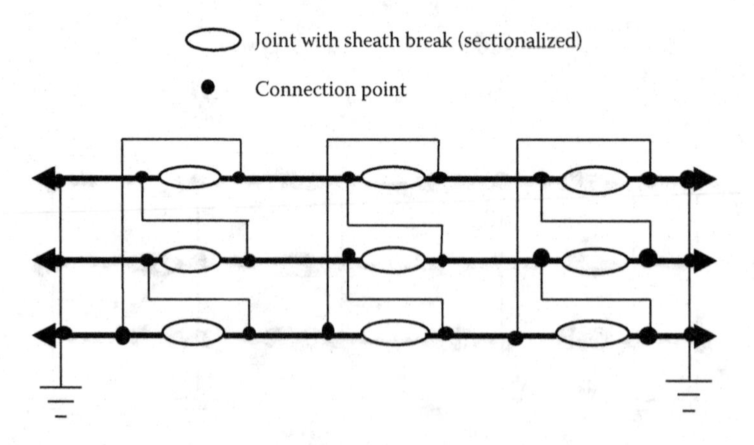

FIGURE 16.9 Cross-bonding connections.

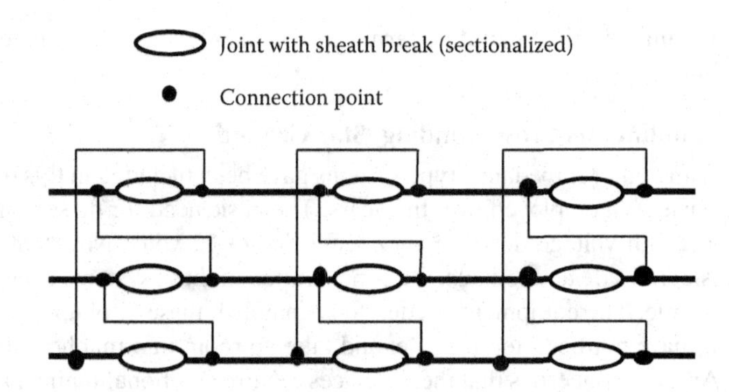

FIGURE 16.10 Continuous cross-bonding.

Longer circuits may require that continuous cross-bonding be used where the sections are divided into sets of three sections of cable. This is shown in Figure 16.10.

16.5.6.4 Auxiliary Cable Bonding

Sheath losses may be controlled by installing sheath insulators in every splice and then grounding one point on each cable to a parallel "auxiliary" cable. An additional duct is required. This arrangement is shown in Figure 16.11.

This system is similar to the continuous cross-bonding method since all the joints must have shield isolation and all shields are bonded at each splice. The unique part of this arrangement is that the shields are connected to each other and to a separate neutral cable that runs along the length of the circuit.

This permits the through fault current to be transmitted on both the shield as well as the parallel neutral cable. A reduction in the amount of shield materials is thus possible. A cable fault must still be cleared by having the fault current of that phase taken to ground at a remote point. This means that you must still put on a

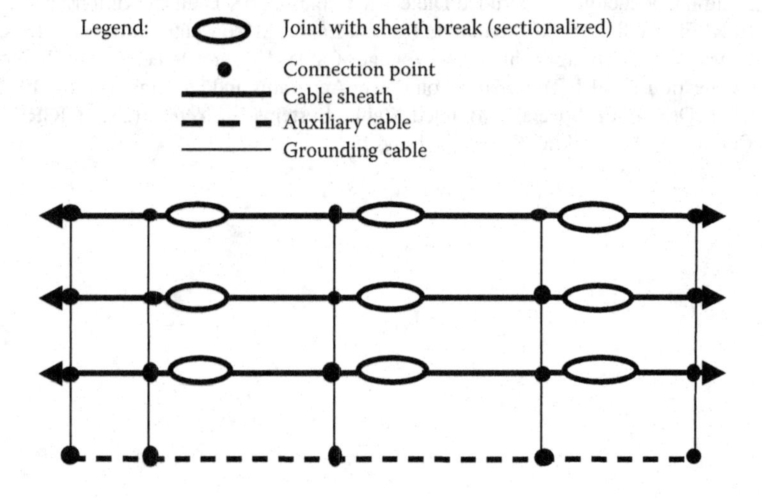

FIGURE 16.11 Auxiliary cable bonding.

sufficient amount of shield metal to permit the breaker, or other backup device, to "see" the fault.

16.5.6.5 Continuous Cross-Bonding, Star Ground

This system and other impedance type systems have been included in this discussion since they have been employed over the years. The basic need for these systems was to keep the sheath voltage down to very low value of 12 volts. Now that over 100 volts is considered safe and reliable, the complication of these systems does not seem to be worthwhile. The equipment needed to accomplish these hookups are not very easy to find, have not been too reliable, and take up room in a manhole that is at a premium. Another concern is that these devices require additional maintenance time to be certain that they remain operational.

REFERENCES

1. IEEE Std. 575-1988, *IEEE Guide for the Application of Sheath-Bonding Methods for Single-Conductor Cables and the Calculation of Induced Voltages and Currents in Cable Sheaths.*
2. *Underground Systems Reference Book,* EEI, 1957.
3. Halperin, H. and Miller, K. W., April 1929, "Reduction of Sheath Losses in Single-Conductor Cables", Transactions AIEE, p. 399.
4. *Sheath Bonding Transformers,* Bulletin SBT 2, H. D. Electric Co., Chicago, IL.
5. AIEE-IPCEA *Power Cable Ampacities,* AIEE Pub. No. S-135-1 and -2, IPCEA P-46-426, 1962.
6. *IEEE Standard Power Cable Ampacities,* IEEE 835-1994.
7. ICEA P-53-426.
8. *Engineering Data for Copper and Aluminum Conductor Electrical Cables,* Bulletin EHB-90, The Okonite Company.
9. IEEE Std. 532-1993 ISBN 1-55937-337-7, *IEEE Guide for Selection and Testing Jackets for Underground Cables.*
10. EPRI EL-3014, RP 1286-2, "Optimization of the Design of Metallic Shield / Concentric Neutral Conductors of Extruded Dielectric Cables Under Fault Conditions."
11. EPRI EL-5478, "Shield Circulating Current Losses in Concentric Neutral Cables."
12. "Sheath Over-voltages in High Voltage Cable Due to Special Sheath Bonding Connections," IEEE Transactions on Power Apparatus and Systems, Vol. 84, 1965.
13. "The Design of Specially Bonded Cable Systems," *Electra,* (28), CIGRE Study Committee 21, Working Group 07, May, 1973.

17 Underground System Fault Locating

James D. Medek

CONTENTS

17.1 INTRODUCTION

Underground systems originally established for concentrated areas such as cities or towns consisted of cable installed in conduit systems and terminated in manholes. The manholes provided a space to join sections of cables and in some cases, a place to install transformers and switches. Cable construction used in these systems consisted of paper insulation, impregnated with an insulating fluid, and covered with a lead sheath. When failures occurred in this type of system, the procedure for finding the failure was to locate the manhole where smoke or flame was coming out of the manhole cover. In the event smoke was not seen, the procedure called for reenergizing the faulted circuit to produce the detectable smoke location. If this did not work, or if there were other operating problems, a device such as a high voltage transformer could be used to breakdown the failure and generate sufficient smoke for the purpose of detecting which manhole contained the failure. If the fault was found in a manhole or if the fault had occurred in a cable between manholes, no further locating procedures were required. This was because the entire section of cable between adjacent manholes had to be replaced.

When cables were installed directly in the earth, location of the faults became more exacting and difficult to find. The location of a failure on a direct buried cable system requires a great deal of accuracy. That is because as small a hole as practical is excavated so that repairs can be made to the cable. On the other hand, the hole must be large enough for the personnel to complete repairs. Instruments have been developed to aid in this exact location, but with these new instruments come the requirement that operators become better trained and proficient in their use. This chapter will introduce the state of the art in fault location and some of the equipment, which is or can be used to find underground failures. There have been modifications made by users so as to better suit their various applications.

While major strides in fault locating have been achieved through technology and field experience, no one instrument has been able to find the approximate location of the field and an exact location for repairs. It should also be considered that training and experience are a major factor in locating underground failures. Without continual use of the equipment or re-training sessions, accuracy of finding the failures and extended time requirements will affect the finding of faults.

17.2 CONDUIT AND MANHOLE SYSTEMS

With the increase of electrical loads and more reliable demands of customers connected to the conduit and manhole systems, the requirement for quicker and safer methods of finding electrical failures became a goal in many companies. As system voltage increased, the use of a high voltage transformer to jump the gap of a fault, and a high current supply to burn the failure into a detectable condition became the approach that many companies utilized. When the failure is reduced to a low value of resistance, a bridge circuit can be used to give an approximate location of the failure. The bridge circuit most widely used is the resistance bridge known as the Murray Loop. This instrument requires a good conductor to complete the circuit, a very low resistance jumper between the faulted phase and a good phase, and an accurate length of the circuit being tested. Accurate circuit lengths are normally known in a conduit system and since the cables in a conduit system usually have three phases, a return conductor is available most of the time to utilize a bridge instrument.

17.2.1 Murray Loop Bridge

The Murray Loop bridge instrument provided an approximate location within the capability of the fault locator or accuracy of the instrument used. When the location found gives a questionable location such as a location near a manhole, the location opens three possibilities. The failure may be in the manhole, or in one of two sections of cable being spliced in the manhole. A way of verifying the correct failure location has been developed and used in the field with success. The instrument used for this purpose consists of a high voltage capacitor that is charged with a direct current supply. The charged capacitor is then connected to the faulted cable for the purpose of discharging at the failure producing a loud noise. The noise may be detected either at the conduit opening at either end of the manhole, or in the manhole itself.

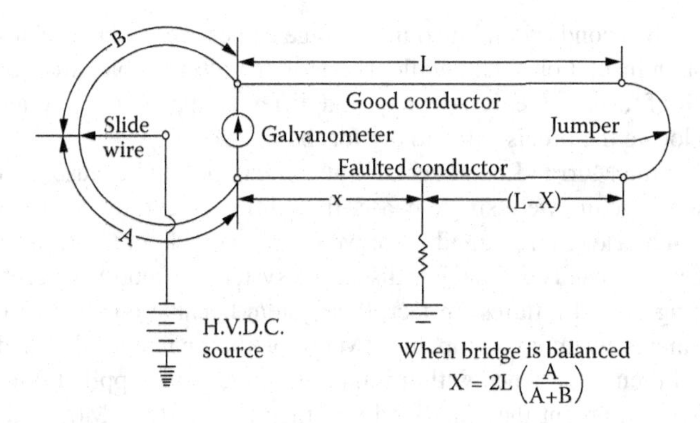

FIGURE 17.1 Murray Loop diagram. *Underground Systems Reference Book*, 1957.

Visual detection of the discharge may also occur to aid in the location of the defective cable or splice.

A bridge circuit that has its best application on three-phase circuits consisting of three-conductor paper-insulated cables in a lead sheath. The reason for this is because it is a bridge circuit that requires a good conductor of the same length and size as the faulted conductor. In these types of circuits, the most common failure is on one of the phase conductors. Since the fault itself is part of the bridge circuit, an operator knows if it is going to work by noting the ability to pass a small current through the fault itself. One technique is to reduce the resistance of the failure to the shield or ground by passing a higher current through the fault prior to using the Murray Loop bridge circuit. The bridge circuit normally can be used with a low voltage direct current source; however, if the lower voltage is insufficient to pass current through the fault, a higher level of voltage can be applied. This assumes that the bridges, as well as the connections, are capable of withstanding the higher voltage. The Murray Loop can be used to find approximate locations on cables that are of shielded or nonshielded design. Accurate records of conductor length and size are a must; if known, this method can give very accurate results (Figure 17.1) [2].

17.3 UNDERGROUND RESIDENTIAL DISTRIBUTION

In the US around 1950, direct buried cable installations were used to a great extent in new residential areas. This type of installation was less obtrusive to the eye and both economical and faster to install. After several years, failures occurred on these installations. Cables were found to have inherent problems because of the various applications and field environments to which they were exposed. Consequently, the need for accurate fault locating equipment became a prime requirement. Since the cables were direct-buried, the ability to remove a section of cable was eliminated. It was now necessary to locate the failures within the area that a crew could dig and find the failure. This was not as easy as it sounds. Not only was the equipment needed, but also the men must be trained and able to operate the instruments and accurately locate failures. The process was compounded since there were different designs of both secondary and medium voltage cables. Secondary cables were usually without an insulation shield while the medium voltage cables were of a shielded design. Both of these cable designs required different approaches and techniques to adequately locate the precise area to dig for the failure.

The use of the Murray Loop became impractical mainly because the majority of the residential circuits were single phase. In addition, accurate lengths of cable are required with a bridge circuit and these were not available in the residential direct-buried cable systems. The capacitor discharge system became very useful in accurately locating buried failures. In fact, this system became one of the most widely used instruments by many companies. Many of the engineers that had served in the war used their experience with military equipment and applied it to cable fault locating. The equipment they modified was the radar system. Modified to connect to a cable or overhead line, these sets were the first attempts to utilize the theory. They were not accurate and were very complicated to the point that an engineer was needed to properly adjust the various settings.

17.4 METHODS OF LOCATING FAILURES

The following are methods of finding failures on underground cable systems. The success of some of these methods is often dependent on the type of system on which it is used, or the skill of an individual operator. You may find that the name used to refer to these methods may change from one company to another, so it is necessary to refer to the basic concepts when investigating new equipment [3].

17.4.1 Cable Locating

Location of the cable route is the most important part of fault finding on a direct-buried system. Once the faulted cable has been identified, the next step is to verify the route. Unlike a conduit system where accurate locations are maintained for conduit runs, manholes, and terminations, the direct-buried system does not have a specific route. Installation problems in the field may cause the construction crew to change the route. These changes are very seldom noted in the original installation drawings. Splices may be installed as part of the initial installation or added when repairs are made on a previous failure. In some cases, extra cable may be buried with the intent of extending the circuit for future loads to be added, or for situations where the cable route is changed to go around unknown objects buried in the ground.

17.4.2 Cut and Try

This method of isolating the failure was originally used when no other instrument was available. Now seldom used, it has an application where cable length exceeds the capability of available instruments. Long, buried cables can be cut in the center of that length and then tested both ways to see which half has the failure. The resulting length of cable to be tested is now reduced by 50%. This method can be repeated until a short section of cable is known to contain the failure and is within the capability of the available instruments. "Cut and try" should only be used when other methods are not available since it may damage otherwise serviceable facilities.

17.4.3 Radar or Time Domain Reflectometry

Shortly after World War II, radar or TDR sets became available on the surplus market. People who had used the radar sets during the war used their experience to modify these sets for use by underground personnel to find failures on underground cables. At first, these sets were cumbersome to use and fairly complicated to adjust. Like many new instruments, improvements have been made through the years. For the people performing fault locating, radar is becoming the instrument of choice. Radar sets send out a pulse of relative low voltage which is reflected back to the set itself when it encounters a change of impedance. The extremes of the change of impedance are the open end of a cable and a short between the conductor and the ground shield. The trace on a screen is a measure of time for the pulse to travel to the failure and back. It is necessary to know the velocity of propagation of a pulse for the type of cable involved and the length of the cable being evaluated, as this length is the length of the trace on the radar instrument. Application of this equipment is restricted to a shielded cable design in which the characteristic impedance is fairly constant throughout the length.

The TDR pulse travels along the cable system at a velocity that is a fraction of the speed of light. Depending on the insulation on the cable, this speed is in the order of 50% to 70% of the speed of light.

17.4.4 Capacitor Discharge (Thumper) Set

The capacitor discharge instrument has been nicknamed "Thumper" because of the audible sound it makes when the energy of the capacitor discharge occurs at the failure in the ground. It was originally used for duct and manhole systems but it is this thumping sound that made it one of the most useful devices for exactly locating the failure in direct buried systems. The capacitor discharge consists of three basic components. A high voltage, direct current power supply; a high voltage capacitor; and a timing device to control the times the charged capacitor is connected to the faulted cable. The timing device may be an adjustable spark gap or a set of contacts that are controlled by a timing device. This instrument requires that the discharge from the capacitor occurs between the phase conductor and a shield or nearby ground. Therefore, it is mostly suited for shielded cable designs.

Experience demonstrated that if the thumper was frequently used over too great a time period, damage to extruded dielectric cable systems ensued [1]. By combining the TDR with the thumper, a fault can be pinpointed with only one or two thumps—thus minimizing possible damage to the system.

Figure 17.2 shows a fault-locating instrument suitable for field application that combines the TDR and thumper. The instrument consists of two testing features. It contains a high-voltage direct current circuit used to energize the capacitor discharge circuit and a TDR that locates the distance to the fault. The locating team can then

FIGURE 17.2 Combination thumper and TDR equipment. Photograph courtesy of Von Corporation.

turn off the set until one member can measure off that distance and wait for another thump to pinpoint the location.

Here are a few of the major improvements that have been made through the years:

- Controls have become simplified so as to become user-friendly.
- Screen traces have become more visible in bright light.
- Comparative traces are available for the purpose of comparing a good circuit with a faulted circuit.
- Velocity of propagation has been narrowed so as to make fewer choices for various cable constructions. It is still necessary to have an idea about what it should be as it effects the accurate distance to the failure. [See Chapter 20 for a discussion of Velocity of Propagation.]
- Taking TDR readings from both ends of a cable improves the accuracy of the measurement.
- Introduction of an electrical discharge at the failure will show the fault location by breaking down the failure and capturing it on the viewing screen.

17.4.5 Earth Gradient

The earth gradient instruments are used on nonshielded cables by applying a signal to the phase conductor of the faulted cable and a return path to a ground connection. The signal travels through the ground returning to the ground connection at the transmitting instrument. The returning signal can be detected above the ground in close proximity of the buried cable with a sensitive meter.

Transmitters normally consist of a direct current signal of sufficient voltage to supply a detectable signal with probes on the surface of the ground. Direct current signal is preferred where there is a chance of stray alternating currents in the vicinity of the faulted cable.

Detectors used with the direct current transmitter consist of: microammeters with a zero center feature; a control for the sensitivity; and a set of probes, which either can be frame mounted or used separately.

Used mostly on direct buried nonshielded cable normally used for secondary main and service cables, this method has an excellent record for finding failures on aluminum conductor cables where the aluminum conductor has corroded. During the corrosion process at the failure, the aluminum powder expands, causing the insulation to break (if it not already broken) and make a connection to ground. This is where the earth gradient starts.

17.4.6 Capacitance Measurements

This method of fault location can best be used on cables where the failure consists of an open circuit of the phase conductor.

A shielded cable has fairly uniform capacitance between the phase conductor and the insulation shield. By knowing the capacitance per foot of the cable, an approximate location of an open-type failure can be calculated. If the capacitance per foot is not known, an alternate method consists of measuring capacitance from both ends of the faulted cable. A ratio of the two readings will provide an approximate location as long as the cable is uniform throughout the run under test (Table 17.1).

17.4.7 Instruments Infrequently Used in North America

The following are methods of fault locating on underground systems that have been used by some fault-locating personnel. They are listed here only for reference and not as a recommendation of widely used practices in the US today. Some of the methods listed here are similar if not identical to others in the list, but have been given different names by users depending on which reference material you are using as a guide:

- Murray/Fisher Loop
- Varley Loop
- Hilborn Loop
- Insulation resistance
- Pulse decay

TABLE 17.1
Capacitance of 175 mil 15 kV Cable

Condutor Size	Insulation Type	Length in Feet	Capacitance Value in Microfarad (μF)
#2 AWG	EPR	500	0.04
#2 AWG	XLPE or TR-XLPE	1,000	0.06
4/0 AWG	EPR	1,000	0.10
1,000 kcmil	EPR	1,000	0.19

- Differential decay method
- Standing wave
- Charging current
- Impulse current differential method

17.5 FAULTED CIRCUIT INDICATORS

Faulted circuit indicators (normally referred to as "fault indicators") do not provide a location of the fault on a circuit, but do reduce the area in which the fault-locating equipment can be applied. This is extremely important when locating failures on long lengths of cable. It is also a valuable tool when locating faults that are installed in areas where there is limited access for the personnel performing the fault locating.

Fault indicators of the type addressed here can be obtained to cover a range of currents depending on the type of circuit to which they are attached. It is important to know what the normal current is in the circuit, and in addition, intermittent increases in current, which may occur on the circuit. Knowing these two factors will minimize the occurrence of a false tripping of the device. The effect of the exact fault current minimizes range-faulted indicators, which will automatically reset when the circuit is energized.

Early applications had limited success in many areas due to the following reasons:

1. Initial fault indicators were of the manual reset type (see Figure 17.3) that required the operating departments return to the previously faulted cable circuit, access each location where they were installed, and manually reset the device. An external tool is required to reset this indicator.
2. Information on the correct placement of the fault indicators was often not available to the installer. Consequently, they were installed on the cable in such a location where they did not work properly. When properly installed and reset after every failure on the circuit, these devices provided a valuable service to the people locating faults.

Fault indicators are designed so that they detect a given level of current passing through the cable to which they are connected. On alternating current circuits, the indicators do not show which direction the fault current is flowing—only that the current has exceeded the rated current of the indicating device. It is necessary for the user to inspect the fault indicators starting at the source of the current into the circuit. This continues until the user finds one indicator that had not experienced the fault current. The user then knows that the circuit failure is between the fault indicator, which showed excessive current, and the one that did not show excessive current (no target).

Magnetic fault indicators are placed over a conductor and detect the magnetic field caused by electric current flowing in the conductor. When the current exceeds the rating of the fault indicator, a warning will appear on the device, as a target, light, or noise showing that current much greater than normal (fault current) has passed through the fault indicator.

FIGURE 17.3 Manual reset fault indicator. Photograph courtesy of E.O. Schweitzer Mfg. Co.

An improvement over the manual reset fault indicator consists of an operating button, which eliminates the need for a separate tool to reset the indicator. It also eliminates the requirement of removing the fault indicator from the circuit as is required in Figure 17.4.

The automatic reset feature shown in Figure 17.5 is one of the most important features to the utility operating department. It was no longer necessary

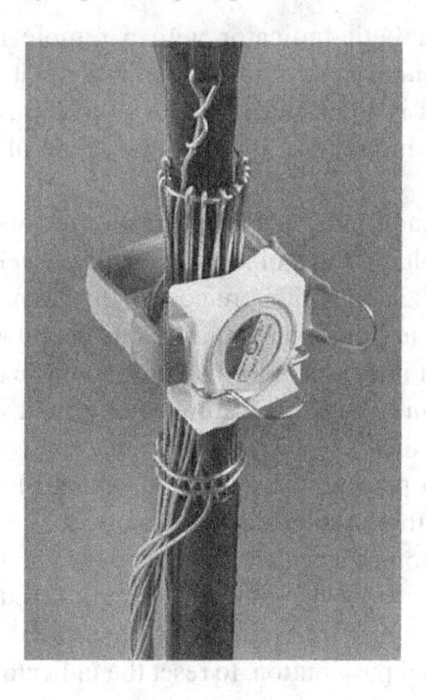

FIGURE 17.4 Fault indicator designed with a reset button. Photograph courtesy of E.O. Schweitzer Mfg. Co.

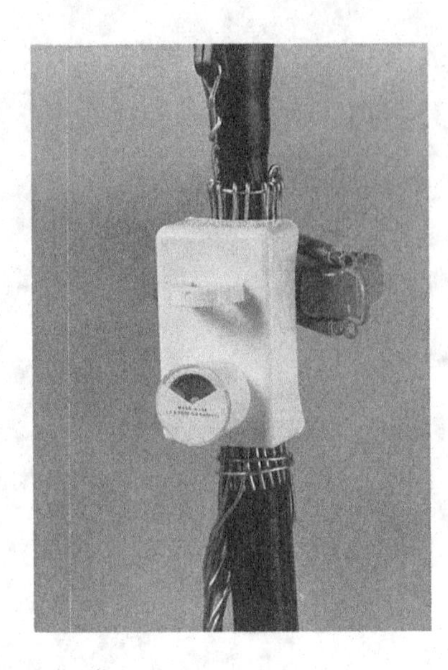

FIGURE 17.5 Fault indicator with automatic reset. Photograph courtesy of E.O. Schweitzer Mfg. Co.

to return to the circuit that had been faulted and repaired, access every fault indicator, and reset the device manually. These indicators are designed so as to take a sampling of current from the repaired and energized line and use that to reset the fault indicator. In other words, all fault indicators return to a no fault indication.

Figure 17.6 shows a fault indicator with a remote target attachment. This allows operating personnel access to the indicator without entering the transformer or switchgear. It is sometimes used in substations on outgoing circuit for a quick evaluation of which phase is in trouble. One of these must be attached to each phase.

Figure 17.7 shows a fault indicator mounted on the voltage test point on a separable connector. The voltage test point was initially developed for the purpose of testing the line or energization without removing the elbow. However, now adapting the fault indicator to the test point has expanded its usefulness.

The voltage test point is available for purchase to those customers who choose to use it. If you have, or plan to buy, this feature when purchasing separable connectors and if you desire to use this type of fault indicator, you will need to specify this at the time of purchase. At present, there is not a standard for the physical size of the voltage test point. Therefore, when using this feature make sure that the fault indicator fits the voltage test point for which it is intended. The separable connector fault indicator uses the energy from the energized circuit through the voltage test point to reset the indicator after the circuit is returned to service. Previous indicators used current, or the mechanical push button, to reset the indicator. This fault indicator has the advantage of not being able to be installed incorrectly as it can only be used with a voltage test point.

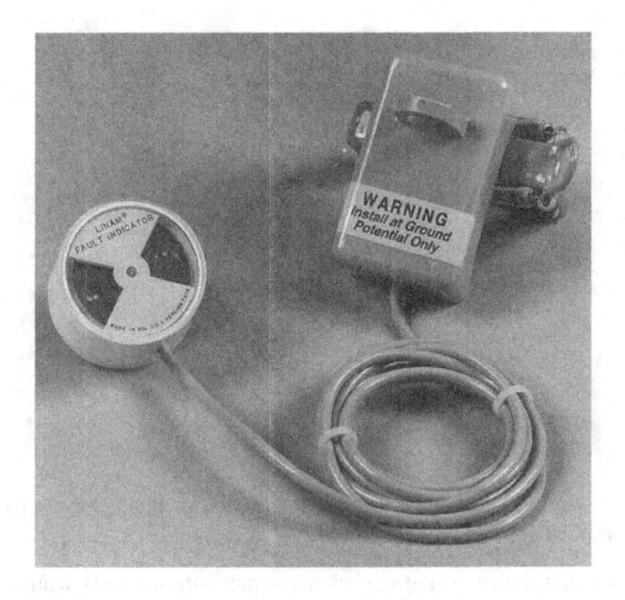

FIGURE 17.6 Fault indicator with remote target. Photograph Courtesy of E.O. Schweitzer Mfg. Co.

FIGURE 17.7 Separable connector with fault indicator mounted on test point. Photograph courtesy of E.O. Schweitzer Mfg. Co.

Figure 17.8 shows an application for a fault indicator on overhead circuits. This one utilizes the energy from the voltage in the conductor to reset the target. While the reset feature is an advantage for underground applications, one major advantage of the manual reset fault indicator on overhead applications is the fact that the indicator does not reset.

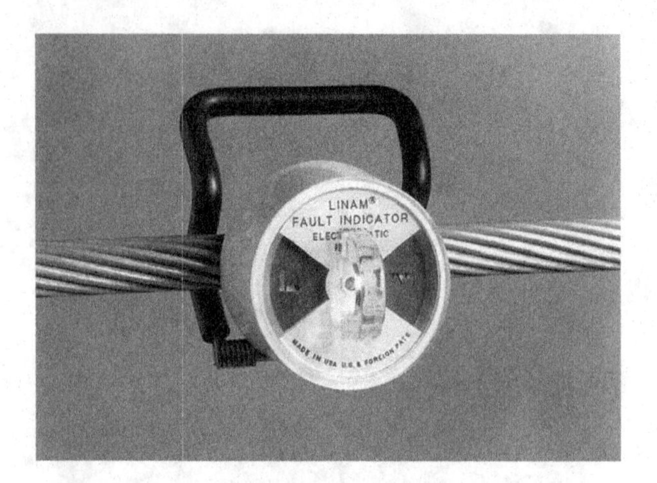

FIGURE 17.8 Fault indicator for overhead application. Photograph courtesy of E.O. Schweitzer Mfg. Co.

This means that a circuit that experiences intermittent faults would normally be difficult to find. However, since the manual fault indicator does not reset, operating personnel can return to the previously faulted circuit and examine the fault indicators to locate the previous intermittent failure.

REFERENCES

1. Hartlein, R. A., Harper, V. S. and Ng, H., April 1994, "Effects of Voltage Surges on Extruded Dielectric Cable Life," IEEE Transactions on Power Delivery, Vol. 9, pp. 611–619, EPRI Project RP-2284.
2. *Underground Systems Reference Book,* 1957, Edison Electric Institute, Publication #55–16, New York, NY.
3. *Underground Cable Fault Location Reference Manual,* November 1995, EPRI TR-105502.

18 Field Assessment of Power Cable Systems

William A. Thue

CONTENTS

18.1 INTRODUCTION

This chapter provides an overview of methods of performing electrical tests in the field on shielded power cable systems [11,13]. It is intended to help the reader select a test that is appropriate for a specific situation of interest. Guides to field testing are contained in the group of Institute of Electrical and Electronics Engineers (IEEE) Standards shown under References [1–5].

[*Editors note: IEEE 400-2001 is an Omnibus document that covers all test methods that were available at the time it was prepared. This document is under revision. The 'point' documents such as 400.1, 400.2, 400.3, and 400.4 cover specific methods in greater detail than in Omnibus 400.*]

Field tests can be broadly divided into the following categories:

1. Acceptance tests
2. Installation tests
3. Maintenance tests

18.1.1 Acceptance Tests

Acceptance tests are field tests made after the cable system is installed—including joints and terminations but before the system is energized. Such tests are made to (a) verify that no damage occurred during shipping, handling, or installation; and (b) identify poor workmanship during repair or jointing/terminating.

These tests are usually conducted at voltages somewhat below factory test levels and durations.

18.1.2 Installation Tests

Installation tests are field tests similar to the acceptance tests described previously except that the tests are conducted after the cable is installed but not jointed or terminated. These tests only verify that no damage occurred during shipping, handling, or installation.

18.1.3 Maintenance Tests

Maintenance tests are field tests made to assess the condition of the cable system at the time of making such a test. A baseline test can be useful in determining deviations during later tests, but most of these tests are conducted after a system has been in service for a period of time. Such tests may be further categorized as follows:

18.1.3.1 Withstand Tests

Withstand tests are maintenance tests that are intended to detect defects in the insulation of a cable system in order to improve the service reliability after the defective part is removed and appropriate repairs performed. These are also referred to as "go/no go" or "pass/fail" tests.

18.1.3.2 Condition Assessment Tests

Condition assessment tests are tests that are intended to provide indications that the insulation system has deteriorated. Some of these tests will show the overall condition of a cable system and others will indicate the locations of discrete defects that may cause the sites of future service failures. A major goal of these tests is to estimate the future performance of the cable system being tested—hence, the term "diagnostic test" is used [10].

The various field test methods that are currently available for testing shielded, insulated power cable systems rated 5 kV through 500 kV are discussed. A brief listing of "advantages" and "disadvantages" is included, but the users should avail themselves of the technical papers that are referenced, the material listed in the references, manufacturers' literature, and recent research results to make decisions on whether to perform a test and which test method (or methods) to use. In making such decisions, consideration should be given to the performance of the entire cable system, including joints, terminations, and associated equipment.

18.2 FUNDAMENTALS

Test equipment for field use must obviously be reasonably small so that it may be conveniently transported to the location. The source of excitation energy can be the normal line equipment (on-line) or a separate source that is generally capable of applying moderately elevated voltages (off-line).

18.2.1 Elements of Testing

There are three building blocks of ALL test methods [1]:

- Voltage
- Current
- Time/Frequency

18.2.2 Excitation Voltage

- Direct voltage (DC)
- Low frequency (about 0.1 Hz)
- Power frequency (50/60 Hz)

- High frequency (100 to 500 Hz)
- Oscillating or decay voltages

18.2.3 Options for Property Measured

- Amperes
- Partial discharge (PD)
- Dissipation factor (tan δ)
- Decay voltage or current
- Inception/extinction voltages

18.2.4 Time Considerations

The time involved in the reading varies considerably. One of the factors that must be considered here is the frequency of the applied source because it is necessary to average several readings—hence, the term "repetition" comes into use. This means that if the frequency is 0.1 Hz and three readings are to be averaged, the source must be applied for at least 30 seconds. If the frequency is 60 Hz, this time can be shortened to just 2 or 3 seconds.

Another consideration for the time of application involves the philosophy of the test. If the goal of the test is to have the failure occur while the circuit is out of service, then an extended time is usually specified (up to 1 hour). If the goal is to only evaluate the circuit, then the time is generally kept very short (a few seconds). Experience is also given consideration in the time decision.

18.2.5 Global Evaluation versus Pinpointing Defects

A broad distinction exists between the test methods in their ability to pinpoint a defect as opposed to looking at the "average" of the total circuit—often referred to as a global evaluation. As will be seen later in this chapter, the PD measurement can lead to the location of one or more defects while dissipation factor gives an average of the circuit under test and hence cannot locate a particular site.

The location of the defect site frequently makes use of a measurement of the time it takes for a signal to travel in the specific cable being tested. Time domain reflectometry (TDR) can measure the time for the signal to go from the test set to the defect and then return (a round trip) or just from the defect to the test set (one way). The velocity of propagation of the wave varies with the insulation material as well as other construction features of the cable.

18.2.6 Need for Testing

While medium and high voltage (HV) power cables are carefully tested by the manufacturer before shipment with alternating current (AC) or DC to ensure conformance with published specifications and industry standards. During transport, installation, and accessory installation, cables are vulnerable to external damage. Therefore, cables may be tested prior to placing them in service to locate any external mechanical damage and to ensure that jointing and terminating work has been satisfactory. Periodic testing of service-aged cables may also be performed with the desire to determine system degradation and to reduce or eliminate service failures.

Additionally, many users find that, with time, these cable systems degrade and service failures become troublesome. The desire to reduce or eliminate those failures may lead cable users to perform periodic or maintenance tests after some time in service. Cable users also need special diagnostic tests as an aid in determining the economic replacement interval or priority for replacement of deteriorated cables.

Research work has shown that certain types of field testing may lead to premature service failures of cross-linked polyethylene (XLPE) cables that exhibit water treeing [6]. This substantiates some field observations that led to concern about field test methods, levels of voltage applied, fault location methods, and lightning surges.

Experience with paper insulated, lead covered (PILC) cable systems that have been tested in the field with DC for over 60 years has shown that testing with the recommended DC voltage does not deteriorate sound insulation or only at such a very slow rate that it has not been detected.

The decision to employ maintenance testing must be evaluated by the individual user taking into account the costs of a service failure, including intangibles, the cost of testing, and the possibility of damage to the system. As proven nondestructive diagnostic test methods become available, the users may want to consider replacing withstand type voltage tests with one or more of these assessment methods.

18.3 OVERVIEW

18.3.1 On-line versus Off-line Testing

Off-line testing has been the usual way of testing cable systems in the field. As the name implies, the circuit to be tested is de-energized and taken out of service. This is necessary with any method that requires higher than normal test voltages or frequencies other than for normal operations. A disadvantage is that there are times when it is not practical to take a circuit out of service in order to test. Switching may also be time-consuming and hence expensive.

Methods are now available that allow diagnostic measurements to be made while the circuit is in normal operation—on-line testing [14,15]. This method is applicable to all system voltages (5 through 500 kV) and all types of cable construction and insulations (paper, XLPE, ethylene–propylene rubber [EPR], etc.). A disadvantage of this method is that sensors must often be placed around the cable or equipment at frequent intervals such as at joints and termination.

The advantage of this approach is that the readings may be taken while the circuit is energized and are not influenced by time of day, year, or load.

18.3.2 Direct Voltage Testing

Testing with DC was accepted for many years as the standard field method for performing HV tests on the cable insulation systems that were found on utility systems—paper-insulated cables [2]. One of the most important uses of testing PILC cable systems prior to placing them in normal service (proof test) was to confirm that the circuit was safe for the workers to energize. This was accomplished very satisfactorily by DC. Testing with DC is still useful in finding gross problems with paper-insulated cable systems either as a proof test or if it is done on a periodic basis (maintenance test). The important consideration is "Are other types of insulation—such as XLPE—in cables on that circuit?"

As systems were added that had extruded insulations, DC was used in the same manner as with paper-insulated systems. Unfortunately, this extrapolation was not based on experience or research.

Research has shown that DC testing tends to be blind to certain types of defects and that it can aggravate the deteriorated condition of some aged XLPE-insulated cables that have water trees. Whenever DC testing is performed, full consideration should be given to the fact that steady-state DC creates an electrical field determined by the conductance of the insulation within the insulation system, whereas under service conditions, AC creates an electric field determined chiefly by the dielectric constant (or capacitance) of the insulation. Under ideal, homogeneously uniform insulation conditions, the mathematical formulas governing the steady-state stress distribution within the cable insulation are of the same form for DC and for AC, resulting in comparable relative values. However, should the insulation contain defects where either the conductivity or the dielectric constant assumes values significantly different from those in the bulk of the insulation, the electric stress distribution obtained with DC will no longer correspond to that obtained with AC. As conductivity is generally influenced by temperature to a greater extent than dielectric constant, the comparative electric stress distribution under DC and AC voltage application will be affected differently by changes in temperature or temperature distribution within the insulation. Furthermore, the failure mechanisms triggered by insulation defects vary from one type of defect to another. These failure mechanisms respond differently to the type of test voltage utilized, for instance, if the defect is a void where the mechanism of failure under service AC conditions is most likely to be triggered by PD, application of DC would not produce the high PD repetition rate that exists with AC. Under these conditions, DC testing would not be useful. However, if the defect triggers failure by a thermal mechanism, DC testing may prove to be effective. For example, DC can detect the presence of contaminants along a creepage interface.

Testing of extruded dielectric, service-aged cables with DC at the currently recommended DC voltage levels can cause the cables to fail after they are returned to service. The failures would not have occurred at that point of time if the cables had remained in service and not been tested with DC [7]. Furthermore, from the work of Bach [8], we know that even massive insulation defects in solid dielectric insulation cannot be detected with DC at the recommended voltage levels.

After engineering evaluation of the effectiveness of a test voltage and the risks to the cable system, high DC may be considered appropriate for a particular application. If so, DC testing has the considerable advantage of being the simplest and most convenient to use. The value of the test for diagnostic purposes is limited when applied to extruded installations, but has been proven to yield excellent results on laminated insulation systems.

18.3.3 Alternative Test Methods

Renewed interest in alternative tests has appeared that stress the insulation as is done in actual operation and may be used in the factory for testing and evaluation purposes.

A serious disadvantage of power frequency AC tests at elevated voltage levels was the requirement for heavy, bulky, and expensive test transformers that may not be readily transportable to a field site. This problem has been mitigated through use of resonant (both series and parallel) test sets and compensated (gapped core) test transformers. They are designed to resonate with a cable at power frequency, the range of resonance being adjustable to a range of cable lengths through a moderate change of the excitation voltage frequency, or a pulse resonant system. Power frequency AC tests are ideally suited for PD location and dissipation factor (tan δ) evaluation.

Some of the practical disadvantages of power frequency tests are reduced while retaining the basic advantages by the use of very low frequency (VLF, or about 0.1 Hz) voltage or by the use of other time-varying voltages such as oscillating waves (OSW).

18.4 DIRECT VOLTAGE TESTING

18.4.1 Introduction

The use of DC has a historical precedent in the testing of laminated dielectric cable systems. Its application for testing extruded dielectric cable systems at HV is a matter of concern and debate. Reference [6] contains information relevant to these concerns.

This section presents the rationale for using DC testing, including the advantages and disadvantages and a brief description of the various DC field tests that can be conducted. These are generally divided into two broad categories, delineated by the test voltage level: low voltage DC testing (LVDC) covering voltages up to 5 kV and high voltage DC testing (HVDC) covering voltage levels above 5 kV.

Testing with a DC voltage source requires that only the DC conduction current be supplied rather than the capacitive charging current. This may greatly reduce the size and weight of the test equipment.

18.4.2 Performing Low Voltage Direct Current Tests

Equipment for producing these voltages is typified by commercially available insulation resistance testers. Some have multivoltage range capability.

Cable phases not under test should have their conductors grounded. Ends, both at test location and remote, should be protected from accidental contact by personnel, energized equipment, and grounds.

Apply the prescribed test voltage for specified period of time. It may be advantageous to conduct the test with more than one voltage level and record readings of more than one time period.

Such test equipment provides measurements of the insulation resistance of the cable system as a function of time. Interpretation of the results, covered in greater detail in old IEEE 400-1991 [2], made use of the change in resistance as testing progresses. A value of polarization index could be obtained by taking the ratio of the resistance after 10 minutes to the resistance after 1 minute.

The Insulated Cable Engineers Association (ICEA) provides values of insulation resistance in its applicable publications.

18.4.3 Performing High Voltage Direct Current Tests

Equipment for producing these voltages is typified by rectification of an AC power supply. Output voltage is variable by adjusting the AC input voltage. Output current into the cable system under test may be measured on the HVDC side or ratio transformation of the AC input. For the latter case, the test equipment leakage may mask the test current and the interpretation of results.

Apply the prescribed test voltage for the specified period of time. Reference [2] provides guidance for the selection of test voltage and time for cable systems having laminated dielectric cables.

The following three general types of test can be conducted with this equipment.

18.4.3.1 Direct Current Withstand Test

A voltage at a prescribed level is applied for a prescribed duration. The cable system is deemed to be acceptable if no breakdown occurs.

18.4.3.2 Leakage Current—Time Tests

Total apparent leakage output current is recorded as a function of time at a prescribed voltage level. The variations of leakage current with time (rather than its absolute value) provide diagnostic information on the cable system.

18.4.3.3 Step-Voltage Test or Leakage Current Tip-up Tests

The voltage is increased in small steps while the steady-state leakage current is recorded, until the maximum test voltage is reached or a pronounced nonlinear relationship between the current and the voltage is displayed. Such departures from linearity may denote a defective insulation system.

18.4.4 Advantages and Disadvantages

Some of the advantages and disadvantages of DC testing are listed as follows.

18.4.4.1 Advantages

- Relatively simple and light test equipment, in comparison to AC, facilitates portability
- Input power supply requirements readily available
- Extensive history of successful testing of laminated dielectric cable systems and well-established database
- Effective when the failure mechanism is triggered by conduction or by thermal consideration
- Purchase cost generally lower than that of non-DC test equipment for comparable kilovolt output

18.4.4.2 Disadvantages

- Blind to certain types of defects, such as clean voids and cuts
- May not replicate the stress distribution existing with power frequency AC voltages
- The stress distribution is sensitive to temperature and temperature distribution
- May cause undesirable space charge accumulation, especially in water-treed cable and at accessory/cable insulation interfaces
- May adversely affect future performance of water-tree-affected extruded dielectric cables
- Relatively high voltages have been used and they may damage the cable system

18.5 ON-LINE POWER FREQUENCY TESTING

18.5.1 Introduction

An on-line condition assessment service is available that utilizes the measurement of a form of PD. The sensitivity is in the order of 0.5 pC. The circuits are energized by their normal source of supply voltage. High frequency inductive sensors are moved to accessible portions of a cable circuit such as cables in manholes, terminations, etc. Novel techniques are used to filter noise in order to obtain these low sensitivity levels. Rather than using the picocoloumb unit for the PD level, the measurements are in millivolts.

18.5.2 Measurements

Sensors are placed at convenient locations along the cable route, such as near a splice, and noise level readings are taken over a frequency spectrum. Frequencies that produce high noise levels are avoided during actual data collection.

Data is collected and analyzed at the service provider's office. Sites are located and ranked as to the severity of their findings—at condition levels of 1 through 5. Level 1 means that no action is required, level 2 means consider testing again in 2 years, level 3 means a low probability of a service failure within 2 years, level 4 indicates a medium probability of failure in 2 years, and level 5 indicates a high probability of failure and that replacement should be considered.

18.5.3 Advantages

- The circuit may be left in service during the assessment
- All types of cable may be analyzed: paper, high molecular weight polyethylene (HMWPE), XLPE, tree-retardant polyethylene (TR-XLPE), etc.
- All voltage levels of cable, joints, and terminations may be analyzed
- Transformers up to 40 kV may be analyzed

18.5.4 Disadvantages

- Attenuation of these small signals that are being recorded is so large that readings may need to be taken in atleast every other manhole

- Wet manholes may require pumping
- Analysis of data requires highly trained engineers

18.6 OFF-LINE POWER FREQUENCY TESTING

18.6.1 Introduction

These methods have the advantage, unique among the off-line test methods described in this chapter, of stressing the insulation exactly as same as the normal operating conditions. It also replicates the most common method of factory testing on new cables and accessories.

There is a practical disadvantage in that the cable system represents a large capacitive load, and in the past, a bulky and expensive test generator was required if the cable system was to be stressed above normal operating levels. This size and bulk can be offset by the use of resonant and pulsed resonant test sources, which are described later.

A further advantage of power frequency testing is that it allows PD and dissipation factor (tan δ) testing for diagnostic purposes. Some other test sources also permit these measurements, but give rise to some uncertainty in interpretation, since the measurements are then made at a frequency other than the normal operating frequency.

It may seem logical that field tests use the same type of test voltage as when the circuit is in normal operation. However, a conventional power frequency transformer required even for full reel tests in the factory is a large and expensive device. Since a power cable may be made up of multiple reels of cable spliced together in the field, an even larger test transformer would be required to supply the heavy reactive current drawn by the geometric capacitance of the cable system.

The size of the transformer can be substantially reduced by using the principle of resonance. If the effective capacitance of the cable is resonated with an inductor, the multiplying effect of the resonant circuit (its Q factor) will allow the design of a smaller test transformer. In the ideal case of a perfect resonance, the test transformer will only be required to supply energy to balance the true resistive loss in the inductor and cable system. A further and significant reduction in size and weight of the test voltage generator can be achieved by use of the pulsed resonant circuit.

18.6.2 Test Apparatus Requirements

The following requirements are common to all three types of line frequency, resonant testing systems:

1. The apparatus is provided with an output voltmeter that responds to the crest of the test waveform. For convenience, this may be calibrated in terms of the root mean square (rms) voltage of the output. (i.e., as 0.707 times the crest voltage.)
2. The output waveform is sinusoidal and contains a minimum of line frequency harmonics and noise. This is of particular importance when diagnostic measurements (PD, power factor, etc.) are performed.

Suggested maximum values for total harmonic and noise are:

- For withstand tests ±5% of the output voltage crest
- For diagnostic tests ±1% of the output voltage crest

It should be noted that certain types of voltage regulators using inductive methods for regulation tend to produce large amounts of harmonic distortion. Line filters to minimize noise introduced from the power line are recommended for diagnostic measurements.

For withstand tests, the detection and indication of breakdown of the point at which breakdown occurs is defined by the overcurrent protective device of the test system. For this reason, it is desirable that a high speed and repeatable electronic circuit be used to operate the system circuit breaker and that the circuit breaker be as fast operating as practical.

18.6.3 Characteristics of Test Systems

The operating characteristics of a conventional test set are similar to that of a power transformer, although there are significant differences in the design of the source equipment.

Resonant systems operate differently than conventional transformers in that they have a specific tuning range for the capacitance of the cable under test. Capacitance outside this range cannot be energized. The minimum that can be energized can be reduced to zero in the series resonant system by using an auxiliary capacitor of appropriate rating in parallel with the test sample. The parallel resonant test system can be energized with no connected capacitance. The maximum value is independent of the current or thermal rating of the test system and cannot be exceeded. A typical tuning range is of the order of 20:1, maximum-to-minimum capacitance.

Both conventional and resonant test transformers provide an output that stresses the cable system under test identically to that under normal operations.

The output of a pulsed resonant test system consists of a power line frequency modulated at a low frequency, such as 1 Hz. The stress distribution in the cable system under test is therefore identical to that under normal operation. The only difference is that the magnitude of the stress varies periodically. The duration of the test must therefore be extended so that the cable under test is subjected to the same volt–time exposure as with a constant amplitude line frequency test.

18.6.4 Test Procedures

18.6.4.1 Testing with System Voltage (Medium Voltage)

It has been a utility practice for field crews to reclose an overhead circuit after a visual patrol is made of the circuit. The visual patrol is important in order to verify that any damaged equipment has been removed, downed lines have been restored, and feeds from alternate circuits have been disconnected. Fusing used in these operations are normally the size and type that was originally found in the switch.

This practice has been carried over to the underground residential distribution (URD) circuits by the same operating crews that switched the overhead system. URD circuits have been reenergized either by the overhead fuse connection or by the use of a separable connector (elbow). In some cases, continual reclosing and sectionalizing have been used to isolate a failure. This practice should be used sparingly since it may be damaging to the underground system. Reclosing in this manner may cause HV transients to be generated, and hence subject the circuit to excessive current surges. Both of those conditions may reduce the life and reliability of the underground circuit.

Devices have been developed to eliminate the need to reenergize a faulted underground circuit. With the use of a standard operating tester, a HV rectifier, the correct adapters, and an AC system source, a test can be applied to the underground circuit that will indicate whether a circuit is suitable for reenergization. A voltmeter phasing tester, in common use for overhead testing, can be modified to test underground circuits with the application of a HV rectifier and proper adapters. The voltmeter indicates the amount of charging current that is on the circuit being tested. Since the underground cable is a good capacitor, an unfaulted circuit would give a high reading when the tester is first connected to the circuit. As the capacitance charges, the reading on the voltmeter will decrease. If the reading fails to decrease, a faulted circuit is indicated.

When any of the above tests are complete, all parts of the circuit should be grounded four times the time it was being tested and the system made secure. The recommended rate-of-rise and rate-of-decrease of the test voltage is approximately 1 kV per second.

The duration of an acceptance test on a new cable system is normally 15 minutes at specified voltage. Maintenance tests may be 5 to 15 minutes. Any diagnostic tests (such as PD) may be performed during this period. The voltage should be maintained at the specified value to within ±1%.

18.7 PARTIAL DISCHARGE TESTING

18.7.1 Introduction

PD measurement is an important method of assessing the quality of the insulation of power cable systems, particularly for extruded insulation materials. A significant advantage of PD is that the site of discharge can be located with considerable accuracy by the use of TDR technology.

This chapter considers PD from two points of view: the measurement of all PDs occurring within the cable system and the location of individual PD sites. Keep in mind that PDs are high frequency events and are attenuated by changes in impedance in the cable system. A significant attenuation problem can occur with medium voltage cables having a tape shield. The metal surfaces that overlap each other can become sufficiently corroded so that the shield functions as an open spiral—not as a tube. This increase in impedance can greatly reduce the PD signal to the sensor, making PD testing not viable. (This condition is quickly assessed by placing a TDR signal on the cable system prior to the actual PD test. If the return signal does not appear at the sensor, it is obvious that PD testing will not work.)

18.7.2 Measurement of Partial Discharge

Perhaps the most significant factory test made on the insulation of full reels of extruded cable is the PD test. This is usually done at power frequency, but can also be carried out at VLF and at some voltage significantly higher than normal working voltage to ground. Experience has shown that this test is a very sensitive method of detecting small imperfections in the insulation such as voids or skips in the insulation shield layer.

It would therefore seem logical to repeat this test on installed cables to detect any damage done during shipping or laying, or any problems created by jointing and terminating the cable. In the past, this has been a difficult measurement to perform in the field due to the presence of other PD signals.

Once the necessary steps are taken to reduce the noise level below the PD level to be measured, the test can provide a great deal of useful diagnostic data. By observing the magnitude and phase of the PD signals and how they vary with increasing and then decreasing test voltage, the results will disclose information on the type and position of the defects and their probable effect on cable life.

Noise reduction methods necessary for field tests of PD usually include the use of an independent test voltage source such as a motor-generator, power line and HV filters, and shielding and sometimes the use of bridge detection circuits.

PDs can also be detected on-line with special sensors connected to splices or terminations using the frequency spectrum of the discharges. Signals generated from the cable system are separated so that it is possible to distinguish signals generated from different sources.

After signal separation is accomplished, identification is performed using proprietary software.

In summary, if the cable system can be tested in the field to show that its PD level is comparable with that obtained in the factory tests on the cable and accessories, it is the most convincing evidence that the cable system is in excellent condition.

18.7.3 Test Equipment

Excitation voltages for PD test sets that are commercially available include both 0.1 Hz and line frequencies. Resonant transformers are used for operating frequencies such as 50 and 60 Hz. Both resonant sets and 0.1 Hz sets are relatively small so that they can fit in a van or trailer.

A TDR measurement is first applied to determine the length of the circuit, the location of splices, and the condition of the neutral.

18.7.3.1 Advantages

- Locates PD sites.
- Measures both inception and extinction voltages of discharge.
- Voltage application (dwell) time is about 2 seconds after an approximate 10 second ramp time.
- PDs at 1.3 to 2.0 times operating voltage are indicative of a near-term service failure in XLPE or TR-XLPE cables.

18.7.3.2 Disadvantages

- PD magnitudes in joints and terminations are much higher than acceptable levels for extruded dielectric cables and have not been correlated with the remaining life the of accessories.
- Cable length limited to about 1 to 2 miles due to attenuation.
- Joints limit length of cable that may be tested because of their added attenuation of the signal.
- Voltages above 2 V_o may cause damage to extruded insulations.

18.8 DISSIPATION FACTOR TESTING

18.8.1 Introduction

Periodic testing of service-aged cables is practiced with the desire to determine system degradation and to reduce or eliminate service failures. Dissipation factor testing describes a diagnostic testing technique for field testing of service-aged shielded cable systems.

18.8.2 Dielectric Loss

Service-aged, shielded cable can be described by an equivalent circuit as shown in Figure 18.1. The cable capacitance per unit length, C, is:

$$C = 7.354/\varepsilon \log_{10}\left(d_i/d_C\right) \tag{18.1}$$

where

ε = dielectric constant of the insulation
d_i = the diameter over the insulation
d_c = the diameter under the insulation (over the conductor shield)

If the space between the coaxial conductors is filled with a conventional insulating material, the cable conductance per unit length, G, is:

$$G = 2\pi k e_v f C \tan\delta \tag{18.2}$$

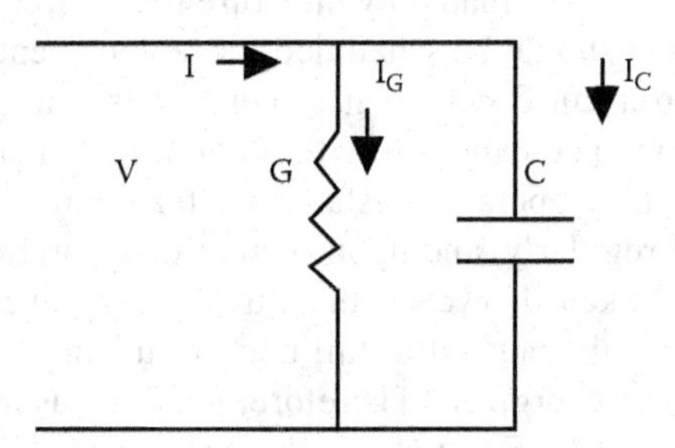

FIGURE 18.1 Equivalent circuit of a lossy portion of a power cable.

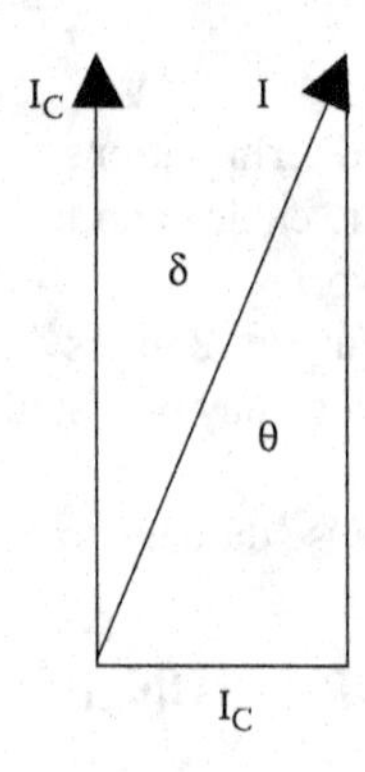

FIGURE 18.2 Phasor diagram for lossy dielectric.

The quantity tan δ is used to designate the lossyness of the insulating dielectric in an AC electric field. This is called loss factor or the tangent of the loss angle δ of the material. Typical values of e_v and tan δ are shown in Table 6.1.

For an applied voltage, V, the current through the loss-free dielectric is I_C and the current due to the lossyness of the material is I_G, see Figure 18.2. The angle formed by the current, $I = I_C + I_G$ and the current I_C, is δ. The angle formed by the current, $I = I_C + I_G$ and the voltage, V, is q and cos q is the power factor. I, I_C, and I_G are phasor quantities.

18.8.3 Method

The tan δ test is a diagnostic test that allows an evaluation of the cable insulation at operating voltage levels. The test is conducted at operating frequency or at the VLF frequency of 0.1 Hz. When the tan δ measurement exceeds a historically established value for the particular insulation type, the cable is declared defective and will have to be scheduled for replacement. If the tan δ measurements are below a historically established value for a particular insulation type, additional tests have to be performed to determine whether the cable insulation is defective.

Tests conducted on 1,500 miles of XLPE insulated cables have established a figure of merit for XLPE, tan $\delta = 4 \times 10^{-3}$. If the cables measured tan $\delta > 4 \times 10^{-3}$, the cable insulation is contaminated by moisture (water trees). The cable may be returned to service, but should be scheduled for replacement as soon as possible. If the cables measured tan $\delta < 10^{-3}$, it is not possible to predict the integrity of the cable insulation. The cable insulation could have many small defects, in which case the cable may operate satisfactorily for many more years. The tan δ should be monitored regularly, and upon further deterioration of the dissipation factor, proper action taken. However, the cable could have only a few isolated large defects, which could cause it to fail upon returning it to service or within days after it has been reenergized. Therefore, if the measured tan $\delta < 4 \times 10^{-3}$, it is recommended that a VLF test at 3 V_o be performed to identify remove, and repair the large defects.

18.8.4 Measurement and Equipment

Bridge type circuits are used to measure cable capacitance and tan δ. The most common are a Schering bridge and transformer ratio arm bridge. Both test sets require an AC HV source and a loss-free capacitor standard. For balanced bridges, the dissipation factor and cable capacitance are:
Schering bridge

$$\tan\delta = 2\pi f C_1 R_1 \tag{18.3}$$

$$C_X = C_N R_1 / R_2 \tag{18.4}$$

Transformer ratio arm bridge

$$\tan\delta = 2\pi f C_1 R_1 \tag{18.5}$$

$$C_X = C_N W_1 / W_2 \tag{18.6}$$

18.8.5 Advantages

- Tan δ measurements are diagnostic tests that permit assessment of the state of aging or damage of the cable insulation.
- Cables are tested with an AC voltage at operating voltage levels.
- The tests are performed at operating or at VLF frequencies.
- The tan δ test can be used on extruded as well as on PILC cables.
- When a cable does not pass the tan δ test, it can still be returned to service until repair or replacement has been scheduled.
- Monitoring of the tan δ will establish a cable history and a deterioration can be observed.

18.8.6 Disadvantages

- When a cable passes the tan δ test, it is not possible to declare the cable insulation sound.
- A VLF or breakdown test should be performed to identify any large defects in the cable system insulation.

For typical values of dissipation factor, see Table 6.1.

18.8.7 Dissipation Factor with Very Low Frequency Sinusoidal Wave Form

18.8.7.1 Method

Bach reported that loss factor (tan δ) measurements at VLF (0.1 Hz sinusoidal) could be used to monitor aging and deterioration of extruded dielectric cables [8]. The 0.1

Hz loss factor is mainly determined by water-tree damage of the cable insulation and not by water along the conducting surfaces. The measurement of loss factor with 0.1 Hz sinusoidal waveform offers comparative assessment of the aging condition of polyethylene, XLPE, and EPR type insulations. The test results permit differentiating between new, defective, and highly degraded cable insulations. The loss factor with a 0.1 Hz with sinusoidal waveform is a diagnostic test. Cables systems can be tested in preventative maintenance programs and returned to service after testing. The dissipation factor measurements at VLF can form the basis for the justification of cable replacement or cable rejuvenation expenditures.

18.8.7.2 Measurement Equipment

A programmable HV VLF test generator with loss factor measurement capability is connected to the cable under test. If for a test voltage of the V_0 $\delta > 4 \times 10^{-3}$, the service aged cables should eventually be replaced. The test voltage should not be raised above V_0 in order to prevent insulation breakdown. If for test voltage of V_0 the $\tan \delta \ll 4 \times 10^{-3}$, the service-aged cables should be additionally tested with VLF at 3 V_0 for 60 minutes. When the cables pass this test, it can be returned to service without reservation. Loss factor measurements at VLF can form the basis for the justification of cable replacement or cable rejuvenation expenditures.

18.8.7.3 Advantages

- The test is a nondestructive, diagnostic test.
- Cables are tested with an AC voltage equal to the phase-to-ground voltage at which they operate.
- Cable system insulation can be graded as excellent, defective, or highly deteriorated.
- Cable system insulations can be monitored and history developed. Cable replacement and rejuvenation priority can be planned.
- Test sets are transportable and power requirements are comparable to standard cable fault locating equipment.

18.8.7.4 Disadvantages

The maximum available test voltage is 36 kV rms and the maximum capacitive load is about 3 μF.

- The test works best after comparative cable system data have been developed.

18.9 VERY LOW FREQUENCY (ABOUT 0.1 HZ) TESTING

18.9.1 Introduction

VLF testing methods can be categorized as withstand or diagnostic. In withstand testing, insulation defects are caused to breakdown (fault) at the time of testing. Faults are repaired and the insulation is retested until it passes the withstand test. The withstand test is considered a destructive test. Diagnostic testing allows the identification of the relative condition of degradation of a cable system and establishes, by comparison with figures of merit, if a cable system can or cannot continue operation. Diagnostic testing is considered as nondestructive [9].

In extreme cases, when the cable system insulation is in an advanced condition of degradation, the diagnostic tests can aggravate the condition of the cable and cause breakdown before the test can be terminated.

The VLF withstand test methods for cable systems are:

- VLF testing with cosine-pulse waveform
- VLF testing with sine waveform
- VLF testing with square wave with programmable slew rate

The VLF diagnostic test methods for cable systems are:

- VLF dissipation factor (tan δ) measurement
- VLF PD Measurement

Field testing techniques frequently employ a combination of diagnostic and withstand test methods. They are selected based on their ease of operation and cost/benefit ratio. The various VLF test methods described are in commercial use and are accepted as alternate test methods in international standards.

18.9.2 Very Low Frequency Withstand Testing

The Electric Power Research Institute (EPRI) and the Canadian Electrical Association (CEA) funded a study to evaluate the advantages and limitations of 0.1 Hz testing as a possible substitute for DC maintenance testing of in-service, aged power cables [9]. The use of DC was intended to confirm the soundness of a cable and to weed out weakened cables.

The subject of appropriate methods for fault location and maintenance testing of extruded dielectric distribution cables has been given considerable attention. This interest was generated by the knowledge that transient voltages and DC testing can reduce the life of XLPE and TR-XLPE insulated cables. There is an unanswered question as to the effect on EPR cables, but the concern exists.

The laboratory phase of this work evaluated a wide range of insulations and ages of cables. During this laboratory evaluation, various voltages were applied to model cables and service-aged XLPE, 15 kV samples. Similar tests were run on full size EPR, PILC, and transition joints. The results of the study concluded that low frequency AC testing can detect cable imperfections in XLPE cable and transition joints at lower voltages than for a DC test, with none of the detrimental effects of DC.

Once the suitable magnitude and application time of the voltage were determined, field tests were conducted on four utility systems. It was concluded that 0.1 Hz is a satisfactory alternative for DC testing. The shape factor of the 0.1 Hz is the same as the operating system and no damage was found as judged by the decrease in magnitude of AC breakdown. If the circuit fails during the test, the lower test voltage reduces the transient voltages that the system must endure.

Procedures were developed for 0.1 Hz testing of 15 kV cable systems at 22 kV for 15 minutes. Other values are suggested for 5 to 35 kV. A large eastern utility has been using this procedure for several years on mixed cable systems and is very satisfied with the results. Absence of long-term effects on service-aged extruded cables after this test remains to be established.

18.9.3 Very Low Frequency Testing with Cosine-Pulse Waveform

18.9.3.1 Method

The VLF cable test set generates a 0.1 Hz bipolar pulse wave, which changes to sinusoidal polarity. Sinusoidal transitions in the power frequency range initiate a PD at an insulation defect, which the 0.1 Hz pulse wave develops into a breakdown channel. Within minutes, a defect is detected and forced to become a failure. It can then be located with standard, readily available cable fault-locating equipment. Cable systems can be tested in preventative maintenance programs or after a service failure. Identified faults can be repaired immediately and no new defects will be initiated during the testing process. When a cable system passes the VLF test, it can be returned to service.

18.9.3.2 Measurement and Equipment

A DC test set forms the HV source. A DC-to-AC converter changes the DC voltage to the VLF AC test signal. The converter consists of a HV choke and a rotating rectifier that changes the polarity of the cable system being tested every 5 seconds. This generates a 0.1 Hz bipolar wave. A resonance circuit, consisting of a HV choke and a capacitor in parallel with the cable capacitance, assures sinusoidal polarity changes in the power frequency range. The use of a resonance circuit to change cable voltage polarity preserves the energy stored in the cable system. Only leakage losses have to be resupplied to the cable system during the negative half of the cycle.

The 0.1 Hz test set is easily integrated in a standard cable fault-locating and cable-testing system by making use of available DC high potential sets. Stand-alone VLF systems should be supplemented by cable fault-locating equipment.

The cable system to be tested is connected to the VLF test set. In five to six steps, the test voltage is regulated to the test voltage level of 3 V_0 (V_0 is phase-to-ground voltage). The recommended testing time is 15 to 60 minutes. When the cable system passes the VLF voltage test, the test voltage is regulated to zero and the cable and test set are discharged and grounded. When a cable fails the test, the VLF test set is turned off to discharge the system. The fault can then be located with standard cable fault-locating equipment.

18.9.3.3 Advantages

- The VLF test uses a 0.1 Hz pulse wave that changes to sinusoidal polarity. The sinusoidal transitions in the power frequency range may initiate a PD at a defect that the 0.1 Hz pulse wave may develop into a breakthrough channel.
- Due to sinusoidal transitions between the HV pulses, traveling waves are not generated.

- Due to continuous polarity changes, dangerous space charges cannot develop. Cables can be tested with an AC voltage up to three times the conductor-to-ground voltage with a device comparable in size, weight, and power requirements to a DC test set.
- The VLF test can be used on extruded as well as oil impregnated paper insulations.
- The VLF test with cosine-pulse waveform works best when eliminating a few singular defects from an otherwise good cable insulation. The VLF test is used to "fault" the cable defects without jeopardizing the cable system integrity.
- When a cable passes the recommended 0.1 Hz VLF test, it can be returned to service.

18.9.3.4 Disadvantages

- When testing cables with extensive water-tree damage or ionization of the insulation, VLF testing alone is often "not conclusive." Additional tests that measure the extent of insulation losses will be necessary.
- Present limitations are the maximum available test voltage of 56 kV.

18.9.4 Very Low Frequency Testing with Sinusoidal Waveform

18.9.4.1 Method

The VLF test set generates sinusoidal changing waves that are less than 1 Hz. When the local field strength at a cable defect exceeds the dielectric strength of the insulation, PD starts. The local field strength is a function of applied test voltage, defect geometry, and space charge. After initiation of PD, the PD channels develop into breakthrough channels within the recommended testing time. When a defect is forced to break through, it can then be located with standard, readily available fault locating equipment. Cable systems can be tested in preventative maintenance programs after failure. Identified faults can be repaired immediately. When a cable passes the VLF test, it can be returned to service.

18.9.4.2 Measurement and Equipment

The VLF test set is connected to the cable or cable system to be tested. The test voltage is regulated to the test voltage level of 3 V_0. The recommended testing time is 60 minutes or less than that found in VLF testing guides. VLF sets that have sufficient capacity to be able to supply and dissipate the total cable system charging energy can be found. When the cable system passes the VLF voltage test, the test voltage is regulated to zero and the test set and cable system are discharged and grounded. When a cable fails the test, the VLF test is turned off to discharge the cable system and test set and the cable fault can then be located with standard cable fault-locating equipment.

In addition to standard 0.1 Hz sinusoidal VLF test sets, which have been in use for many years for VLF testing of electrical machines, several variations are also available to meet specific cable system test requirements:

1. VLF, less than 0.1 Hz, high-voltage generator with programmable test voltage waveforms for cable systems with mixed insulation:
   a. Sine wave test voltage
   b. Bipolar pulse wave with defined slew rate
   c. Regulated DC test voltage with positive and negative polarities
   d. Programmable step test for all voltage waveforms
2. VLF, 0.1 Hz, HV generator with dissipation factor (tan δ) measurement capability
3. PD free, VLF, 0.1 Hz HV generator for PD testing
4. PD free, VLF, bipolar pulse with defined slew rate, HV generator for PD testing

18.9.4.3 Advantages

- Cables are tested with an AC voltage up to three times the conductor-to-ground voltage. After initiation of a PD, a breakthrough channel at a cable defect develops.
- Due to continuous polarity changes, dangerous space charges do not develop in the cable insulation.
- Test sets are transportable and power requirements are comparable to standard cable fault-locating equipment.
- The VLF test can be used on extruded as well as paper-type cable insulations. The VLF test with sinusoidal waveform works best when eliminating a few defects from an otherwise good cable insulation. The VLF test is used to "fault" the cable defects without jeopardizing the cable system integrity. When a cable passes the recommended 0.1 Hz VLF test, it can be returned to service.
- VLF test sets with 0.1 Hz loss factor measurement capability for diagnostically identifying cables with highly degraded cable insulations are available and can be used with a 0.1 Hz withstand test.

18.9.4.4 Disadvantages

- When testing cables with extensive water-tree damage or ionization of the insulation, VLF withstand testing alone is often "not conclusive." Additional tests that measure the extent of insulation losses will be necessary.
- Limitations are the maximum available test voltage of 36 kV rms and the maximum capacitive load of approximately 3 μF at 0.1 Hz (30 μF at 0.01 Hz). The total charging energy of the cable has to be supplied and dissipated by the test in every electrical period. This limits the size of the cable system that can be tested. A long testing time must be seen as an inconvenience rather than a limitation.

18.9.5 Tan δ Test with VLF Sinusoidal Waveform

18.9.5.1 Method

Bahder [18] first used dissipation factor (tan δ) measurements to monitor aging and deterioration of extruded dielectric cables. Bach reported a correlation between an increasing 0.1 Hz dissipation factor and a decreasing insulation breakdown voltage level at power frequency. The 0.1 Hz loss factor is mainly determined by water-tree damage of the cable insulation and not by water along the conducting surfaces. The measurement of the loss factor with a 0.1 Hz sinusoidal waveform offers comparative assessment of the aging of polyethylene, XLPE, and EPR type insulations. The test results permit differentiation between new, defective, and highly degraded cable insulations. The loss factor with a 0.1 Hz sinusoidal waveform can be used as a diagnostic test. Cables can be tested in preventative maintenance programs and returned to service after testing. The loss factor measurements at VLF can be used to justify cable replacement or cable rejuvenation expenditures.

18.9.5.2 Measurement and Equipment

A programmable, HV VLF, 0.1 Hz test generator with dissipation factor measurement capability is connected to the cable system under test. The dissipation factors of tan δ at V_0, tan δ at 2 V_0, and the differential loss factor Δ tan δ (tan δ at 2 V_0 minus tan δ at V_0) are measured. The measured values are used as figures of merit to grade the condition of the cable insulation as good, defective, or highly deteriorated.

For example, XLPE insulation is tested at a 0.1 Hz test voltage of V_0 and the tan $\delta > 2 \times 10^{-3}$, the service-aged cables should be replaced. The test voltage should not be raised above V_0 in order to prevent an insulation breakdown. If for a 0.1 Hz test voltage of V_0, and the tan $\delta >> 1.2 \times 10^{-3}$, the service-aged cables should additionally be tested with VLF $3 \times V_0$ for 60 minutes. When cable passes this test, it can be returned to service.

If for a 0.1 Hz test voltage at V_0 and 2 V_0, the loss factors are $1.2 \times 10^{-3} <$ tan $\delta < 2 \times 10^{-3}$ and the differential loss factor Δ tan $\delta < 0.6$, the cable should be returned to service but monitored semiannually.

If for a 0.1 Hz test voltage at V_0 and 2 V_0, the loss factors are $1.2 \times 10^{-3} <$ tan $\delta < 2 \times 10^{-3}$ and the Δ tan $\delta > 1.0$, the cable should be replaced.

It must be understood that, for different insulations, installations, and cable types, tan δ figures of merit can vary significantly from the values listed in the previous paragraph. The test gives the best results when comparing present measurements against established historical figures of merit for a particular cable.

18.9.5.3 Advantages

- This test is a diagnostic, nondestructive test. Cable systems are tested with an AC voltage equal to the conductor-to-ground voltage.
- Cable system insulation can be graded between good, defective, and highly deteriorated.
- Cable system insulation condition can be monitored over time and a cable system history be developed. Cable replacement and cable rejuvenation priority and expenditures can be planned.
- Test sets are transportable and power requirements are comparable to standard cable fault-locating equipment.

18.9.5.4 Disadvantages

- For a 0.1 Hz VLF test set, the maximum available test voltage is 36 kV rms and the maximum capacitive load in approximately 3 μF.
- The test becomes useful after historical comparative cable system data have been accumulated.

The suitability, practicality, and effectiveness of these testing methods for service-aged power cables with extruded dielectric insulation should be determined based on experience.

It is known that DC testing of extruded dielectric insulated cables is not very useful. In fact, it may cause cables to fail after having been returned to service. At this time, VLF test techniques are effective alternatives for testing of service-aged power cables with extruded dielectric insulation.

18.10 OSCILLATING WAVE TEST

18.10.1 Introduction

OSW testing, also known as damped alternating current (DAC) testing [5], was selected by a CIGRE task force "alternative tests after laying" as an acceptable compromise using the following criteria:

- The ability to detect defects in the insulation that will be detrimental to the cable system under service conditions, without creating new defects or causing aging
- The degree of conformity between the results of tests and the results of 50 or 60 Hz tests
- The complexity of the testing method
- The commercial availability and costs of the testing equipment

The purpose of the OSW testing method is to detect defects that may cause failures during service life without creating new defects that may threaten the life of the cable system.

Although OSW testing does not have a wide reputation with respect to cable testing, it is already used for testing in metal-clad substations and is being recommended for gas insulated cable testing.

18.10.2 General Description of Test Method

The test circuit basically consists of a DC voltage supply, which charges a capacitance C_1, and a cable capacitance C_2. After the test voltage has been reached, the capacitance is discharged over an air coil with a low inductance. This causes an oscillating voltage in the kilohertz range. The choice of C_1 and L depends on the value of C_2 to obtain a frequency between 1 and 10 kHz.

18.10.3 Advantages

- The OSW method is based on an intrinsic AC mechanism.
- The principal disadvantages of DC (field distribution, space charge) do not occur.
- The method is easy to apply.
- The method is relatively inexpensive.

18.10.4 Disadvantages

- Effectiveness of the OSW test method in detecting defects is better than with DC but worse when compared with AC (60 Hz).
- In particular for medium voltage cable systems, the factor f* OSW/60 Hz voltage approaches 1, indicating the mutual equivalence.
- For HV cable systems f* OSW/60 Hz is significantly higher (1.2 to 1.9), which means that OSW is less effective than 60 Hz.
- For both HV and MV cable systems, f* OSW/DC is low (0.2 to 0.8), indicating the superiority of OSW over DC voltage testing.
- Since the capacitance of a cable is dependent on length, each length of otherwise identical cable, for instance, oscillates at a different frequency. This difference in frequency creates a change in measured properties. [*Note:* f* OSW/60 Hz is the ratio of breakdown values for a dielectric containing a standard defect when using OSW voltage and 60 Hz voltage, respectively.]

18.10.5 Test Apparatus

The cable is charged with a DC voltage and discharged through a sphere gap into an inductance of appropriate value so as to obtain the desired frequency. The voltage applied to the cable is expressed by:

$$V(t) = V_1 e^{\alpha t/\sqrt{LC}} \cos 2\pi f t \tag{18.7}$$

where

V_1 = charging voltage provided by the generator
α = damping ratio
$C = C_1 + C_2$

$$f = 1/2\pi\sqrt{LC}$$

Other test circuits are possible and give alternative solutions using different circuit configurations.

18.10.6 Test Procedure

Many tests carried out so far are of an experimental nature. Artificial defects like knife cuts, wrong positions of joints and holes in the insulation were created and subjected to different testing procedures including the OSW testing method.

These test procedures were intended to obtain breakdown as a criterion for comparison. The general testing procedure is as follows:

- Start to charge the cable with a DC voltage of about one or two times the operating voltage.
- Increase with steps of 20 to 30 kV.
- Produce 50 shots at each voltage level.
- Time interval between shots 2 to 3 minutes.
- Proceed until breakdown occurs.

In one case, the Dutch testing specification for HV extruded cables, the OSW method is mentioned as a withstand test to be used as an after-laying (installation) test. The test procedure is as follows:

- Charge the cable slowly, using the DC power supply.
- After reaching the value of 3 V_0, the DC source will be disconnected and the rapid recloser activated.
- The cable circuit will be discharged through a reactor, causing the OSW testing voltage.
- This procedure should be repeated 50 times.

In the Netherlands, the OSW testing method is applied as an after-laying test for HV extruded cable systems.

18.10.7 Further Development Work

Since the effectiveness of the OSW testing method is not as high as would be desired, it might be very attractive to combine OSW with PD site detection as an additional source of information. In the literature, details are given of an automatic PD measurement system, enabling statistical analysis and generating phase, time, and amplitude resolved PD fingerprints. Compared to AC 50/60 Hz generated PD fingerprints, additional information results from the decreasing voltage amplitude of each OSW pulse. For medium voltage cable systems according to [12], this measuring system looks feasible.

18.11 DIELECTRIC SPECTROSCOPY

Dielectric spectroscopy is one of several test methods to measure dielectric losses in a cable system [16,17]. The proponents believe that PD activity from internal cavities and surface discharges are the main cause of aging. This measurement can provide information on the global effects caused by water treeing and oxidation of the insulation.

The general approach is to separate any "noise" from the PD patterns. The measurement is based on the response of the insulation when subjected to ever increasing voltages over a wide frequency range:

- If the response is linear, there is little or no aging.
- If the response is nonlinear, the cable is aged and deteriorated.

The significance of the deterioration is determined from the magnitude of response. The frequency range is from 0.0001 kHz to 1 kHz and the voltage for XLPE is from 0 to 20 kV peak.

REFERENCES

1. IEEE Std 400-2001, "Guide for Field Testing and Evaluation of Shielded Power Cables", SH94972.
2. IEEE Std 400.1-2007 (previously IEEE 400-1991), "Guide for Field Testing Laminated Dielectric, Shielded Power Cable Systems Rated 5 kV and Above With High Direct Current Voltage".
3. IEEE Std 400.2-2005, "Guide for Field Testing of Shielded Power Cable Systems Using Very Low Frequency".
4. IEEE Std 400.3-2007, "Guide for PD Testing of Shielded Power Cable Systems in a Field Environment".
5. IEEE Std 400.4 (under development), "Guide for Field Testing of Shielded Power Cabled Systems Rated 5 kV and Above with Damped Alternating Current Voltage (DAC)."
6. Srinivas, N. H., et al., June 1991, "Effect of DC Testing on Aged XLPE Insulated Cables with Splices", JICABLE 91, Paris, France.
7. Eager, G. S., Jr., et al., July 1992, "Effect of DC Testing Water Tree Deteriorated Cable and a Preliminary Evaluation of VLF as an Alternative", IEEE Transactions on Power Delivery, Vol. 7 (No. 3).
8. Bach, R., 1989, "Quervergleich Verschiedener Spannungarten zur Pruefung von Mittelspannungs Kabelanlagen", (in German), Technical University of Berlin, Annual Report of Research Activities.
9. Report TR-110813, "High Voltage, Low Frequency (0.1 Hz) Testing of Power Cables", EPRI, Palo Alto, CA and Canadian Electrical Association, Montreal, Quebec, July 1998.
10. AEIC CG7-2005, "Guide for Replacement and Life Extension of Extruded Dielectric 5–35 kV Underground Distribution Cable", Second Edition.
11. IEEE 510-1992, *IEEE Recommended Practices for Safety in High-Voltage Power Testing*, reaffirmed 1992.
12. IEEE 4-2001, *IEEE Standard Techniques for High-Voltage Testing*.
13. IEEE 48-2009, *IEEE Standard Test Procedures and Requirements for High-Voltage Alternating-Current Cable Terminations*.
14. Ahmed, N. and Srinivas, N., 1998, "On Line Partial Discharge Detection in Cables", IEEE Dielectrics and Electrical Insulation, Vol. 5 (No. 2).
15. Srinivas, N., Nishioka, T., Sanford, K., and Bernstein, B., 2006, "Non Destructive Condition Assessment of Energized Cable Systems", IEEE/PES Transmission and Distribution Conference, Dallas, TX.

16. Werelius, P., Tharning, P., Erikson, R., Holmgren, B. and Gafvert, U., February 2001, “Dielectric Spectroscopy for Diagnosis of Water Treed Deterioration in XLPE Cables”, IEEE Dielectrics and Electrical Insulation, Vol. 1, pp. 27–41.
17. Reiter, M. Gockenbach, E. and Bursi, H., “Dielectric Spectroscopy of Multi-stress Aged XLPE Insulation of Synergistic Effects”, *Proceedings of the 2005 International Symposium of Electrical Insulation Materials*, Vol. 1, ISEM 2005.
18. Baher, G., et al., 1977, “In Service Evaluation of Polyethylene and Cross-Linked Polyethylene Insulated Cables Rated 15 to 35 kV”, IEEE *Transactions Power Apparatus and Systems*, PA&S-96, pp. 1754–1766.

19 Treeing

William A. Thue

CONTENTS

19.1 INTRODUCTION

Treeing in extruded dielectric cable insulation is the term that has been given to a type of electrical prebreakdown deterioration that has the general appearance of a tree-like path through the wall of insulation. This formation is radial to the cable axis and hence is in line with the electrical field.

Trees form in insulations such as polyethylene, cross-linked polyethylene (XLPE), and ethylene–propylene rubber (EPR) cables and are considered as two distinct types:

- Water trees (also known as electrochemical trees)
- Electrical trees

They are differentiated by the distinctions shown next and other parameters that will be discussed in detail later in this chapter.

Water Trees

- Diffuse, indistinct small voids separated by somewhat deteriorated insulation
- Moisture is required
- Slow growth (months, years)
- Must be stained to see them. This may be from chemicals in or around the cable or be stained while the cable is examined
- May exist without immediate electrical failure

Electrical Trees

- Distinct hollow channels
- Water not required
- Rapid growth (seconds, minutes, hours)
- Electrical failure occurs very quickly

19.2 BACKGROUND

The phenomenon known as treeing in dielectrics was first described by Rayner in 1912 [1]. He had been investigating electrical breakdown in the presence of discharges in *paper-insulated cables.* The tree-like appearance of Lichtenberg figures was well known during the 1920s. These "trees" are totally different from what is seen in extruded dielectric cables because those older trees were carbon paths burned into the paper insulation that proceed concentrically around the insulating wall. In Figure 19.1, six adjacent tapes are shown together.

Treeing in extruded dielectric cables was described by Whitehead [2] in 1932 in his work on electrical breakdown. The development of corona detection equipment in 1933 by Tykociner, Brown, and Paine [3] made quantitative studies possible. Kreuger [4] thoroughly described the methods for detection and measurement of discharges in 1965.

The announcements by Vahlstrom and Lawson [5,6] in 1971 and 1972 that direct buried high molecular weight polyethylene (HMWPE) cables installed in the field

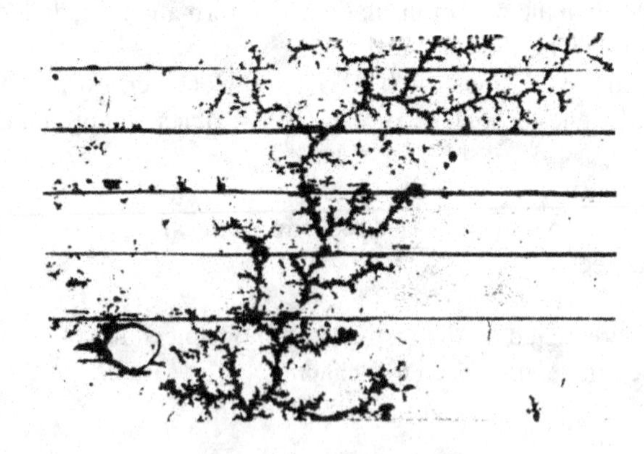

FIGURE 19.1 Tree-like deterioration in paper insulation. Courtesy of University of Wisconsin–Madison.

in underground residential distribution (URD) systems contained water trees made a significant impact on the cable industry. This was the first time that this type of treeing had been reported in a typical utility environment. Previously reported results, especially by the Japanese [7] that they called "sulfide trees," now became required reading.

Up until that time, mechanical damage was considered to be the predominant mechanism of failure for cables in a utility installation. The remaining 10% were thought to be due to unknown causes that were not necessarily due to electrical deterioration in the cable.

19.3 WATER TREES

Water (also called electrochemical or chemical) trees form at a slow rate that may take many years to propagate and grow. Water trees can occur in all solid (extruded) dielectric materials. The tree-like appearance can be described by reference to many natural shapes that are sometimes obvious upon cutting wafers from aged cables. Examples are shown later in this chapter.

The visibility of water trees stems from the staining of the interior of the tree wall by some form of chemical staining. Nonstained water trees disappear when the sample is dried. Staining techniques are discussed later in this chapter.

Water treeing is influenced by the following:

Moisture (an essential ingredient!)
Voltage stress
Voids
Contaminants
Ionic impurities
Temperature
Temperature gradient
Aging time
PH

19.3.1 Mechanisms of Water Treeing

There are three stages in the development of water trees in extruded insulation:

- Inception
- Propagation [12,13]
- Conversion/electrical breakdown

19.3.1.1 Treeing Mechanisms

Laboratory investigation of water tree inception has demonstrated that several probable mechanisms are involved:

1. Voltage stress concentration at a very small void or contaminant can be orders of magnitude higher than the average electrical stress at that location. Although very sensitive measuring equipment has been utilized (about

0.05 pC sensitivity), no partial discharge has been detected during this stage of inception or growth. Electron bombardment in such a cavity is capable of erosion of the dielectric in the cavity.
2. Mechanical stress can be produced by very high electrical fields around an electrical stress concentration. This may result in fatigue and the development of stress cracks in the dielectric. Partial discharges in such cracks may lead to local electrical breakdown and tree formation.
3. Electron bombardment can be sufficiently high to cause a cavity to form due to ionization of the dielectric that causes chain scission and decomposition.
4. Local intrinsic breakdown of the dielectric.

None of these, by itself, can explain the results of laboratory studies, so it is probable that initiation of a water tree involves more than one mechanism. The effect of high electrical stress is universally accepted, however.

19.3.1.2 Treeing Propagation

Propagation or growth of water trees seems as elusive as inception. Only the presence of an alternating current (AC) electrical field seems to be a unifying conclusion. Although direct current (DC) can produce water trees, it is at a much slower rate and with a different shape. It is well established that polymers are readily permeable by moisture and that the presence of moisture in these insulations is deleterious.

Two significant forces to explain the penetration of moisture through extruded dielectrics are:

1. Electrophoresis is a term used to describe the movement of charged particles in an electric field. Particles with a positive charge tend to move toward a negative electrode, while negative ions tend to move toward the positive electrode. A practical application of this theory when DC testing a cable is to place the negative charge on the conductor. This pulls moisture in toward the center of the cable and makes the test more definitive.
2. Dielectrophoresis describes the movement of uncharged but polarized particles or molecules in a divergent field (Figure 19.2). In the example of an otherwise uniform single-conductor electrical cable, the field increases as

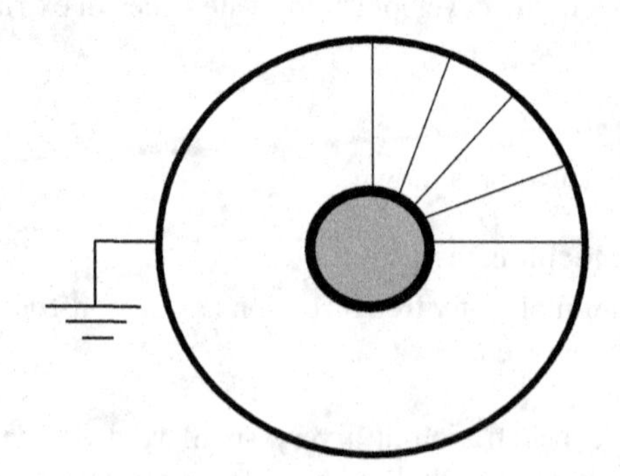

FIGURE 19.2 Dielectrophoresis.

> a particle or molecule gets closer to the conductor. An uncharged particle will be polarized at any given point of time so that it will have a negatively charged dipole with its negative side, for instance, toward the conductor that is positive at that instant. Since the negative side of this dipole exists in a stronger field than the positive side, the particle will be attracted toward the field of greatest field intensity. In an AC system, as the conductor becomes negatively charged, the polarization process is reversed. This means that the particle is still attracted toward the conductor with its higher electric field.

The practical effect of dielectrophoresis is that moisture will be drawn to the higher dielectric field regions even in an alternating field. This high stress point may be at the conductor or a small void that was formed during manufacturing. A void that was initially filled with gas now becomes water filled. While this does not fully explain the formation of the water tree, it does shed some light on the growth of such trees and the dispersion of moisture in an energized cable [14].

Water tree growth is a slow mechanism of deterioration in extruded dielectrics when exposed to both moisture and voltage stress. They initiate from sites that create high electrical stress such as voids, contaminants, protrusions, and loose or rough shields. Such trees may grow from either shield and are known as vented trees. These are often known as fans, broccoli, etc. When water trees grow from imbedded voids or contaminants, they are often called bow ties and are usually not vented to the atmosphere.

Water trees have fine structures when seen with low magnification. Under high power magnification, they may be seen as discreet ellipsoids that basically line up with the electrical field. The insulation between these ellipsoids may be somewhat deteriorated through oxidation or other means, but they remain as adequate electrical insulation—but may be sites for electron entrapment. (There will be more discussions on this later.)

In contrast to electrical trees, water trees are characterized by:

1. Slow growth. They may grow for many years before they contribute to breakdown.
2. Water trees are initiated and grown at much lower electric fields than do electrical trees.
3. They can be so large that they extend from one shield to another without resulting in breakdown.
4. Some moisture is required.
5. Temperatures above 70°F accelerate tree growth.
6. If partial discharge is present, we are not able to detect it at that time.
7. They disappear when allowed to dry unless stained. They reappear when placed in boiled water.

19.3.1.3 Examples of Water Trees

Some examples of water trees are shown in Figures 19.3 through 19.6 along with the names that describe their shape.

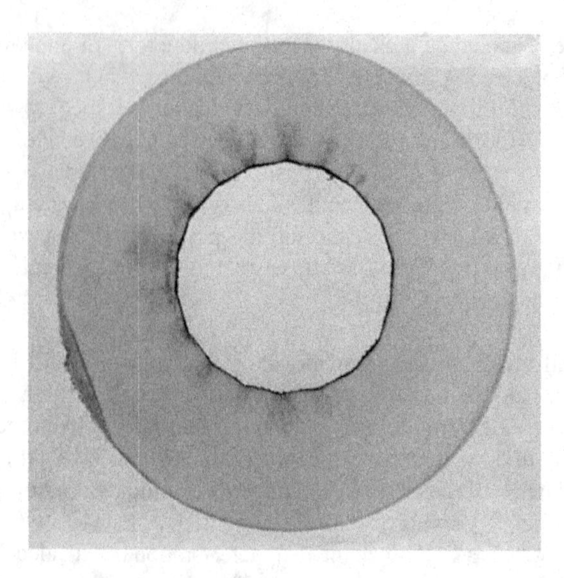

FIGURE 19.3 Vented fan trees growing from high electrical stress points in a HMWPE cable with taped strand shielding and a staining antioxidant. Photo by W. A. Thue.

19.3.1.4 Methods to Minimize Water Treeing

The most effective, if not economic, method to avoid the formation of water trees is to keep the insulation absolutely dry. This can be accomplished with an impervious metallic sheath such as lead, copper, or aluminum. Cables with these sheaths have been in service for over 40 years with no known deterioration. Of course, the problem is to keep the metal from corroding or being damaged mechanically. (This concern also applies to paper-insulated cables!)

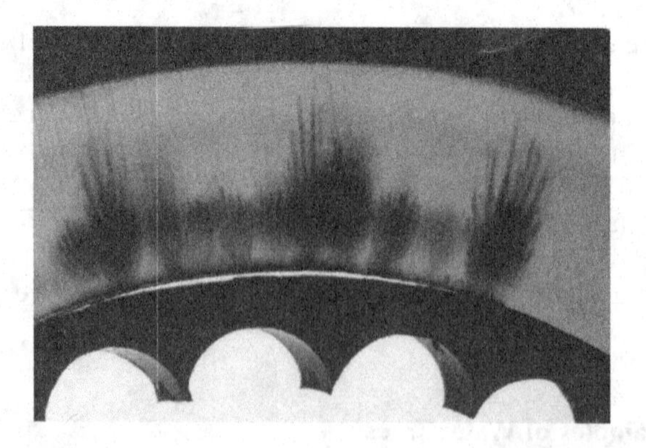

FIGURE 19.4 Vented water trees growing from a loose extruded strand shield in a 175-mil wall of XLPE cable.

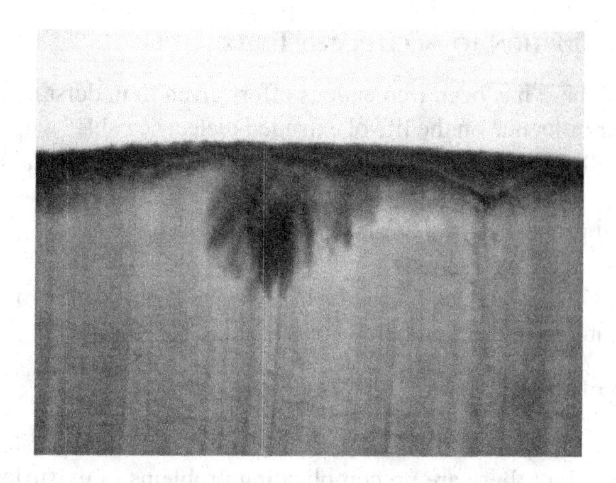

FIGURE 19.5 Water trees growing from taped insulation shield. Photo by W. A. Thue.

Other possible deterrents to water tree formation include:

1. Smooth electrical interfaces at the shields—especially at the conductor shield where the electrical stresses are the highest
2. Tightly bonded insulation shields that must be removed mechanically
3. Semiconducting compounds with small amounts of ionic impurities. It has been shown that ions from the shields are concentrated at water tree sites in the insulation
4. Solid conductors
5. Strand blocking
6. Internal pressure, such as nitrogen, to keep out moisture
7. Reduction or elimination of voids and contaminants

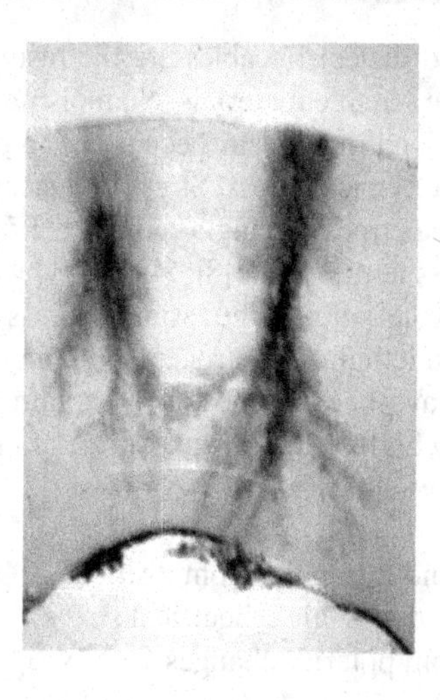

FIGURE 19.6 Two large water trees across a 220-mil HMWPE cable made with staining antioxidant. (Note that there is no electrical failure showing that even these water trees are capable of holding line voltage.) Photo by W. A. Thue.

19.3.2 Introduction to Accelerated Testing

Over the years, there has been tremendous effort given to understanding water trees because of their influence on the life of extruded dielectric cables with the subsequent loss of life [15–18]. Tests in laboratories can be separated into several methods:

1. Electrical properties of material specimens
2. Miniature cable evaluations
3. Typical full sized cables—usually 1/0 AWG, 175-mil wall cable
4. Actual cable construction being considered

The testing of specimens:

1. Testing of samples is the least expensive and is best suited for screening purposes. Here there are no complicating problems of extrusion and other manufacturing conditions.
2. Miniature cables are the least expensive way to introduce the manufacturing and processing concerns, but have not shown to be in good correlation with full sized cables.
3. The 1/0 AWG, 175-mil cable has become the most commonly used full sized cable because it is a very common size among the utilities in North America, and because of its modest cost of production. Manufacturing and processing are factored into the results. These tests are still very costly with a typical AEIC Qualification Test requiring $50,000 or more and 12 months to accomplish.
4. Tests on the actual cable under consideration are even more costly than that for the 1/0 cable described above.

19.4 ELECTRICAL TREES

Electrical trees in extruded dielectric cables are the result of internal electrical discharges that decompose the organic materials. No moisture is needed for this process. Overvoltages are required and may occur because of imperfections in the structure that cause high, localized electrical stress. Stress enhancement calculations from the tip of a sharp conductive electrode having a radius of 2.5 microns vary from 240 to 480 times the average stress at that point [8]. Since a typical distribution cable has an average stress of about 35 volts per mil, the resultant stress level is considerable. Other overvoltage situations include lightning strokes, switching surges, and test voltages.

Several mechanisms have been proposed to explain the initiation of electrical trees. Each requires a very high electrical stress to attain the energy level that would be required. Some of the possibilities are as follows:

1. Localized heating and thermal decomposition
2. Mechanical damage due to high electrical stress
3. Fatigue cracking from polarity changes
4. Small voids
5. Air inclusions around contaminants
6. Electron injection

Partial discharges that decompose the organic material in insulations are generally considered the common factor in the formation of electrical trees. The intrinsic electrical strength of the commonly used material is many times higher than the electrical stresses that are encountered in actual service. How can these excellent materials fail at such low stresses? The presence of internal voids, contaminants, and external stress points leads to electrical stress enhancements that are sufficiently high to originate water trees.

Impulses, surges, and DC stresses seem to create hollow channels through the insulation that we know as electrical trees. When seen in wafers, electrical trees are distinct and opaque. The waters usually do not have to be stained those trees to see them, but staining is certainly a recommended practice. Electrical trees require high stress but not water, and they grow relatively fast—in minutes to hours. Examples of electrical trees are shown in Figures 19.7 through 19.9.

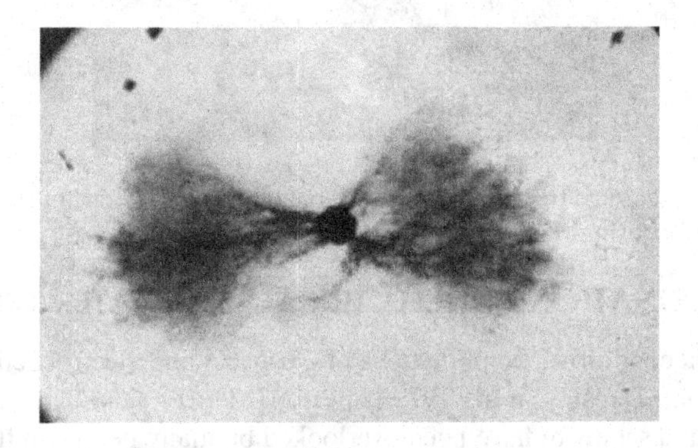

FIGURE 19.7 Bow tie growing from a contaminant (hence nonvented). Photo by W. A. Thue.

FIGURE 19.8 Bush tree growing from a shield (hence vented). Photo by W. A. Thue.

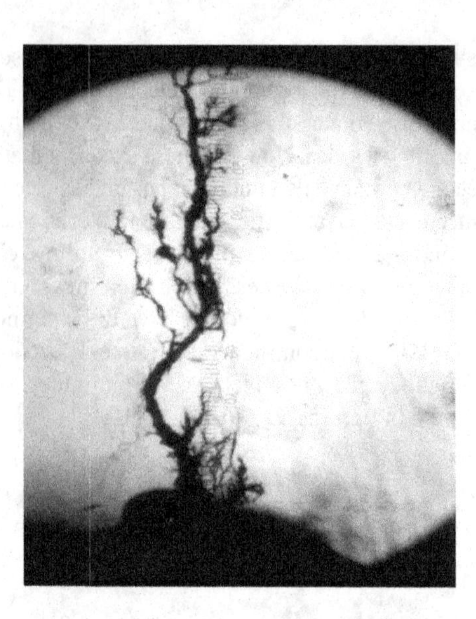

FIGURE 19.9 Electrical tree growing from a contaminant on conductor shield. Courtesy of University of Wisconsin–Madison.

19.5 COMBINATIONS OF ELECTRICAL AND WATER TREES

"Under certain conditions, combinations of electrical and electrochemical trees can occur" said John Densley in his 1979 paper [11]. Thirty years have passed and yet this combination seems to have been overlooked by many people in their attempt to have only two categories of trees in extruded dielectrics.

The fact is that field samples have shown this condition for an equal number of years. They generally appear as water trees with small sections of dark, obvious electrical trees in the branch. Another form of this condition is found when installed cables are tested after service aging using diagnostic testing procedures. Wafers show the presence of a large, vented water tree with an electrical tree growing from the leading edge.

The importance of these observations is that they most likely show the transition of water trees (benign as they may be) to partial electrical trees—where the end of life of the cable is at hand. This may explain why water trees are deleterious to extruded dielectric cables.

One concept of the mechanism where water trees are converted to sections having electrical trees is referred to as trapped space charges. Electrons with high energy levels are used to crosslink polyethylene in what is called radiation curing. Cross-linkable polyethylene is applied to a conductor, the conductor is grounded, and electrons are beamed through the dielectric cable with sufficient energy that they reach the conductor.

If the energy level is not high enough to penetrate the entire wall, electrons may be trapped in the insulation. Likely sites for such trapping are oxidized material (such as are found in water trees in service-aged cables). In a new cable, these trapped electrons can stay in a dormant condition for months. The mechanical effects of installation or increase in temperature can give these electrons sufficient energy to start moving. As they move toward the metallic member (conductor or shield), these electrons drill a channel through the insulation. This is not called an electrical tree, but it has many of the same characteristics. If the electron is trapped near the conductor for instance, the channel only extends from the site to the conductor—not in both directions.

Trapped charges can affect installed cables. A high-energy lightning stroke can deposit electrons at sites in the insulation wall—just as described previously. Again, these electrons may stay dormant until some additional energy is applied such as a temperature increase. When the energy level is sufficiently high, the electron will move. It travels may be to one of the metallic components or merely to one of the adjacent sites such as a void. The result of this action is to breech a portion of the good insulating wall. Another lightning stroke may do additional damage or the owner might use high-energy direct voltage or capacitor discharge equipment on the cable. All of these sources can provide electron entrapment with subsequent additional deterioration of the insulating wall.

It is very important to know that water trees can be the *sites* for the trapped charges. This means that after being in service for a period of time, cables having water trees are more prone to trapped charges because the sites for this entrapment have been developed. These sites may be in the microvoids or in the somewhat deteriorated insulation between those microvoids.

These combination water and electrical trees have also been seen on service-aged cables that have been tested with modestly elevated voltages during diagnostic evaluations. A partial discharge site may be located with or without a failure in the field. A typical wafer examination of this area shows extensive water treeing patterns; however, one will have an electrical tree growing from the water tree. The opinion is that a partial discharge test will locate a water tree, but during the test, an electrical tree will begin to grow and hence be detected by the test.

Figure 19.10 shows a tree site with both water and electrical trees. Note that there is one water tree from a shield at the left of the picture and that the electrical tree has grown from the same site as the earlier water tree.

FIGURE 19.10 Combination of water and electrical trees. Courtesy of University of Wisconsin–Madison.

Another combination of water and electrical trees is shown in Figure 19.11.

Another combination of electrical and water trees at the same site is shown in Figure 19.12. Note that the water tree has burned sections indicating the transition from a water tree to an electrical tree.

An example of a field failure pattern may help illustrate this phenomenon. Service failures of extruded dielectric cables peak during lightning seasons but, even with the sophisticated lightning counter facilities now available, the actual field failure

FIGURE 19.11 Water tree and electrical tree. Photo by W. A. Thue.

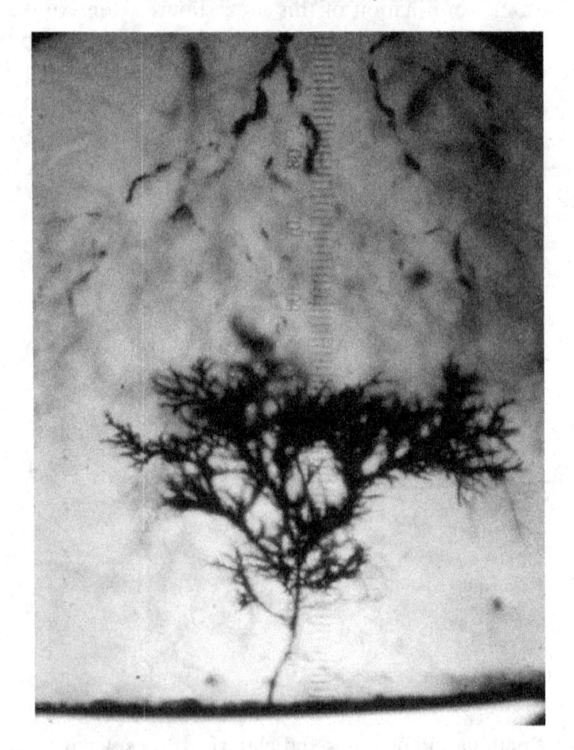

FIGURE 19.12 Electrical tree in a water tree. Photo by W. A. Thue.

seldom occurs just as the lightning strikes in the vicinity of the cable. The failure is delayed by days or weeks, but the failure location and the lightning strike point are very close. It seems highly likely that the lightning strike converted a water-treed area to a partial electrical tree that continued to grow until the insulation was no longer capable of serving its function under normal operating voltage.

Conversion of a water tree to an electrical tree seems to be an important factor. This occurs by the formation of an electrical tree from the tip of a water tree as well as inside a water tree where a portion of the "branch" is an electrical tree that has not progressed sufficiently that it has caused complete electrical breakdown. This conversion can certainly be accelerated by the effect of trapped charges.

19.6 LABORATORY TESTING

Treeing was considered to be a laboratory "trick" until the 1970s. Some of the earliest work was done by Simplex Wire & Cable. Hunt, Pratt, Ware, Kitchin, and others [19–21] reported on the work done with one needle embedded in small slabs of polyethylene beginning in 1956. From this work, they developed the first commercial tree retardant HMWPE insulation. They reported in 1958 that moisture was an inhibitor to tree growth. What was *not* known at that time was that they were looking only at electrical trees. They confidently predicted in 1958 that "HMWPE may last more than 40 years in water at operating stress up to 45 volts per mil." They were not aware of the existence of water trees as we now understand them nor did they repeat that statement made in that first paper that "...at the end of 40 years, half the lengths of cable will have failed."

Other researchers in that same time period began using two embedded needles. See Figure 19.13 for the test geometry of a double needle test. The "standard defect" consists of a specially sharpened sewing needle that is inserted into one of the small faces of a block of compression molded plastic that is 1" × 1" × 0.25" and the point of the sharpened needle is 0.5" away from an ordinary (dull) needle. These needles are inserted under carefully controlled annealing conditions. McMahon and Perkins came up with similar conclusions and reported in 1960 that "corona life of a specimen of HMWPE in air is a strong function of humidity. A relative humidity of 95

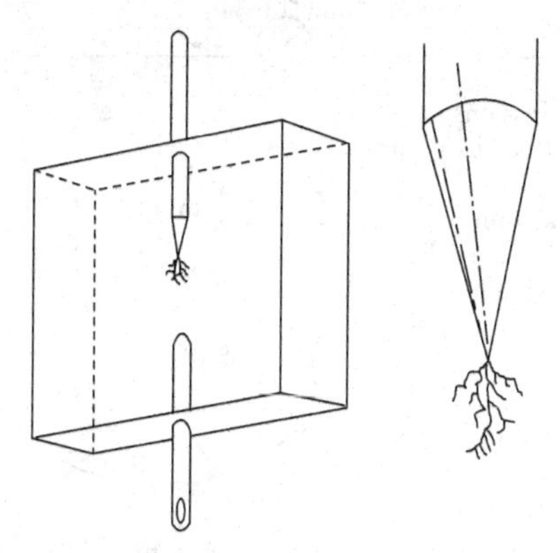

FIGURE 19.13 Double needle test specimen. Courtesy of University of Wisconsin–Madison.

to 100% gives approximately 15 times longer life than dry air." They were also only looking at electrical trees.

After the reported findings of Lawson and Vahlstrom [5,6] and the reports by Tabata et al. in 1972 of "sulfide trees" [7] in cables removed from the field, laboratory work moved toward wet testing of insulating materials such as the test developed by McMahon and Perkins. Specimen testing for water treeing is generally done in this "pie plate." This is a compression molded, dish-shaped specimen that has 24 conical depressions molded into the bottom (see Figure 19.14). The specimen dish is placed into a grounded bath that contains an electrolyte—usually made by adding 0.01 N NaCl in distilled water. A platinum wire connects a power supply to the upper portion of the dish. From 2 to 8 kV AC is applied. A low-density polyethylene (LDPE) specimen will develop water trees at the tips of the cones from 120 to 240 μm in length in 24 hours at 5 kV. The length of these water trees is then measured.

Figure 19.15 is a photograph of a water tree growing from one of the indents in a pie plate.

By 1975, the Association of Edison Illuminating Companies (AEIC) had developed an accelerated water-treeing test (AWTT) on actual full sized cable samples placed in water-filled pipes. This test is part of the qualification testing requirement to meet AEIC specifications.

Tank tests were developed at the Marshall, Texas, facility of Alcoa by Lyle and Kirkland in 1981 [9]. Testing in tanks has continued for more than 20 years at that facility even with all of the changes in names that have occurred!

The test protocol is based on samples of full sized cables formed into coils and placed in tanks filled with water. Water is also kept in the strands of the 30-feet long cable samples. Voltage is supplied continuously while the current flows for 8 hours every day. Twelve samples are energized until one fails. That failed sample is replaced with a dummy cable while the remaining active samples are energized until the next one fails. After eight of the twelve samples have failed, the remaining four samples are broken down with AC step-rise voltage application.

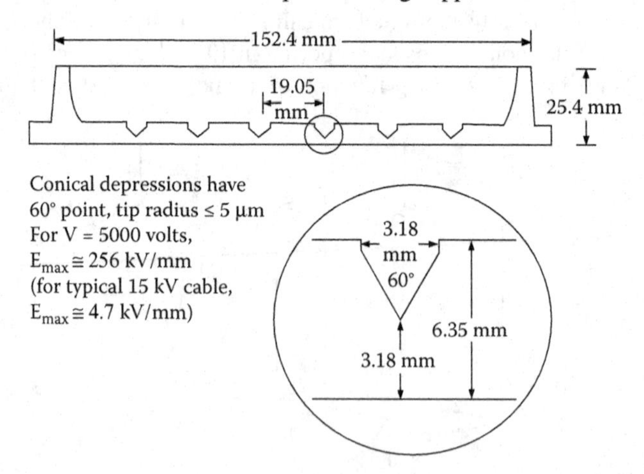

FIGURE 19.14 Water tree test specimen—"pie plate". Courtesy of University of Wisconsin–Madison.

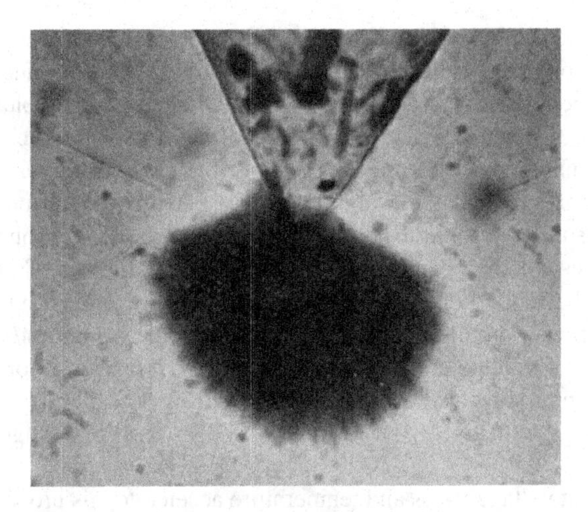

FIGURE 19.15 Water tree from a pie plate indent. Courtesy of University of Wisconsin–Madison.

In the original work, nine different voltage stress/temperature combinations were applied: 60°C, 75°C and 90°C; two, three, and four times rated voltage to ground. The results of these nine conditions are mathematically analyzed to obtain a model of a life curve that is used to demonstrate life at other stress/voltage levels such as would be found in typical operation.

This time-to-failure accelerated laboratory aging information that is developed is used to predict the life of that cable construction and processing. In this manner, other insulations, stress relief materials, and constructions can be compared directly. The significant advantage of the method is that twelve samples give a much higher confidence level in the results than a smaller number and that time to breakdown is more nearly like real life than if an AC step breakdown is used.

19.7 TECHNICAL DISCUSSION OF TREEING

Treeing has been demonstrated as one of the most important factors involved in loss of life for medium voltage cables. Electrical trees are considered to be associated with the final cable failure and do not exist for a long period of time. Water trees are the slower growing variety. They can extend from one electrode to the other without a service failure. Once they have formed, water trees seem to be converted to electrical trees for part or all of their length by DC, surges, and impulses. Conclusions in recent research work show that treed cables that are subjected to DC, surges, or impulses have shorter life in service after that application than cables not subjected to those stresses.

There are several possible explanations for this "conversion" of a water tree to an electrical tree, but the more commonly accepted explanation is that charges are trapped in the insulation wall. When these trapped charges are disturbed by heat or mechanical motion, they can literally bore a hole through the insulation wall. A likely scenario is that the trapped charges bore a tunnel from one void or contaminant to the next one. The insulation between these voids may be in a deteriorated condition, thus speeding up the damage from the trapped charges. This continues until the wall has virtually been destroyed and the cable cannot hold even line voltage.

Inception of water trees is likely to be the result of voltage enhancements at voids, contaminants, or other imperfections in the cable. Another significant factor is that the presence of ionic impurities is especially deleterious to cables. At one time it was thought that the source of these ions was from ground water or the like. It is now established that the source of these impurities is often the materials in the cable—basically contaminants in the older semiconducting shield materials. Microscopically small "chunks" of sand make the insulation/shield interface another source of voltage enhancement. Growth or propagation of the water tree is apparently quite slow—several years in a well-made cable. Bow tie trees may stop propagating as they grow large enough to decrease the voltage stress at their extremities.

We know that voltage stress and temperature accelerate this propagation of water trees. XLPE and thermoplastic polyethylene are adversely affected by temperatures above about 75°C—as demonstrated by laboratory aging studies [10].

As briefly mentioned previously, there is only a small distinction between water and electrochemical trees that result from a "natural" staining of the interior or the voids. Pre-1970 HMWPE insulation was formulated with a staining antioxidant. These cables did not require any dying to see the trees. The change to nonstaining antioxidant around 1970 resulted in water trees that could not be seen unless the wafers were put in a dye solution. In the transition period, it was thought that possibly the staining antioxidant had caused the trees! The dying procedure is given at the end of this section.

Trees also exist and are visible in EPR insulated cable but they can only be seen at the surface of the cut. A similar dying procedure is used for EPR but the staining time must be increased considerably. There are also proprietary methods for staining EPR cable samples. Tree counts in EPR are lower than for the nonopaque types because of not being able to see down into the material, but they may be lower also because they simply do not tree the same way as XLPE cables.

Trees positively initiate at defects within the cable such as at discontinuities between the interfaces of the insulation and the two shields, and at voids and contaminants—metal particles, threads, oxidized bits of insulation (ambers), and even at chunks of undispersed antioxidant.

Trees that have one of their points of origin at the insulation/shield interface are called "vented" trees. They always show up as the dangerous trees as compared with the ones that stay completely within the wall of insulation—the nonvented tree. The probable explanation here is that pressure can build up within the nonvented tree, and this would suppress the partial discharge.

19.8 METHYLENE BLUE DYING TECHNIQUE

Prepare the specimen for dying by removing the conductor to obtain a sufficient length of insulation for examination. Using a very sharp tool (lathe, microtome, or trimmer), make slices of the insulation that are about 25 mils thick.

In a 500 ml beaker with watch glass cover, place:

1. 250 ml distilled water
2. 0.50 gm methylene blue
3. 8 ml concentrated aqueous ammonia

Heat to boiling with continuous stirring. Use a fume hood or other adequate ventilation. Place the specimens to be stained in the solution using a wire for installation and removal. Remove specimens from hot solution from time to time to be certain that the staining is neither too light nor too dark. The time will be dependent on the thickness of the sample as well as the type of insulation.

When the specimens are adequately stained, remove from the hot solution, rinse in hot water, and wipe dry.

A thin film of oil on the surface of the sample makes observation with a microscope much less confused by scratches. A 30-power lens should be used for initial observation purposes.

Record the number of bow tie trees (originating within the insulation wall) and the vented trees (originating from the conductor or insulation shields) that are greater than 2 mils in length. The length for bow ties shall be the peak-to-peak length. For vented trees, the length shall be the maximum distance from the surface of the shield to the tip of the tree. List the numbers in size ranges and by type.

19.9 OBSERVATION IN SILICONE OIL

An excellent method of observing several inches of insulation at one time is to place a 1-foot sample on the insulated cable (the semiconducting insulation shield must be removed!) in a glass beaker with silicone oil that has been heated to about 130°C. At about this temperature, all of the crystallinity is gone and the insulation becomes quite clear. The surface of the conductor shield can be observed for smoothness. Voids or contaminants in the insulation wall can be readily seen. Note: "voids" can be created during the test by moisture in the insulation resulting from service conditions. Such "voids" will grow in size during the observation period so that they are obviously caused by the moisture. A way to eliminate this possibility is to dry the field sample in an oven prior to the observation.

REFERENCES

1. Rayner, E. H., 1912, British JIEE 49, p. 3.
2. Whitehead, S., 1932, "Breakdown of Solid Dielectrics", Earnest Bend, London.
3. Tykociner, J. T., Brown, H. A. and Paine, E. B., 1933, University of Illinois Engineering Experiment Station, Bulletins 259 and 260.
4. Kreuger, F. H., 1965, "Discharge Detection in High Voltage Equipment", American Elsevier Publishing Co., New York.
5. Vahlstrom, W., Jr., 27 September–1 October 1972, "Investigation of Insulation Deterioration in 15 and 22 kV Polyethylene Cables Removed from Service – Part I", IEEE Underground Distribution Conference, Detroit, MI.

6. Lawson, J. H. and Vahlstrom, W., Jr., 27 September–1 October 1972, "Investigation of Insulation Deterioration in 15 and 22 kV Polyethylene Cables Removed from Service – Part II", IEEE Underground Distribution Conference, Detroit, MI.
7. Tabata, T., Nagai, H., Fukuda, T. and Iwata, Z., 18–23 July 1971, "Sulfide Attack and Treeing of Polyethylene Insulated Cables – Cause and Prevention", IEEE Summer Power Meeting.
8. "Treeing Update", *Kabelitems,* Part 1, Vol. 150, pp. 8–9, 1977.
9. "Treeing Update", *Kabelitems,* Parts I, II, III, Vols. 150, 151 and 152, Union Carbide, 1977. (There are 162 references in these three volumes.)
10. "Electrochemical Treeing in Cable", EPRI EL-647, Project 133, January 1978.
11. Densley, R. J., June 1979, "An Investigation into Growth of Electrical Trees in XLPE Cable Insulation", IEEE Vol. EI-14 (No. 3).
12. Sletbak, J., August 1979, "A Theory of Water Tree Initiation and Growth," IEEE Vol. PAS-98, #4.
13. Lyle, R. and Thue, W. A., 1983, "The Origin & Effect of Small Discontinuities in Polyethylene Insulated URD Cables", IEEE 83 WM 002-3.
14. Nunes, S. L. and Shaw, M. T., December 1980, "Water Treeing in Polyethylene – A Review of Mechanisms", IEEE Vol. EI-15 #6.
15. Lyle, R. and Kirkland, J. W., "An Accelerated Life Test for Evaluating Power Cable Insulation", IEEE 81 WM 115-5.
16. Sletbak, J. and Ildstad, E., "The Effect of Service and Test Conditions on Water Tree Growth", IEEE 83 WM 003-1.
17. Lyle, R., 1986, "Effect of Testing Parameters on the Outcome of the Accelerated Cable Life Test", IEEE 86 T&D 577-1.
18. Walton, M. D., Smith, J. T. III, Bernstein, B. S. and Thue, W. A., "Accelerated Aging of Extruded Dielectric Power Cables Parts I, II, III", IEEE Trans. PD Vol. 7, April, 1992 and 93 SM 559-5, 1992 and 1993.
19. Hunt, G. H., Koulopoulos, M. J., and Ware, P. H., April 1958, "Dielectric Strength and Voltage Life of Polyethylene Part III", AIEE Transactions, Vol. 77, pp. 25–28.
20. Kitchin, D. W. and Pratt, O. S., June 1958, "Treeing in Polyethylene as a Prelude to Breakdown", AIEE Transactions, Part III, Vol. 77, pp. 180–186.
21. Kitchin, D. W. and Pratt, O. S., June 1962, "An Accelerated Screening Test for Polyethylene High Voltage Insulation", AIEE Transactions, Pt III, Vol. 81, pp. 112–121.
22. McMahon, E. J. and Perkins, J. R., 1963, "Volume Discharges and Treeing – A Primer", 5th Electrical Insulation Conference on Materials and applications.

20 Lightning Protection of Distribution Cable Systems

William A. Thue

CONTENTS

20.1 INTRODUCTION

Distribution cable systems have peak failure rates during the summer months throughout North America. Research work has shown that impulse surges to cables shorten their service life [1,2]. Both temperature and rainfall may peak during the same time period. All of these factors may influence this failure rate. It is also well documented that water trees reduce the impulse level of extruded dielectric insulated cables and contribute to the failure situation. Most of the effort that has been spent in the past on lightning protection of distribution system components has been on overhead transformers. This is logical when you consider that the companies that build transformers are also the ones that sell arresters [3].

The older paper-insulated cables were manufactured with an inherently high impulse level and that level was maintained over the 50 plus years of life of the system. Today, the extruded dielectric insulated cables that are used so extensively in underground systems exhibit a dramatic drop in electrical strength in just a few months of service. It is important to note that cross-linked polyethylene (XLPE) cables start with a much higher impulse level than ethylene propylene (EPR) or paper cables. EPR cables have initial impulse strengths less than the others, but their impulse level does not drop as quickly and levels out. With time, both XLPE and EPR have impulse levels that are much nearer the basic impulse level (BIL) of the system than for paper cables. Because of this, lightning protection is a significant consideration for these newer cables.

20.2 SURGE PROTECTION

20.2.1 Protective Margin

This is defined as:

$$\frac{\text{Insulation withstand level}}{\text{Arrester protection level}} \times 100$$

Another form of this equation for protective margin is:

$$\left[\frac{\text{Equipment BIL in kV}}{\text{Arrester discharge voltage in kV} + \text{discharge voltage of arrester leads in kV}}\right]^{-1} \times 100$$

A minimum protection margin of 20% over BIL is usually recommended for transformers.

20.2.2 Voltage Rating

Voltage rating of a metal oxide varistor (MOV) arrester is based on its duty-cycle test. The duty-cycle test defines the maximum permissible voltage that can be applied to an arrester and allow it to discharge its rated current. Another way to consider this is that it is the voltage level at which power follow-current can be interrupted after a surge discharge has taken place. At voltage levels above this, power follow-current interruption is doubtful. The safe arrester rating is usually determined by the highest power voltage that can appear from line to ground during unbalanced faults and shifting of the system ground [4–7].

20.2.3 Highest Power Voltage

The highest power voltage can be calculated by multiplying the maximum system line-to-line voltage by the coefficient of grounding at the point of arrester placement.

20.2.4 Coefficient of Grounding

This is defined as the ratio, expressed in percent, of the highest root mean square (rms) line-to-ground voltage on an unfaulted phase during a fault to ground. Systems have historically been referred to as being effectively grounded when the coefficient of grounding does not exceed 80%.

20.2.5 Sparkover

This refers to the initiation of the protective cycle that occurs when the surge voltage reaches the level at which an arc develops across the device's electrodes to complete the discharge circuit to ground. In terms of voltage across gapped arresters, this is somewhat indefinite since sparkover of a simple gap structure is a function of both the wave front and the voltage of the incoming surge.

The essential requirement of a proper sparkover level is the speed of response to steep fronts, such as natural lightning, yet gives a consistent response to waves with slower rates of rise, which are typical of indirect strokes and system generated surges.

Sparkover of an arrester should not be confused with "flashover." Flashover refers to the exterior arcing that may occur, for instance, when surfaces become contaminated.

20.2.6 Surge Discharge

Surge discharge refers to the situation where the arrester must handle the power frequency line current as well as the momentary surge current. This power follow-current continues to flow until the arrester can extinguish the arc.

20.2.7 Insulation Resistance Discharge Voltage

The insulation resistance (IR) discharge voltage of an arrester is the product of the discharge current and the resistance or inductance of the discharge path. While the resistance may be very low, the discharge current can be very high and the IR discharge voltage can reach levels that equal or exceed the arrester sparkover voltage. The inductance of the combined line and ground leads must be kept as short as possible. This is accomplished by placing the arrester as close as practical to the cable termination and *always* connecting the arrester closer to the incoming line than the termination (see Figure 20.5).

20.3 WAVE SHAPE AND RATE OF RISE

Natural lightning must be simulated in the laboratory to test and evaluate lightning protection devices and equipment. This is accomplished with a surge generator. A group of capacitors, spark gaps, and resistors are connected so that the capacitors are charged in parallel from a relatively low voltage source and then discharged in a series arrangement through the device being tested.

The terms used to describe both natural and artificial lightning are "wave shape" and "rise time" (Figures 20.1 and 20.2). The wave crest is the maximum value of

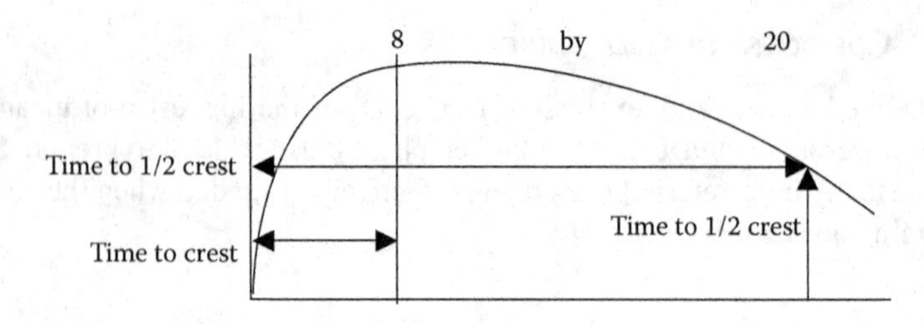

FIGURE 20.1 Wave shape.

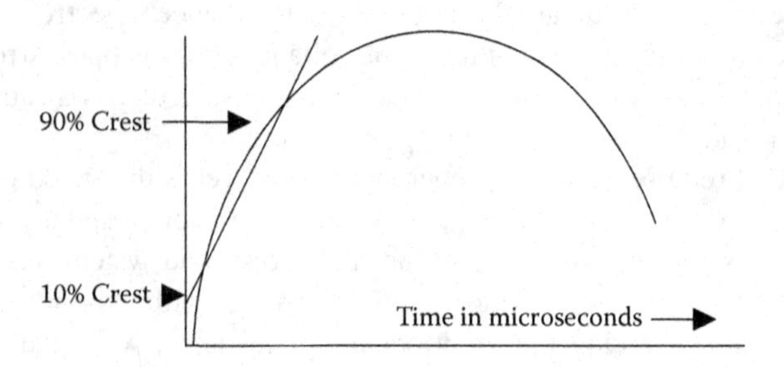

FIGURE 20.2 Rate of rise.

voltage reached. Wave shape is expressed as a combination of the time from zero to crest value for the front of the wave and the time from zero to one-half crest of the wave tail. Both values of time are expressed in microseconds. The rate of rise is determined by the slope of a line drawn through points of 10% and 90% of crest value.

Testing of surge arresters has historically been done with an 8 × 20 microsecond wave, but more recent work has been done at 4 × 10 even though a direct stroke of natural lightning is nearly 1 × 1,000. See Figure 20.1 as to how these times are defined.

20.4 OPERATION OF A SURGE ARRESTOR

20.4.1 Air Gaps

The original surge arrester was a simple air gap. They were made of a simple rod or spheres installed between the line and the ground that were far enough apart to keep the line voltage from sparking over but close enough to discharge when a surge occurred. Air gaps have the disadvantage of allowing system short circuit current to continue to flow until the breaker, fuse, or other backup device operates.

Air gaps have another disadvantage. Electrically speaking, they are sluggish and their response varies as stated above. Sparkover may not occur until a considerable portion of a rapidly rising lightning surge has been impressed on the system. The short gap spacing necessary to provide adequate protection against steep front lightning waves may result in frequent and unnecessary sparkovers on minor power frequency disturbances.

20.4.2 Valve Arresters

Nonlinear resistance can best be considered as resistance that varies inversely with applied voltage (Figure 20.3). Under normal voltage conditions, the resistance is high; under unusual high voltages, the resistance is low.

The material that, in the past, has been used extensively in valve arresters is silicon carbide. It is blended with a ceramic binder, pressed into blocks under high pressure, and fired in kilns at temperatures of over 2,000°F. This component is the valve block. The number of valve blocks used in an arrester is determined by resistance requirements for the rating of the system.

For silicon carbide blocks, it is essential that an air gap be in series with the blocks since they would conduct a large amount of current at operating voltage. This gap must ionize the atmosphere in the arc chamber to break down that gap before the blocks encounter any voltage. After the air gap breaks down, the valve blocks begin to conduct the combination of surge current and power current. The high voltage of the lightning surge decreases the resistance of the valve blocks and the current flows to ground. The voltage now across the blocks is approximately the line-to-ground voltage of the system. The valve blocks revert to their normal high resistance. This forces the power flow current to be reduced to a value that the series blocks can interrupt at the next system current zero.

20.4.3 Metal Oxide Varistor (MOV)

MOVs (variable resistors) became available for distribution systems in 1978. Their first use on distribution systems was on terminal poles, hence the term riser pole arrester.

Gaps are not required because the material is extremely nonlinear. The lower half of the schematic shown in Figure 20.4 represents a MOV arrester. A voltage increase of just over 50% results in a conduction current change of 1 to 100,000. The absence of gaps allows these devices to operate much faster than the older gapped silicon carbide arresters. The absence of gaps is a major factor in allowing MOVs to be used in load break elbow arresters.

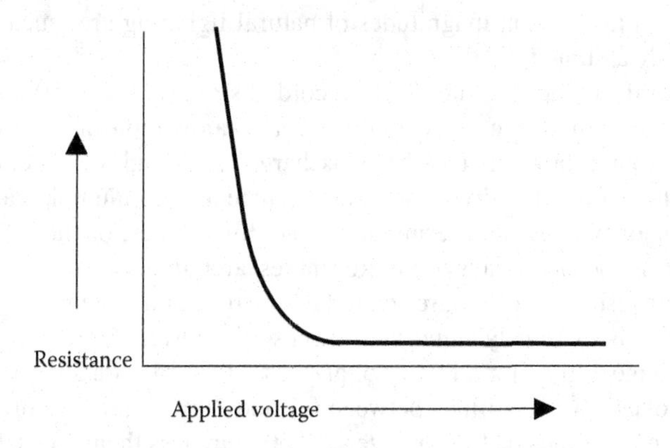

FIGURE 20.3 Nonlinear resistance.

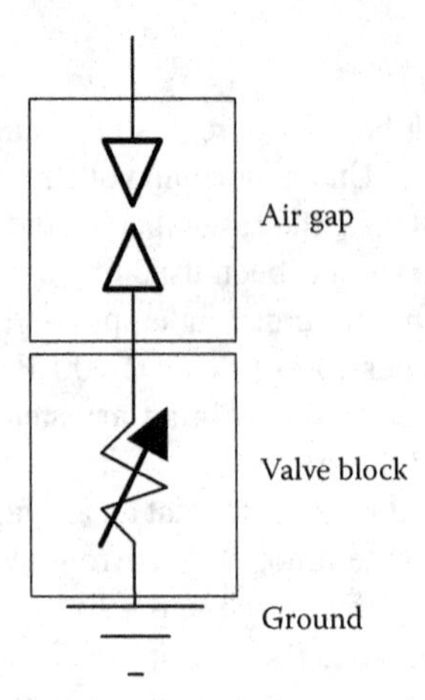

FIGURE 20.4 Schematic of a silicon carbide arrester.

Grounding resistance/impedance must be treated more seriously now that the underground residential distribution (URD) systems are using conduit and/or jacketed neutral wires. With bare neutral wires, the stroke energy was dissipated along the cable run. The insulation provided by the jacket or conduit makes low resistance grounds at the terminals an essential factor. Customer's grounds helped older systems have low resistance before those pipes became plastic. Even gas pipes are now plastic!

20.5 NATURAL LIGHTNING STROKES

The understanding of natural lightning has increased tremendously since the early 1980s. Electric Power Research Institute (EPRI) efforts led to the construction of antennas throughout the US, and beyond, to record lightning strokes. These systems are now capable of pinpointing the time, location, magnitude, and polarity of strokes that occur between the clouds and the ground. What has been determined is that the rate of rise and the current magnitudes of natural lightning are much more severe than previously assumed.

From this information, we now have recorded strokes of over 500,000 amperes. Although these high stroke currents do occur, examination of arresters removed from service do not show that they have discharged such high values of current. One possible explanation is the division of stroke currents into multiple paths. Another is that the majority of strokes terminate to buildings, trees, or the ground without directly striking the electrical system. Recent research indicates that indirect strokes may be the biggest cause of failures on today's distribution systems.

Rate of rise is extremely important because the faster the rate, the higher the discharge voltage will be for all types of arresters. Recorded data shows that natural lightning strokes have rise times between 0.1 and 30 μs with 17% of the recorded strokes having rise times of 1 μs or faster and 50% are less than 2.5 μs. For the same wave shape, the average rate of rise increases with the crest magnitude. Using the "standard" 8 × 20 microsecond wave and a 9 kV-gapped arrester, the discharge voltage is about 40 kV. For the same 20 kA stroke but rising to crest in 1 microsecond, the arrester would have a 54.4 kV discharge or a 36% increase. Metal oxide arresters (without gaps, of course) commonly exhibit a 12% to 29% increase under similar circumstances.

The inductance (hence, length and shape) of the arrester leads becomes more pronounced with the faster rate of rise. Applying the generally used value of 0.4 microhenries per foot, the lead voltage is 8 kV per foot of total lead length at 20 kA per microsecond and 16 kV per foot at 40 kA. Assuming new arresters and 2 feet of total lead length, the total voltage at 20 kA and 40 kA would be 70 and 96 kV, respectively. In other words, a stroke having a 40 kA per microsecond rate of rise would add 32 kV to the arrester discharge voltage given in a typical manufacturer's literature.

Prudent engineering suggests that the level of protection should be calculated for a family of possible values of current and rates of rise for the anticipated lightning activity in the service area under study. This suggests currents such as 40 kA for parts of central Florida, but only 10 kA or lower for California. Rates of rise of 1 to 3 microseconds are commonly used in calculations.

For an interesting note, these systems are of use to many organizations. Lightning stroke information is used by the forest service to warn of fires rather than the old fire towers or airplanes. Antennas near Anchorage, Alaska, warn of volcanic eruptions that produce lightning.

20.6 TRAVELING WAVES

Whenever a lightning stroke encounters an electric system, energy is propagated along the circuit from the point of origin in the form of a traveling wave. The current in the wave is equal to the voltage divided by the surge impedance of the circuit. Surge impedance is approximately equal to the square root of the ratio of the self-inductance to the capacitance to ground of the circuit. Both the inductance and capacitance are values per given unit length making the surge impedance of a circuit independent of the actual length of the circuit.

A traveling wave will keep moving without change in a circuit of uniform surge impedance except for the effects of attenuation. As soon as the wave reaches a point of change in impedance, reflections occur.

A wave reaching an open circuit is reflected without change in shape or polarity. The resultant voltage at the open end will be the vector sum of the incident wave and the reflected wave. This is the source of the voltage doubling circumstance. If an arrester is located at the open point, this doubling does not occur *after the arrester begins to discharge.*

When a wave arrives at a ground or other value of impedance that is lower than the surge impedance of the circuit, the incident wave is reflected without change in shape but with a reversal in polarity.

No reflections will occur on a circuit that is connected to ground through a resistance/impedance that is equal to the surge impedance of the circuit since there is no change in impedance.

It is convenient to think of traveling waves as having square shapes to illustrate the points just mentioned, but since real surges have a finite time to crest, the results of the superposition of the actual wave shapes are quite different than the square waves, which are the worst-case scenarios.

20.7 VELOCITY OF PROPAGATION

For practical purposes, a traveling wave on an overhead line travels at the speed of light—984 feet per microsecond. The velocity of propagation of a traveling wave in cables commonly used today is about half the speed of light, or 500 to 600 feet per microsecond. This can be derived from the fact that, in an insulated and shielded cable, the speed is reduced depending on the specific inductive capacity, or permittivity, of the insulating material.

$$V = \frac{984.3\ \text{ft}/\mu\text{sec}}{\sqrt{LC}} \quad \text{or} \quad V = \frac{32.8\ \text{m}/\mu\text{sec}}{\sqrt{LC}} \tag{20.1}$$

This calculates out to 659 ft/μsec for tree-retardant cross-linked polyethylene (TR-XLPE) and 577 ft/μsec for an EPR. The use of time domain reflectometry (TDR and also known as "radar") for finding faults, neutral corrosion, and so forth, uses this velocity to locate the problem. A reasonably accurate location will be obtained using this formula, but it must be noted that there are other factors involved such as the type of shield. A cable with a longitudinally corrugated shield will have a velocity of propagation that is slightly slower than what would be calculated using Equation 20.1, for instance. A cable with an overlapped flat tape shield that has corroded may slow the time down by several percentage points or be so severe that the wave is so greatly attenuated that it never reflects back to the sending point.

Velocity of propagation becomes important to the protection of distribution cables because the travel time from the junction arrester to the end of the cable run is very short as compared with the conduction time of the arrester. Consider a typical 5,000-feet long loop that is open at the midpoint. At 500 feet per microsecond, the travel time is only 5 microseconds to the end and 10 microseconds for the round trip. The arrester conduction time for an 8 × 20 microsecond wave is about 50 microseconds. This means that the junction arrester still has 90% of its conduction time left when the wave has traveled to the end of the cable. If the end does not have an arrester, the reflected wave will travel back toward the junction point and add to the incoming voltage wave throughout the length of the cable. Thus, the entire cable is exposed to the "doubled" wave. The amount of time the incoming wave is maintained becomes an important consideration as to the exposure of the cable to this full doubling of voltage.

Attenuation due to losses in the insulation has a negligible effect on the reflected voltage because the low loss insulations that are in use today do not attenuate the wave appreciably in the relatively short runs used for distribution systems. Attenuation due to shield type and configuration can be a serious problem. Helically applied flat tapes can deteriorate and corrode to the point that for a pulse having a fast rate of rise, the shield system looks like an open spiral. A TDR pulse cannot travel very far without being attenuated to unreadable levels. The good thing about this fact is that a time domain reflectometry (TDR) pulse can be used to find open shields as well as alert the user to limitations on what method of assessment testing can be utilized.

20.8 PROPER CONNECTION OF ARRESTERS

There are several extremely important installation rules for arresters:

- Keep both the line and ground side leads as short and straight as possible. (It is the sum of the two lead lengths that must be used in the calculations.)
- The lead from the line should go to the arrester FIRST—then to the termination.
- The ground resistance should be as low as practical. This means 10 ohms if the cable has an insulating jacket or is in a conduit.

20.8.1 Lead Lengths

The issue of lead length on the voltage that will be impressed on a cable has been discussed earlier in this chapter. All of that is correct. There is, however, one more issue here. Does that lead carry the lightning current? If the lightning current flows in that lead, its length is a factor. If, on the other hand, the lead does not carry lightning current, its length and impedance are not factors. In the real world, the current generally flows through all the paths that are available. The amount of current times the length of each lead establishes the voltage that is impressed on the cable. The practical point is that the circuit must be analyzed in its entirety, which includes both the hot lead and the ground lead.

20.8.2 Route of Current Flow

In the beginning of this section, it was stated that the lead from the incoming line should first be attached to the arrester—then to the termination. Wait a minute. This is not the way we have always done it! Are you certain of that?

Yes. If we can visualize the flow of lightning current as a flood of water, we can easily recognize that we would be much better off if we could divert that flood around our house—not through it. That is why the arrester is the first connection point. The bulk of the current flows through the arrester to and through its ground. The termination lead length is not very significant because it is not carrying that much current.

20.8.3 Ground Resistance/Impedance

Why is the ground resistance/impedance important? We are concerned about voltage, and voltage is the product of current and impedance (length). Almost all of the current that goes through the arrester must flow to the ground at the arrester location. Remember that the impedance of an overhead line (the neutral for our purposes) is about 50 to 60 ohms. If the ground at the arrester is very high, then all of

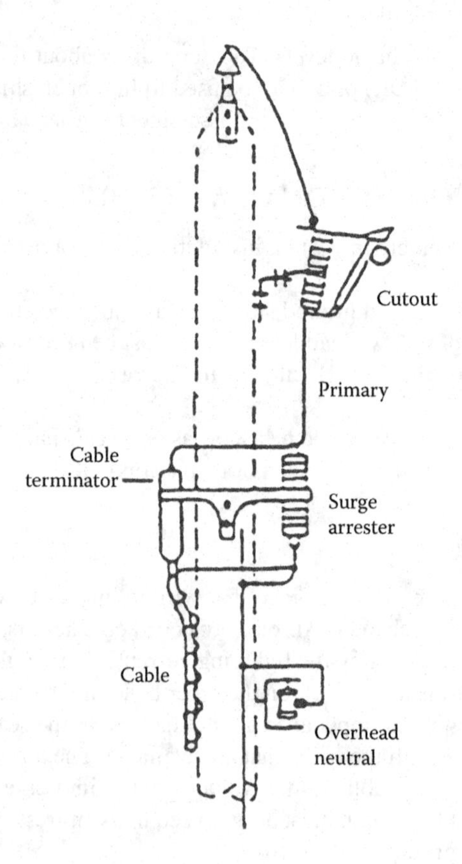

FIGURE 20.5 Correct arrester connection diagram.

that lightning current must flow along those neutrals. This means that the "footing" resistance is 60 ohms. The voltage that is developed is the current multiplied by 60 ohms. Even if there are two directions for the ground current to flow, this can be a very high voltage.

The voltage buildup through the arrester is increased by the voltage buildup in the ground circuit (Figure 20.5).

REFERENCES

1. *Underground Distribution System Design and Installation Guide,* NRECA, 1993, Section 5.
2. "Effect of Voltage Surges on Solid Dielectric Cable Life", EPRI 6902-1990.
3. "Surge Behavior of URD Cable Systems", EPRI EL-720, April 1978.
4. Westrum, A. C., 1979, "State of the Art in Distribution Arresters", Thirty-Second Annual Time to 1/2 Crest Power Distribution Conference, University of Texas–Austin.
5. Hopkinson, R. H., 1983, "Better Surge Protection Increases Cable Life", *Electric Forum.*
6. Sakshaugh, E. C., 1977, "Influence of Rate of Rise on Distribution Arrester Protective Capability", *Electric Forum.*
7. *Lightning Protection for Rural Electric Systems,* 1993, NRECA Pub. #92-12.

21 Cable Performance

William A. Thue

CONTENTS

21.1 INTRODUCTION

Cable failure reporting in the US began by action of the Edison Electric Institute (EEI) and its predecessor the National Electric Light Association. A significant, early report covered the performance of paper-insulated, lead-covered cables (PILC), splices, and terminations beginning in 1923. Failure rates of cables were reported in units per 100 installed miles for a variety of causes. Splice and termination reports were based on failures per 1,000 units that were in service. These reports continued through 1966 and served both as a useful performance guide as well as a barometer of the effectiveness of the cable specifications in effect. The National Electric Light Association prepared the first U.S. paper cable specification in 1920 for cables rated up to 15 kV. The Association of Edison Illuminating Companies took over this responsibility and upgraded this to 45 kV in 1930 [1].

The advent of underground residential distribution (URD) systems with the extensive use of extruded dielectric cable convinced the U.S. utility group to become involved in specifications for this evolving type of cable. Later usage in conventional urban duct and manhole systems to take the place of the backbone paper-insulated cables finalized this requirement.

During the early 1970s, isolated reports of early cable failures on the extruded dielectric systems began to be documented in many parts of the world. "Treeing" was reintroduced to the cable engineer's vocabulary, but with an entirely new meaning from the paper cable use of the word.

The EEI's last attempt to report distribution cable failures was in 1973. A vacuum was therefore developed in the US for distribution cable failure reporting and no system for similar reporting exists in the world.

21.2 CABLE FAILURE DATA

The data coming from a few U.S. utilities and work funded by the Electric Power Research Institute (EPRI) in 1975 [2] showed that thermoplastic polyethylene (PE) insulated cables were failing at a rapidly increasing rate and that cross-linked polyethylene (XLPE) and ethylene–propylene rubber (EPR) cables had a lower failure rate.

The next compilation of data began in 1976 with 16 and later 21 utilities in North America, reporting their failure rates on an annual basis for both polyethylene and XLPE insulated cables [3].

Failure data kept by the utilities was rather meager. It was decided to only request data based on the type of insulation, number of failures for each year, and the total amount of each of those insulation types in service at the end of the year. It was also decided to only ask for failures of known electrical causes, such as defective cable, insulation deterioration, lightning, etc., and then include all "unknown" causes since treeing analysis was not easily obtainable.

21.3 PERFORMANCE

Comparable data from EEI for paper-insulated, medium voltage power cables installed in the US is included as Figure 21.1 for the years 1923 through 1966—when the data was no longer collected. Similar data showing the electrical failures of polyethylene and XLPE for 21 North American utilities is shown in Figure 21.2. The Association of Edison Illuminating Companies (AEIC) then began to collect and report similar data in 1984 except that data was requested from all utilities. A major future step was to request information on jackets, ducts, voltage stress levels, etc. The old 21-company base was not separately recorded, however. They also began to collect data for tree-retardant XLPE (TR-XLPE) as well as EPR (see Figures 21.3 through 21.6).

AEIC strongly suggests that this data be carefully analyzed and understood. This is important since the ages of the cables were not known and could skew the results.

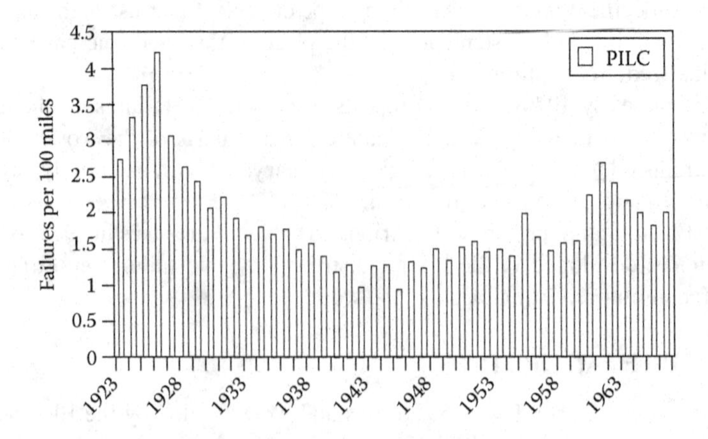

FIGURE 21.1 PILC cable failures in the US.

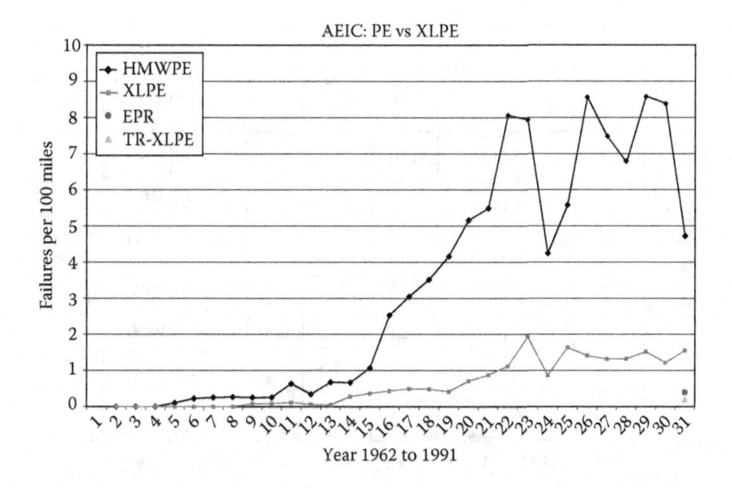

FIGURE 21.2 Electrical failures of extruded dielectric cables in North America.

For instance, jacketed cable is *probably* newer than nonjacketed cable and hence the failure rate of the older cables may not be entirely the result of a jacket.

The European community also began to collect data and their results were published as UNIPEDE-DISCAB that represents most of the European countries. Their data includes polyethylene, XLPE, EPR, and polyvinyl chloride (PVC).

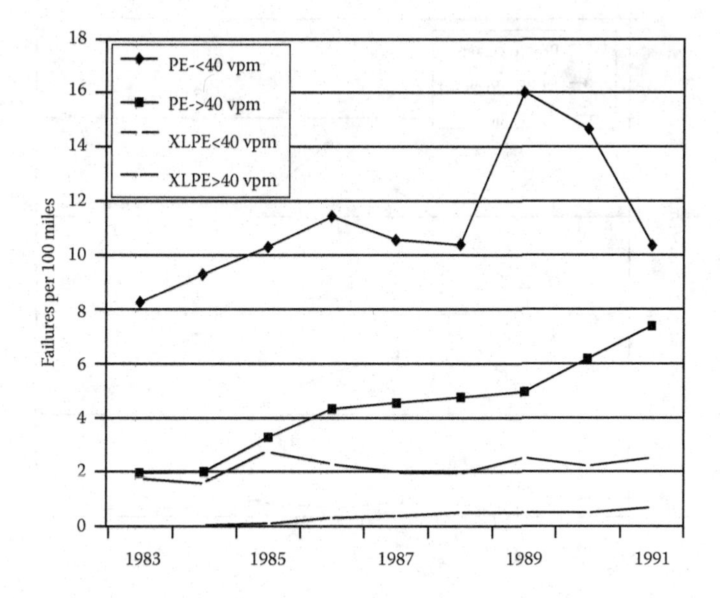

FIGURE 21.3 AEIC cable failure data, high and low electrical stress.

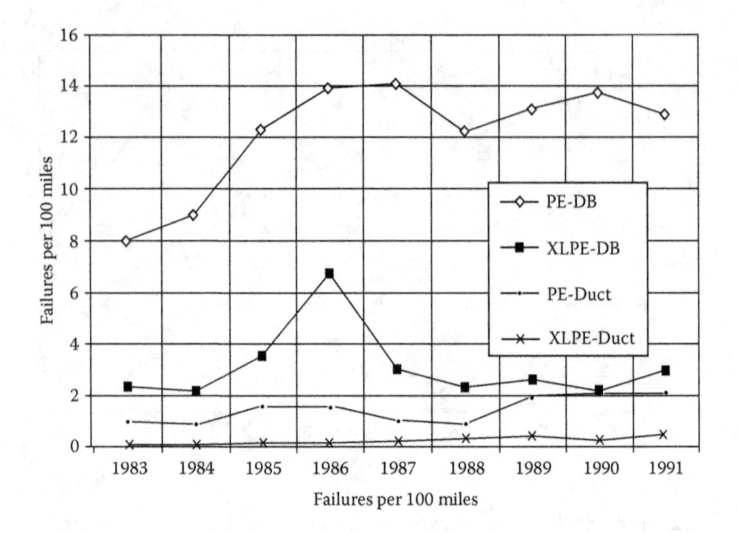

FIGURE 21.4 AEIC cable failure data, duct versus direct buried.

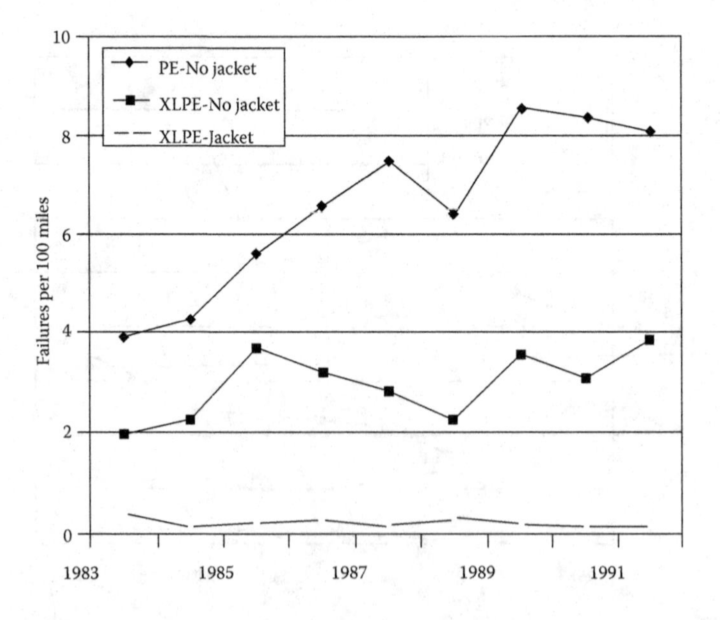

FIGURE 21.5 AEIC cable failure data, jacket versus nonjacket construction.

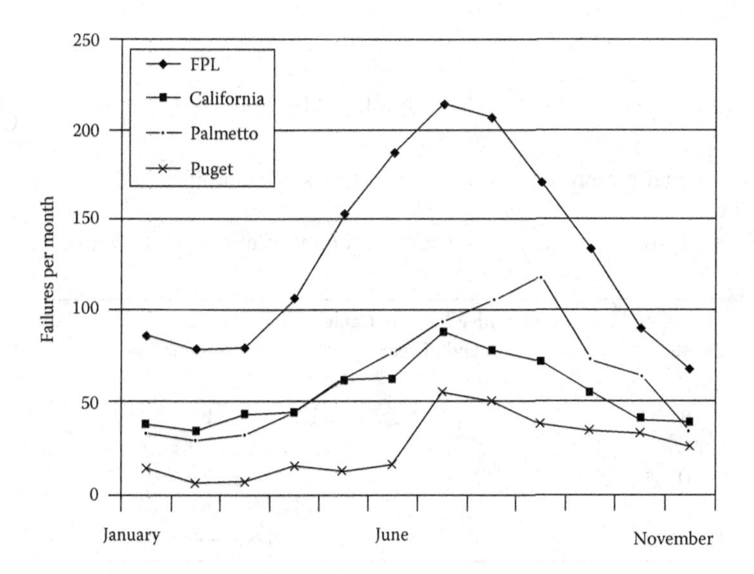

FIGURE 21.6 Cable failure data, seasonal pattern.

21.4 ANALYSIS OF DATA

Cable failure rates in the US have historically been calculated on the basis of failures per 100 miles of installed cable. The rest of the world reports failures per 100 kilometers. All data is shown as rates per 100 miles for ease of comparison.

The most frequently used form of the data shows the number of failures per 100 miles for each year. The disadvantage of such a depiction is that older cable is looked at in the same light as the new one. This data is more readily available, but a preferred method is to take into account the years in service for all cables. This is accomplished by integrating the miles installed with the years of service. The expression is:

$$\text{Service Index } A_j = \Sigma i\left(M_i N_j - 1\right) \tag{21.1}$$

where

A = system age in year j in terms of service mile-years
M_i = the number of miles of cable installed in year i
N_{j-i} = the number of years from i to j.

Years	Miles of Cable Multiplied by Age				
A	a				
B	2a	b			
C	3a	2b	c		
D	4a	3b	2c	d	
E	5a	4b	3c	2d	e
F	6a	5b	4c	3d	2e f, etc.

At the end of year F, the age of the system is:

$$A_F = 6a + 5b + 4c + 3d + 2e + f \quad (21.2)$$

where a, b, and c represent the number of miles of cable installed in A, B, and C, respectively.

This analysis can be shown as a summation of failures per 1,000 mile-years.

Years	Cumulative Miles of Cable at End of Year	Summation
A	a	a
B	b	a + b
C	c	a + b + c
D	d	a + b + c + d
E	e	a + b + c + d + e
		Σ = 5a + 4b + 3c + 2d + e

It is only necessary to add the miles of cable installed each year to the summation of cable installed in all previous years to obtain updated mile—years from this equation.

21.5 PRESENT SITUATION

There is no new data to report regarding cable failure statistics. This is the result of two factors:

- New cable performance is very good. The few problems with the new cable makes collecting data seem unnecessary.
- There are not enough people to do the essential work.

The last North American performance data was collected in 1991 by AEIC. The data shows an extremely low failure rate for TR-XLPE and EPR. The XLPE rate has not escalated to a troublesome level. The European collection of data has also been discontinued.

This is certainly an indication of the effort that has been directed toward improved cables—from both the material suppliers' standpoint as well as the cable manufacturers.

REFERENCES

1. Thue, W. A., October, 2001 adapted from class notes, "Power Cable Engineering Clinic," University of Wisconsin–Madison.
2. EPRI RP 133: "Electrochemical Treeing in Solid Dielectric Cable," 1 EPRI Report EL-647, 1976.
3. Thue, W. A. and Bankoske, J., 1980, "Operating and Testing Experience on Solid Dielectric Cables," CIGRE.

22 Concentric Neutral Corrosion

William A. Thue

CONTENTS

22.1 INTRODUCTION

In nature, metals are usually found in combinations such as oxides or sulfides, not as pure metals. Nature wants to change those pure metals back to their original state after we have refined them to almost pure metals. That process is known as corrosion [1–3].

One hundred years ago, an interesting article was published [4]: "Practically the only factor which limits the life of metal is oxidation, under which name are included all the chemical processes whereby the metal is corroded, eaten away, or rusted. In undergoing this change, the metal always passes through or into a state of solution, and as we have no evidence of metal going into aqueous solution except in the form of ions, we have really to consider the effects of conditions upon the potential difference between a metal and its surroundings. The whole subject of corrosion is simply a function of electromotive force and resistance of the circuit."

Corrosion may be defined as the destruction of metals by chemical or electrochemical reaction with the environment. The fundamental reaction involves a transfer of electrons where, in a moist or wet environment, some positive ions lose electrical charges. These positive charges are acquired by the metallic member, and a portion of the metal surface goes into solution, hence is corroded. The entire process may be divided into an anodic reaction (oxidation) and a cathodic reaction (reduction). The anodic reaction represents acquisition of charges by the corroding metal, and the cathodic reaction represents the loss of charges by hydrogen ions that are discharged. The flow of electricity between the anodic and cathodic areas may be generated by local cells set up either on a single metallic surface, or between dissimilar metals.

22.2 ELECTROMOTIVE SERIES

The tendency for metals to corrode by hydrogen ion displacement is indicated by their position in the electromotive series shown in Table 22.1. To achieve these

TABLE 22.1
Electromotive Series (Anodic End)

Metal	Ion	Volts
Magnesium	Mg + 2e*	−2.34
Aluminum	Al + 3e	−1.67
Zinc	Zn + 2e	−0.76
Chromium	Cr + 3e	−0.71
Iron	Fe + 2e	−0.44
Cadmium	Cd + 2e	−0.40
Nickel	Ni + 2e	−0.25
Tin	Sn + 2e	−0.14
Lead	Pb + 2e	−0.13
Hydrogen	H + e	Arbitrary 0.00
Copper	Cu + 2e	+0.34
Silver	Ag + e	+0.80
Palladium	Pd + 2e	+0.83
Mercury	Hg + 2e	+0.85
Carbon	C + 2e	+0.90
Carbon	C + 4e	+0.90
Platinum	Pt + 2e	+1.2
Gold	Au + 3e	+1.42
Gold	Au + e	+1.68
(Cathodic End)		

* "e" stands for electrons (negative charges).

precise voltages, the metals must be in contact with a solution in which the activity of the ion indicated is 1 mol per 1,000 grams of water and at 77°F (25°C). Different values of voltage will be obtained in other solutions.

Metals above hydrogen displace hydrogen more readily than do those below hydrogen in this series. A decrease in hydrogen ion concentration (acidity) tends to move hydrogen up relative to other metals. An increase in the metal ion concentration tends to move the metals down relative to hydrogen. Whether or not hydrogen evolution will occur in any case is determined by several other factors besides the concentration of hydrogen and metallic ions.

22.2.1 Electrochemical Equivalents

The electrochemical equivalent of a metal is the theoretical amount of metal that will enter into solution (dissolve) per unit of DC transfer from the metal to an electrolyte. Table 22.2 shows that theoretical amount of metal removed in pounds per year with 1 ampere of DC flowing continuously from the material.

22.2.2 Hydrogen Ion Concentration

A normal solution is one that contains an "equivalent weight" (in grams) of the material dissolved in sufficient water to make 1 liter of the solution. The equivalent weight of hydrogen is 1 and therefore 1 gram of hydrogen ions in a liter of water is a normal acid solution. The hydroxyl ion has an equivalent weight of 17 (1 for the hydrogen and 16 for oxygen). Therefore, 17 grams in a liter is equal to the normal alkaline solution.

TABLE 22.2
Electrochemical Equivalents

Metal	Pounds Removed per Amp per Year
Carbon (C++++)	2.16
Carbon (C++)	4.23
Aluminum	6.47
Magnesium	8.76
Chromium	12.5
Iron	20.1
Nickel	21.1
Cobalt	21.2
Copper (Cu++)	22.8
Zinc	23.5
Cadmium	40.5
Tin	42.7
Copper (Cu+)	45.7
Lead	74.2

TABLE 22.3
Significance of Hydrogen Ion Concentration

pH	Hydrogen Ion Concentration	Hydroxyl Ion Concentration	Reaction
0	1.0×10^{-0}	1.0×10^{-14}	Acidic
1	1.0×10^{-1}	1.0×10^{-13}	Acidic
2	1.0×10^{-2}	1.0×10^{-12}	Acidic
3	1.0×10^{-3}	1.0×10^{-11}	Acidic
4	1.0×10^{-4}	1.0×10^{-10}	Acidic
5	1.0×10^{-5}	1.0×10^{-9}	Acidic
6	1.0×10^{-6}	1.0×10^{-8}	Acidic
7	1.0×10^{-7}	1.0×10^{-7}	Neutral
8	1.0×10^{-8}	1.0×10^{-6}	Alkaline
9	1.0×10^{-9}	1.0×10^{-5}	Alkaline
10	1.0×10^{-10}	1.0×10^{-4}	Alkaline
11	1.0×10^{-11}	1.0×10^{-3}	Alkaline
12	1.0×10^{-12}	1.0×10^{-2}	Alkaline
13	1.0×10^{-13}	1.0×10^{-1}	Alkaline
14	1.0×10^{-14}	1.0×10^{-0}	Alkaline

Since acids produce hydrogen ions when dissolved in water, the concentration of the hydrogen ions is a measure of the acidity of the solution. The hydrogen ion concentration is expressed in terms of pH. Stated mathematically, the pH value is the logarithm of the reciprocal of the hydrogen ion concentration in terms of the normal solution. A change of 1 in pH value is equivalent to a change of 10 times in concentration.

In any aqueous solution, the hydrogen ion concentration multiplied by the hydroxyl ion concentration is always a constant. When the concentrations are expressed in terms of normal solution, the constant is equal to 10^{-14}. It follows that a solution having a pH equal to 7 is neutral, less than 7 has an acidic reaction, and more than 7 has an alkaline reaction (Table 22.3).

22.3 MECHANISM OF CORROSION

The basic nature of corrosion is almost always the same, a flow of electricity between certain areas of a metal surface through a solution capable of conducting electric current. This electrochemical action causes the eating away of the metal at areas where metallic ions leave the metal (anodes) and enter the solution. This is the critical step in the corrosion process.

The first basic requirement for corrosion is the presence of an electrolyte and two electrodes—an anode and a cathode. These electrodes may consist of two different kinds of metal, or may be different areas of the same piece of metal. In any case, there must be a potential difference between the two electrodes so that current will flow between them. A wire or some path is necessary for the flow of electrons that are negatively charged particles moving in the wire from the negative to the positive.

At the anode, positively charged atoms of metal detach themselves from the solid surface and enter into solution as ions while the corresponding negative charges, in the form of electrons, are left behind in the metal. For copper:

$$Cu^{o} \rightarrow Cu^{++}2e^{-}$$

The detached positive ions bear one or more positive charges. The released electrons travel through the metal or other conducting media to the cathode area. The electrons reaching the surface of the cathode through the metal meet and neutralize some positively charged ions (such as hydrogen) that have arrived at the same surface through the electrolyte. In losing their charge, the positive ions become neutral atoms again and may combine to form a gas.

$$2H^{+} + 2e \rightarrow H_2$$

The release of hydrogen ions results in the accumulation of OH^- ions that are left behind and increase the alkalinity at the cathode—hence making the solution less acidic. Other common reactions occurring at the cathode are shown as follows. Their occurrence depends on such factors as pH, type of electrolyte, etc.

$$4H^{+} + O_2 + 4e^{-} \rightarrow 2H_2O$$

$$O_2 + 2H_2O + 2e^{-} \rightarrow H_2O_2 + 2OH^{-}$$

$$O_2 + 2H_2O + 4e^{-} \rightarrow 4OH^{-}$$

For corrosion to occur, there must be a release of electrons at the anode and a formation of metal ions through oxidation or disintegration of the metal. At the cathode, there must be a simultaneous acceptance of the electrons by a mechanism such as neutralization of the positive ions or formation of negative ions. Actions at the cathode and anode must always go on together, but corrosion occurs almost always at areas that act as anodes.

22.4 TYPES OF CORROSION

There are numerous types of corrosion, but the ones that are discussed here are the ones that are most likely to be encountered with underground power cable facilities.

In this initial explanation, lead will be used as the reference metal. Copper neutral wire corrosion will be discussed as a separate topic later.

22.4.1 Anodic Corrosion (Stray Direct Current Currents)

Stray DC currents come from sources such as welding operations, flows between two other structures, and—in the days gone by—street railway systems.

Anodic corrosion is due to the transfer of DC from the corroding facility to the surrounding medium—usually earth. At the point of corrosion, the voltage is always positive on the corroding facility. In the example of lead sheath corrosion, the lead provides a low resistance path for the DC current to get back to its source. At some area remote from the point where the current enters the lead, but near the inception point of that stray current, the current leaves the lead sheath and is again picked up in the normal DC return path. The point of entry of the stray current usually does not result in lead corrosion, but the point of exit is frequently a corrosion site.

Clean sided corroded pits are usually the result of anodic corrosion. The products of anodic corrosion such as oxides, chlorides, or sulfates of lead are carried away by the current flow. If any corrosion products are found, they are usually lead chloride or lead sulfate that was created by the positive sheath potential that attracts the chloride and sulfate ions in the earth to the lead.

In severe anodic cases, lead peroxide may be formed. Chlorides, sulfates, and carbonates of lead are white, while lead peroxide is chocolate brown.

22.4.2 Cathodic Corrosion

Cathodic corrosion is encountered less frequently than anodic corrosion—especially with the elimination of most street railway systems.

This form of corrosion is usually the result of the presence of an alkali or alkali salt in the earth. If the potential of the metal exceeds −0.3 volts, cathodic corrosion may be expected in those areas. In cathodic corrosion, the metal is not removed directly by the electric current, but it may be dissolved by the secondary action of the alkali that is produced by the current. Hydrogen ions are attracted to the metal, lose their charge, and are liberated as hydrogen gas. This results in a decrease in the hydrogen ion concentration and the solution becomes alkaline.

The final corrosion product formed by lead in cathodic conditions is usually lead monoxide and lead/sodium carbonate. The lead monoxide formed in this manner has a bright orange/red color and is an indication of cathodic corrosion of lead.

22.4.3 Galvanic Corrosion

Galvanic corrosion occurs when two dissimilar metals in an electrolyte have a metallic tie between them. One metal becomes the anode and the other the cathode. The anode corrodes and protects the cathode as current flows in the electrolyte between them. The lead sheath of a cable may become either the anode or the cathode of a galvanic cell. This can happen because the lead sheath is grounded to a metallic structure made of a dissimilar metal and generally has considerable length. Copper ground rods are most often a source of the other metal in the galvanic cell.

The corrosive force of a galvanic cell is dependent on the metals making up the electrodes and the resistance of the electrolyte in which they exist. This type of corrosion can often be anticipated and avoided by keeping a close watch on construction practices and eliminating installations having different metals connected together in the earth or other electrolyte.

22.4.4 Chemical Corrosion

Chemical corrosion is damage that can be attributed entirely to chemical attack without the additional effect of electron transfer. The type of chemicals that can disintegrate lead are usually strong concentrations of alkali or acid. Examples include alkaline solutions from incompletely cured concrete, acetic acid from volatilized wood or jute, waste products from industrial plants, or water with a large amount of dissolved oxygen.

22.4.5 Alternating Current Corrosion

Until about 1970, AC corrosion was felt to be an insignificant, but possible, cause of cable damage [6]. In 1907, Hayden [7], reporting on tests with lead electrodes, showed that the corrosive effect of small AC currents was less than 0.5% as compared with the effects of equal DC currents.

Later work using higher densities of AC current has shown that AC corrosion can be a major factor in concentric neutral corrosion, see Section 22.5.3.

22.4.6 Local Cell Corrosion

Local cell corrosion, also known as differential aeration in a specific form, is caused by electrolytic cells that are created by a nonhomogeneous environment where the cable is installed. Examples include variations in the concentration of the electrolyte through which the cable passes, variations in the impurities of the metal, or a wide range of grain sizes in the backfill. These concentration cells corrode the metal in areas of low ion concentration.

Differential aeration is a specific form of local cell corrosion where one area of the metal has a reduced oxygen supply as compared with nearby sections that are exposed to normal quantities of oxygen. The low oxygen area is anodic to the higher oxygen area and an electron flow occurs through the covered (oxygen starved) material to the exposed area (normal oxygen level).

Differential aeration corrosion is common for underground cables, but the rate of corrosion is generally rather slow. Examples of situations that can cause this form of corrosion include a section of bare sheath or neutral wires laying in a wet or muddy duct or where there are low points in the duct run that can hold water for some distance. A cable that is installed in a duct and then goes into a direct buried portion is another good example of a possible differential aeration corrosion condition.

Differential aeration corrosion turns copper a bright green.

22.4.7 Other Forms of Corrosion

There are numerous other forms of corrosion that are possible, but the most probable causes have been presented. An example of another form of corrosion is microbiological action of anaerobic bacteria, which can exist in oxygen-free environments with pH values between 5.5 and 9.0. The life cycle of anaerobic bacteria depends on the reduction of sulfate materials rather than on the consumption of free oxygen. Corrosion resulting from anaerobic bacteria produces sulfides of calcium or hydrogen and may be accompanied by a strong odor of hydrogen sulfide and a buildup of a black slime. This type of corrosion is more harmful to steel pipes and manhole hardware than to lead sheaths.

22.5 CONCENTRIC NEUTRAL CORROSION

This section will concentrate on the corrosion mechanisms associated with concentric neutral, medium voltage power cables [11]. The most probable causes of concentric neutral corrosion include:

- Differential aeration
- Stray DC current flow
- DC current generated through AC rectification
- AC current flow between neutral and earth
- Galvanic influence with semiconducting layer (unjacketed cables)
- Galvanic influence of alloy coating and copper neutral wires and other action from dissimilar metals
- Soil contaminants

Electric power systems had used copper directly buried in the ground for over 60 years without problems being experienced. Most of the applications consisted of butt wraps under poles and substation ground grids. The successful operation led to complacency when underground residential distribution cables (URD) began to be installed in vast quantities after 1965.

Although the number of cable failures caused by neutral corrosion were very small, when these cables did fail for other reasons, it became clear that neutral corrosion was taking place in situations that were not anticipated.

22.5.1 Research Efforts

The Electric Power Research Institute (EPRI) funded a series of projects to study the problem and to suggest remedies [9–15]. The subjects include mechanisms of corrosion, cathodic protection methods, procedures for locating corrosion sites, and step-and-touch potential data for jacketed as well as unjacketed cable.

22.5.2 Composition of Soils

Both the physical and chemical characteristics of soils are important although it is difficult to separate these effects. The significant properties of a soil include:

1. Its water retaining capacity
2. Its power of adsorption on the metal surface
3. Its conductivity
4. Dissolved matter in the soil
5. Concentration variations
6. Grain size

The soluble constituents of the soil water include dissolved gases; inorganic acids, bases, and salts; organic compounds (including fertilizers) and other related substances.

22.5.3 Mechanisms of Concentric Neutral Corrosion

Differential aeration is one specific type of local cell corrosion and is probably the most common cause of neutral corrosion. Fortunately, this is a relatively slow form of attack. This type of corrosion is caused when a metal is exposed to soils or water having a difference in oxygen content. Examples of this are:

- Soils with different grain sizes
- Cable going from a direct buried environment to a conduit
- A conduit run that has a section with standing water and another section that has an unlimited supply of oxygen
- Jacketed cable spliced to unjacketed cable

The key concept here is the dissimilar environment and oxygen supply for a run of cable. It can occur in a small crevice made by a large grain of sand or stone in contact with the copper neutral conductor. Areas of low aeration change to an area that is well aerated. This form of corrosion is frequently caused when special backfills are brought in to replace the native soil. The native soil usually has a consistent grain size while the imported material may have quite a different grain size. Pockets are thereby formed.

Another very frequent cause of this form of corrosion is where an unjacketed cable leaves a conduit (such as under a street) and enters the earth. The same sort of cell is created by having a low section of conduit that is filled with water while the adjacent section is in a dry conduit. The use of an overall jacket (either insulating or semiconducting) eliminates this condition.

Stray DC current flow problems are very similar to the lead sheath condition previously described. This situation is frequently encountered when an anode that is used to protect a gas pipeline is installed in close proximity to an unjacketed cable. This damage occurs very rapidly.

Stray DC current causes dissolution of the copper where anions are present that contribute to the reaction. The rate of dissolution may not follow Faraday's law precisely because of other electrochemical oxidation reactions that occur in parallel.

DC current flow can be generated through AC rectification across a film of copper oxide. Copper neutral wires quickly develop an oxide coating. This coating provides a rectification boundary so that AC current is restricted from flowing back to the neutral wires.

AC corrosion was not recognized as a serious problem in the initial URD systems. The opinion was that while AC current flow might take off metal during one half cycle, the other half cycle would bring it back. The concept of rectification was commonly discussed as a possible explanation for AC corrosion in the 1960s. It was not until the 1970s that AC corrosion was recognized as a major concern for copper neutrals. The Final Report of EPRI EL-4042 [13] published in 1985 stated that the effect of high AC current density was creating this rapid corrosion mechanism on bare URD and underground distribution (UD) cables.

Above some threshold of AC current density, the positive cycle tends to dissolve more metal than the negative cycle can plate back. Especially in cables with large conductors that are heavily loaded (such as feeder cables), the amount of current that can flow off the neutral wires at one point and then back on at another is quite large. Another explanation of this flow of current off and then back on the neutral wires is that shifts in potential exist along the cable length due to the differences in the current densities.

Galvanic influence with the semiconducting insulation shield material and bare or tinned copper is another form of concentric neutral corrosion. A voltage differential exists between the carbon in the semiconducting layer and the neutral wires. Corrosion, although not a widespread cause of failure, must be considered.

Dissimilar metal corrosion is probable if plating is used on the neutral wires and no jacket is applied over these wires. Areas of bare copper may exist during the factory plating process or are created by mechanical scraping during handling and field installation efforts. The result is local cell corrosion due to the two different metals.

Research has shown that bare wires outperform plated wires in the field. When jackets are used, bare copper wires are recommended and are almost always specified.

Soil contaminants and other direct chemical action are another source of problems for URD cables. Examples of this are high quantities of chemicals in the soil such as from fertilizers, peat, cinders, and decaying vegetation. Decaying vegetation produces hydrogen sulfide that reacts rapidly to deteriorate copper.

Combinations of the previously discussed corrosion mechanisms do occur in the real world. Multiple sources of corrosion accelerate the problem.

22.6 JACKETS

An overall jacket is the preferred construction for new cable. Both insulating and semiconducting jackets have demonstrated their ability to virtually eliminate corrosion of the neutral wires. An encapsulated jacket made with linear low density polyethylene is the type most frequently specified. See Chapter 8 for a complete discussion of jackets.

22.7 CATHODIC PROTECTION

Cathodic protection [5,8] can be applied to the copper neutral wires of existing cable that did not have a jacket or where the jacket may be damaged. An obvious place where cathodic protection should be considered is where a bare neutral cable goes from a direct buried environment to a conduit as under a road. Another is where a long section of jacketed cable is spliced onto a short section of existing bare neutral cable. Here the short section will be even more vulnerable to corrosion.

Noteworthy efforts have been expended toward solving this concern and the reference section contains excellent sources of advice regarding design and installation recommendations.

22.8 LOCATION OF CORROSION SITES

The existence of deteriorated copper neutral wires is an unwelcomed fact. How to identify their existence and then locate the precise site of the corrosion has shown great advancement in recent years. Several technologies are currently available.

22.8.1 Resistivity Measurements of Neutral Wires

Resistive techniques are used to measure the resistance of neutral wires of installed cables. Instruments are available for testing the resistance of the neutral wires while the cable is energized. This value is compared with the original resistance for that length of cable and the size and number of original wires. A new or undamaged cable would show a resistance ratio of 1 while a cable that has half of the wires remaining would have a ratio of 2.

22.8.2 Location of Deteriorated Sites

If the reading of the resistance ratio warrants, the precise location of the corroded area can be obtained by using a form of time domain reflectometry (TDR). This equipment is similar to that used for locating cable faults [16].

Damage to a neutral wire must be great enough so that a reflection can be seen on the screen of the TDR. This may mean that a wire with only pitting may not be identifiable, and the cable would appear to be sound. The reflection of that wave from the end of the cable has a lower amplitude than intact wires. A cable with uniform corrosion may not be seen, but discrete sites cause reflections and are readily detected.

REFERENCES

1. *Underground Systems Reference Book,* National Electric Light Association, 1931.
2. *Underground Systems Reference Book,* Edison Electric Institute, 1957.
3. Frink, J., September 1947, "Cathodic Protection of Underground Systems", *American Gas Journal.*
4. Whitney, W. R., 1903, "The Corrosion of Iron", *Journal of the American Chemical Society*, #22.
5. *Relieving Underground Corrosion of Multi-grounded Rural Electric Lines,* Rural Electrification Administration, May 1963.
6. Kulman, F. E., 1960, "Effects of Alternating Currents in Causing Corrosion", American Gas Association, Operating Section Distribution Conference.
7. Hayden, J. L. R., 1907, "Alternating Current Electrolysis", AIEE Transactions, Vol. 26, Part I.
8. *Underground Distribution System Design and Installation Guide,* Section 7, National Rural Electric Cooperative Association, Catalogue RERP9087T, 1993.

9. EPRI EL-619, Vol. 1, Phase 1 "Study of Semiconducting Materials Useful for Cable Jackets".
10. EPRI EL-619, Vol. 1, Phase 2, "Cable Neutral Corrosion".
11. EPRI EL-362, "Status Report on Concentric Neutral Corrosion".
12. EPRI EL-1970, Vols. 1, 2 and 3, "Cathodic Protection of Concentric Neutral Cables".
13. EPRI EL-4042, "Corrosion Mechanisms in Direct Buried Concentric Neutral Systems".
14. EPRI EL-4961, "Methods for Mitigating Corrosion of Copper Concentric Neutral Wires in Conduits".
15. EPRI EL-4448, "Methodology for Predicting Corrosion of URD Cables Using Modeling Techniques".
16. "Technologies Locate Corrosion Before It's Too Late", *Transmission and Distribution World*, July 1997, pp. 20–22.

23 Armor Corrosion of Submarine Cables

William A. Thue

CONTENTS

23.1 INTRODUCTION

Submarine cables generally have a metallic armor consisting of one or two layers wound around the outside of the cable. The main function of the armor is to provide mechanical strength during the installation and recovery of the cable since the conductor alone may not have sufficient tensile strength for this purpose.

The armor provides another function as the path for conduction of the return current. Single-conductor submarine cables are frequently spaced a considerable distance apart to preclude the chance of more than one cable being mechanically damaged as a result of a single incident. When the spacing is about 100 feet—about 30 meters—on centers (300 to 500 times the diameter of a single cable), the return current approaches the current in the central conductor. The effect of an increase in spacing above 100 feet (30 meters) is negligible.

Paths that may be available for this return current are:

- The metallic armor wires
- Specially designed return path under the armor
- A metallic sheath
- Reinforcing tapes used in pressurized cables
- Shielding tapes
- Sea and soil surrounding the cables

The choice of a three-conductor or single-conductor cables for large power cables is an important decision. For this moment, we will assume that single-conductor cables have been chosen. We will deal with that situation since it may be of critical importance in the cable design selection.

When the line length is more than 2 miles (3 kilometers), consideration must be given to connecting the components in the cable together. Frequently, a metallic sheath is protected by an insulating jacket to reduce the possibility of corrosion. To prevent electrical breakdown of the insulating jacket, electric connections are usually made at suitable intervals between the sheath, the armor, and the sea/soil. This may be done during the manufacturing process so that no visible signs of interconnection are available or they may be made externally.

23.2 CURRENTS IN ARMOR SYSTEM

This chapter will concentrate on the effects of alternating current (AC) corrosion as a result of return current flow in the "armor system"—the parallel paths formed by the metallic armor wires, any supplemental return conductor or reinforcing tapes, a metallic sheath, and/or a metallic shield. If all of the current stayed in the metallic path or paths, there would not be AC corrosion. The fact is that the current flows off the metal and into the sea or seabed and later returns. It is this process of current leaving and returning to the surface that results in the deterioration. Unfortunately, the metal that is removed by the current does not plate back as the current returns; the greater the current, the greater the damage. This is called AC corrosion (see Chapter 22).

The amount of current will now be considered for single-conductor cables on a system where the phase rotation is A, B, and C and the cables are arranged in a flat configuration as shown in Figure 23.1.

In the case of these three cables with a balanced load three-phase system, the vector sum of the three currents is zero. That is, the currents in each phase all have the same effective value and are 120 degrees out of phase with one another.

Hence,

$$I_A + I_B + I_C = 0 \tag{23.1}$$

We are now going to make the very reasonable assumption that the armor system is connected to ground at more than one location. Even though two grounding points might be a few feet apart, the currents that will be described will flow between these points. The more realistic situation is that the armor system is grounded for the entire route and hence, these currents will flow everywhere in the armor system circuit.

A schematic of the system is shown in Figure 23.2. Only two of the possible components in the armor system are shown to keep the system somewhat simple.

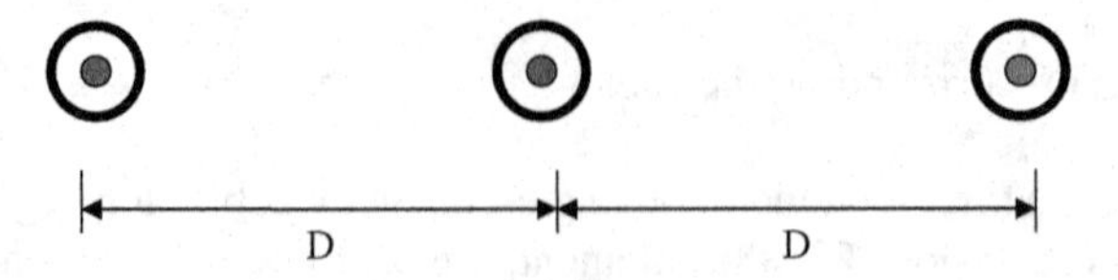

FIGURE 23.1 Single-conductor cables spaced apart.

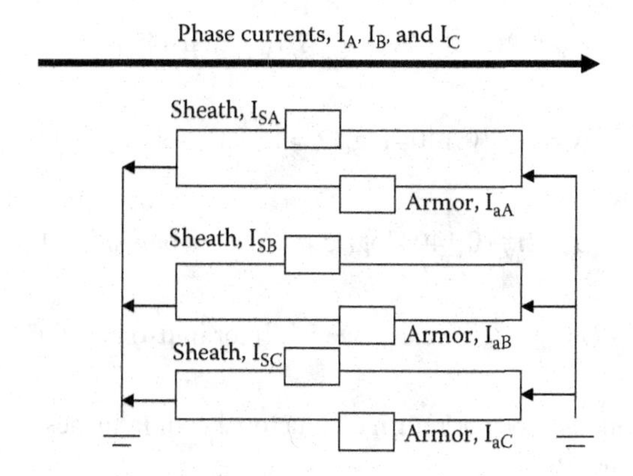

FIGURE 23.2 Parallel currents in spaced single conductor cables.

The shield loss for each phase of a multigrounded cable may be calculated from the following formulas:

A and C phases:

$$W_{LS(\mathrm{A,C})} = I_{\mathrm{AC}}^2 R_S \left(\frac{\left(P^2 + 3Q^2\right) \pm 2\sqrt{3}\left(P - Q\right) + 4}{4\left(P^2 + 1\right)\left(Q^2 + 1\right)} \right) \tag{23.2}$$

B phase cable:

$$W_{LS(\mathrm{B})} = I_{\mathrm{B}}^2 R_S \left(\frac{1}{Q^2 + 1} \right) \tag{23.3}$$

where W_{LS} = micro watts per foot.

Total loss:

$$W_{LS(T)} = 3I^2 R_S \left(\frac{P^2 + Q^2 + 2}{2\left(P^2 + 1\right)/\left(Q^2 + 1\right)} \right) \tag{23.4}$$

$$P = R_S / Y$$

$$Q = R_S / Z$$

and

$$Y = X_M + \mathrm{A}$$

$$Z = X_M - \mathrm{A}/3$$

$$X_M = 2\pi f\left(0.1404 \log_{10} S/r_M\right) \text{ in } \mu\text{-ohms/foot} \tag{23.5}$$

$$A = 2\pi f\left(0.1404 \log_{10} 2 = 15.93 \ \mu\text{-ohms/foot}\right) \tag{23.6}$$

$$B = 2\pi f\left(0.1404 \log_{10} 5 = 36.99 \ \mu\text{-ohms/foot}\right) \tag{23.7}$$

$$R_S = \rho/8r_M = \text{resistance of armor in } \mu\text{-ohms/foot} \tag{23.8}$$

t = thickness of metal tapes or return conductor/sheath in inches
f = frequency in hertz
D = distance (spacing) between centers of cables in inches
r_M = mean radius of shield in inches
I = conductor current in amperes
ρ = apparent resistivity of shield in ohms-circular mils per foot at operating temperature

Effective values in ohms-circular mils per foot:

Overlapped copper tape	30
Overlapped bronze tape 90-100	47
Overlapped Monel tape	2,500
Overlapped Cupro-nickel tape 80-20	350
Lead sheath	150
Aluminum sheath	20
Aluminum interlocked armor	28
Galvanized steel armor wire	102
5052 aluminum alloy	30
Galvanized steel interlocked armor	70
Stainless steel SS304	43

Scc the discussion of ampacity and of metallic shield losses for additional insight into the general problem in Chapter 14. The fact is that the conductor and shield arrangement in a multigrounded system is a one-to-one transformer so the greater the current in the phase conductor of most cable systems, the greater the current induced in the armor system—and hence the chances of corrosion escalate. This is a catch 22 situation in that the desire for additional armor strength and protection runs contrary to the need to reduce sheath currents and hence power losses.

Another way to consider this situation is that the more the metal in the armor system, the closer the current flow comes to the full phase current in the central conductor.

Three-conductor cables have negligible current in the armor system as long as the load is reasonably balanced. An unbalanced load does induce current in the armor system and may create some of the same concerns about armor corrosion as single-conductor cables. The same is true if the three single-conductor cables are tightly laid together—in the form of a triangle. This defeats the concept of spacing them apart to avoid the same ship or anchor breaking all of them at one time.

The point is that the corrosion of a three-conductor or three tightly spaced single-conductor submarine cables is considerably less of a factor than for the spaced single-conductor cables. An unbalanced load here may become significant.

23.3 CORROSION

Although the armor is occasionally insulated, this discussion will assume that the armor is bare and hence in contact with the sea for its entire length. Even in the situation of insulated armor wires, it is prudent to anticipate some damage to the insulating layer during installation or service-inflicted conditions. One single break in the insulating layer could be a very serious corrosion point unless other measures are taken.

The contact between the metallic armor and the sea, in the presence of an electric field, requires a thorough study of the corrosion factors that could be possible. Only a brief overview of the problems will be addressed here. Three types of metals are used for the armor wires of submarine cables:

- Galvanized steel
- Copper
- Aluminum alloys
- All of the above with coverings for each wire

23.3.1 Corrosion Under Direct Current (DC) Conditions

The corrosion properties of metals in seawater are frequently analyzed in the laboratory by performing weight loss studies. An example of such a test is shown in Table 23.1.

The test setup consisted of samples of cold drawn wire having diameters of 4 to 6 mm. DC was circulated between two parallel lengths of the same material about 100 mm long and maintained at a separation of 80 mm in a solution of 3.5% NaCl.

Aluminum, aluminum alloys, and steel corrode according to their electrochemical equivalent and to the transfer of electric charge when subjected to DC in anodic circumstances. Weight loss on anodic copper was diminished due to the formation of a Cu_2O layer that stopped the current flow with these test conditions. Very low weight loss was observed in cathodic conditions.

TABLE 23.1
Weight Loss Under DC and 1,000 Amperes/meter^2 for 30 Minutes

Metal	Weight Loss for Anode, in kg/m^2	Weight Loss for Cathode, in kg/m^2
Aluminum alloy	0.21	0.004
Copper	0.14	0.0012
Galvanized steel	0.65	0.0008

TABLE 23.2
Weight Loss Under DC and 1,000 Amperes/meter^2 for 1,000 Hours

Metal	Weight Loss for 10 A/m^2, kg/m^2	Weight Loss for 100 A/m^2, in kg/m^2	Weight Loss for 1,000 A/m^2, in kg/m^2	Weight Loss for 7,000 A/m^2, in kg/m^2
Aluminum Alloy	3.5×10^{-2}	6.0×10	2.0×10^3	11.0×10^3
Copper	2.5×10^{-2}	7.0×10^{-1}	1.5×10^2	0.4×10^3
Galvanized Steel	1.5×10^{-1}	1.2×1	0.7×10	1.5×10^2

23.3.2 Corrosion Under AC Conditions

Weight loss tests for the same materials were performed under AC conditions for different current densities but for the same 1,000 hour time and in accordance with ASTM G1-72.

There is a well-defined threshold of AC density for these materials. Below this threshold, weight loss is not dependent on current density and follows the rules for chemical weight loss—hence, they are dependent on their electrochemical potentials.

23.4 FIELD EXPERIENCE

Examination was made of this same aluminum alloy armor from a submarine cable with 5 years of service. Almost all of the corrosion was due to chemical corrosion since there was negligible AC current in the cable and the weight loss of 0.054×10^{-3} kg/m^2 compared closely with the data from Table 23.2.

Copper and galvanized steel show a uniform pattern of corrosion in an AC environment while aluminum alloys demonstrate pitting and localized corrosion. An example of steel wire corrosion after 10 years of service is shown in Figure 23.3.

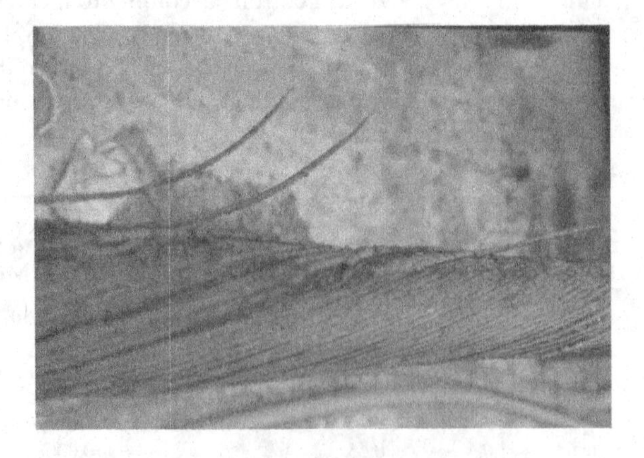

FIGURE 23.3 Steel armor wires corroded by AC current.

23.5 MITIGATION OF ARMOR CORROSION

EPRI Project EL-4042 demonstrated that 1 mA of current made AC corrosion a serious problem for copper concentric neutrals on underground residential distribution (URD) cables. There is no sufficient test data to clarify the amount of AC current in the armor wires of submarine cables to cause rapid deterioration, but field experience with single-conductor cables in a salt water environment has shown that in 10 years corrosion has seriously damaged the cable.

The reverse side of this experience has shown that such single-conductor cables that have a designed return circuit under the armor with a conductivity equal to that of the central conductor have had a new appearance after 12 years of service.

REFERENCE

1. *Corrosion of the Concentric Neutrals of Buried URD Cables.* EPRI EL-4042, Project 1144-1, Final Report May 1985.

24 Glossary and Acronyms

James D. Medek

CONTENTS

24.1 GLOSSARY

Note: This glossary contains many of the cable terms used throughout this book and is furnished as an aid to understand the text. The reader is encouraged to utilize the more complete definitions that may be found in the *IEEE Standard Dictionary of Electrical and Electronics Terms*, IEEE Standard 100-1996.

Abrasion Resistance: Ability to resist surface wear.

Accelerated Life Test: Subjecting a product to test conditions more severe than normal operating conditions, such as voltage and temperature, to accelerate aging and thus to afford some measure of probable life at normal conditions or some measure of the durability of the equipment when exposed to the factors being aggravated.

Acceptance Test: A field test made after cable system installation to demonstrate the degree of compliance with specified requirements or a test demonstrating the quality of the units of a consignment. The term "conformance test" is recommended by ANSI to avoid any implication of contractual relations.

Aging: The irreversible change of material properties after exposure to an environment for an interval of time.

Ampacity: The current carrying capacity of a cable, expressed in amperes. The current that a cable can carry under stated thermal conditions without degradation.

Ampere: The basic SI unit of the quantity of electric current. The constant current that if maintained in two straight parallel conductors of infinite length, or negligible cross section, and placed 1 meter apart in vacuum, would produce a force equal to 2×10^{-7} Newton per meter of length.

Amplitude: The maximum value of a sinusoidally varying waveform.

Annealing: The process of removing or preventing mechanical stress in materials by controlled cooling from a heated state, measured by tensile strength.

Anode: An electrode to which negative ions are attracted.

Antioxidant: A chemical additive incorporated into polyolefins to prevent degradation of materials exposed to oxygen during extrusion in the process that converts insulation materials into cable insulation over conductors.

Asymmetrical: Not identical on both sides of a central line; not symmetrical.

Attenuation: The decrease in magnitude of a signal as it travels through any transmitting medium.

Backfill: The materials used to fill an excavation, such as sand in a trench.

Basic Impulse Insulation Level (BIL): The impulse voltage that electrical equipment is required to withstand without failure or disruptive discharge when tested under specified conditions of temperature and humidity. BILs are designated in terms of the crest voltage of a 1.2 × 50 microsecond full-wave voltage test.

Bedding: A layer of material that acts as a cushion or interconnection between two elements of a device, such as the jute or polypropylene layer between the sheath and the wire armor in a submarine cable.

Bending Radius: The inner radius of a cable, such as when it is trained or being installed.

Braid: An interwoven covering having a flat configuration usually of fiber or metal.

Branching: Term used to describe the portion of the polymeric insulation that "hangs off" the main chain like a "T."

Breakdown: Disruptive discharge through an insulation.

Bridge: A circuit that measures by balancing a number of resistances or impedances through which the same current flows.

Butt Lap: Complete turn of tape where the adjacent layers are next to each other but do not overlap.

Cable, Aerial: An assembly of one or more insulated conductors that are lashed or otherwise fastened to a supporting messenger.

Cable, Belted: A multiconductor cable having a layer of insulation over the assembled but unshielded insulated conductors.

Cable, Spacer: An aerial cable system made of covered conductors supported by insulating spacers; generally for wooded areas.

Cable, Submarine: A cable designed for crossing under bodies of water; having mechanical strength for installation and removal, and limited protection from anchors, debris, and other mechanical damage.

Cable, Tray: A rigid structure to support cables. A type of raceway normally having the appearance of a ladder. May be open at the top (or side) to facilitate changes, or be covered with a ventilated or solid cover.

Cable, Triplexed: A helical assembly of three covered or insulated conductors; sometimes with one bare conductor used as a neutral.

Cambric: A fine weave of linen, cotton, or other fiber that is used as an insulation base.

Capacitance: The storage of electricity in a capacitor. The opposition to voltage change, measured in farads.

Capacitor: Any device having two conductors separated by insulation, with the conductors having opposite electrical charges.

Capstan: A rotating drum used to pull cables or ropes by friction as they are wrapped around the drum.

Carbon Black: Elemental carbon in the form of spherical particles and aggregates, manufactured by thermal decomposition of various hydrocarbons that impart conductivity to semiconducting shields.

Catalyst: A material used to induce polymerization of monomers to convert them into polymeric insulation materials. These may be peroxides or metallocenes.

Catenary: The natural curve assumed by a completely flexible material hanging freely between two supports. A cable curing tube that has a catenary curvature.

Cathode: An electrode to which positive ions (cations) are attracted, the positive pole of a source of direct voltage.

Cellulose: A natural polymer derived from wood that is used to manufacture paper for cables.

Charge: The quantity of positive or negative ions in or on an object; unit: coulomb.

Charging Current: Current that flows in an energized cable as a result of the cable capacitance.

Circular mil: A circular mil has the equivalent area of a circle whose diameter is 0.001 inch.

Coefficient of Friction: The ratio of the tangential force needed to start or maintain relative motion between two contacting surfaces to the perpendicular force holding them in contact.

Concentric Neutral: A neutral conductor that is helically wrapped around the insulation shield of a power cable.

Conduit Fill: The percentage of cross-sectional area used in a conduit as compared with the cross-sectional area of the conduit.

Continuous Vulcanization: A system utilizing heat, and frequently pressure, to vulcanize materials after extrusion onto a conductor.

Conversion of English to Metric:

Length:	1 mil = 0.0254 millimeters
	1 inch = 1,000 mils
	1 inch = 2.54 centimeters
	1 inch = 25.4 millimeters
	1,000 feet = 304.8 meters
	1 mile = 1,609.344 meters
Area:	1 square inch = 6.4516×10^{-4} square meters
	1,000 circular mils = 0.5067×10^{-9} square millimeters
Weight:	1 ounce = 0.028349 kilograms
	1 pound = 0.453592 kilograms
Stress:	1 kilovolt/mil = 25.4 volts/millimeter, 25.4 V/mm

Copolymer: An insulation material composed of more than one type of monomer. Examples of copolymers are ethylene–propylene rubbers, ethylene–ethyl acrylate, and ethylene–vinyl acetate (both used as shields).

Corona: An electrical discharge caused by ionization of a gas by an electrical field.

Corona Extinction Voltage (CEV): The voltage at which partial discharge is no longer detectable within the dielectric structure when measured with instrumentation having specific sensitivity, following the application of a higher voltage to achieve corona inception.

Corona Inception Voltage (CIV): The voltage at which partial discharge is initiated within the dielectric structure with instrumentation having specific sensitivity.

Corrosion: The deterioration of a substance (usually a metal) as a result of a chemical reaction with its environment.

Cross-linking: A molecular structure where different polymer insulation material chains are joined to form a three-dimensional network. This is in contrast to where the chains are entangled together, but not joined. Cross-linking improves mechanical and physical properties.

Cross-linking Agent: A chemical that causes different polymer insulation material chains to join together. The most common type is an organic peroxide, such as dicumyl peroxide.

Crystallinity: Refers to the tendency that certain insulation materials have to "align" and form ordered regions (instead of being random). Insulation materials such as polyethylene (or polypropylene) are considered to be "semicrystalline" as portions of their polymer chains have this tendency.

Curing: A term used interchangeably with "cross-linking," but generally within the context of processing.

Current, Charging: The current needed to bring a cable, or other capacitor, up to voltage; determined by the capacitance of the cable. After withdrawal of voltage, the charging current returns to the circuit. For AC circuits, the charging current will be 90 degrees out of phase with the voltage.

Current, Induced: Current in a conductor due to the application of a time-varying electromagnetic field.

Current, Leakage: Small amount of current that flows through insulation whenever a voltage is present. The leakage current is in phase with the voltage and is a power loss.

Density (physics): The ratio of mass to volume at a specified temperature. Term also used to describe the "tightness" of the packing of the polymer molecules in an insulation; the tighter the "packing", the higher the density.

Dielectric Absorption: The storage of charges within an insulation (dielectric); evidenced by the decrease of current flow after application of DC voltage.

Dielectric Constant: The capacitance of a dielectric in comparison with the capacitance of a vacuum where both capacitors have identical geometry. Also referred to as permittivity or specific inductive capacitance (SIC).

Dielectric Loss: The time rate at which electrical energy is transformed into heat when a dielectric is subjected to a changing electric field; the power created in a dielectric as the result of friction produced by molecular motion in an alternating electric field.

Dielectric Strength: The maximum voltage that an insulation can withstand without breaking down; usually expressed as a gradient—volts per mil or kilovolts per millimeter.

Direction of Lay: The longitudinal direction in which the components of a cable (strands) run over the top of the cable as they recede from an observer looking along the axis of a cable; expressed as left-hand or right-hand lay.

Dissipation Factor: The energy lost when voltage is applied across an insulation due to reactive current flow. Also known as power factor and tan δ (delta).

Drain Wires: A group of small gauge wires helically applied over a semiconducting insulation shield that is designed as a path for leakage current return—as opposed to fault current or a system neutral.

Eccentricity: A measure of the centering of an item within a circular area. The ratio, expressed as a percentage, of the difference between the maximum and minimum thickness (or diameter) of an annular area.

Eddy Currents: Circulating currents induced in conducting materials by varying magnetic fields; usually considered undesirable because they represent loss of energy and create heat.

Elastomer: A term referring to rubber-like materials.

Electrical Treeing: An aging-induced defect that occurs in electrical insulation materials resulting from the application of voltage stress in the absence of water. These trees develop over a period of time significantly longer than do water trees.

Elongation: The fractional increase in length of a material as it is stressed under tension. The amount of stretch of a material in a given length before breaking.

EMI: Electromagnetic field interference.

Endosmosis: The penetration of water into a cable insulation by osmosis. Aggravated and accelerated by DC and AC voltages across the insulation where it is also known as electroendosmosis or dielectrophoresis.

Ethylene Propylene Rubber (EPR): An insulation material composed of a copolymer of ethylene and propylene.

Extrusion: The process that converts insulation material, generally in pellet form, into cable insulation over the conductor and conductor shield.

Filler: An inorganic material such as clay that may be added to a compound to improve physical properties. Fillers may be reinforcing or nonreinforcing.

Flame Retardant: A chemical additive that enhances the flame resistance of a compound.

Gel Fraction: The cross-linked portion of "cross-linked" polyethylene or EPR.

Hard Drawn: A relative measure of temper; drawn to obtain maximum tensile strength.

Hardness: Resistance to plastic deformation; stiffness or temper; resistance to scratching, abrasion, or cutting.

Heat Shrink: A term employed for certain types of joints that will shrink upon heating after they are applied in the field. The "shrinking" process is a result of the polymeric material (polyolefin) being expanded after manufacture and provided to the user in the expanded form.

High Molecular Weight: Molecular weight is a function of the number of polymer molecules in an insulation material. For example, polyethylene comprises many ethylene molecules. High molecular weight means that there are "many" such molecules, rather than a few. A low molecular weight polyethylene (not used as an insulation) would have few molecules.

Hybrid Cable System: A cable system consisting of both extruded dielectric insulation and laminated insulation.

Hypalon: Trade name for chlorosulphonated polyethylene.

Impedance (Z): The opposition to current flow in an AC circuit; impedance consists of resistance, capacitive reactance, and inductive reactance.

Installation Test: A field test conducted after cable installation but before jointing or terminating.

Insulated: Separated from other surfaces by a substance permanently offering a high resistance to the passage of energy through that substance.

Insulation Level: The thickness of insulation for circuits having ground fault detectors that interrupt fault currents within 1 minute (100% level), 1 hour (133% level), or over 1 hour (173% level).

Insulation Resistance: The measurement of DC or AC resistance of a dielectric at a specified temperature. May be either volume or surface resistivity.

Intercalated Tapes: Two or more tapes applied simultaneously so that each tape overlays a portion of the other. Example: copper and carbon shielding tapes in paper-insulated cables.

Interstices: A space between strands of a conductor or between individual phases of a multiconductor cable.

Ionization: (1) The process or the result of any process by which a neutral atom or molecule acquires charge. (2) A breakdown that occurs in gaseous phases of an insulation when dielectric stress exceeds a critical value without initiating a complete breakdown of the insulation system.

Ionization Factor: The difference between dissipation factors at two specified values of electrical stress. The lower of the two stresses is usually selected that the effect of the ionization on the dissipation factor is negligible.

Irradiation: The process of inducing cross-linking of a polymer by exposure to high energy electrons, or by gamma radiation such as Cobalt-60.

Jacket: A nonmetallic polymeric protective covering over cable insulation or shielding.

Jamming: The wedging of three or more cables in a conduit such that they can no longer be moved during cable pulling.

Jam Ratio: The ratio of the overall diameter of one cable to the inner diameter of the conduit in which they are being pulled. For three cables in a conduit, the critical jam ratio is between 2.8 and 3.2.

Lay: The axial length of one turn of the helix of any component of a cable.

Lay Length: Distance along the axis for one turn of a helical component.

Lignin: The material in wood that holds the cellulose and other components together; it is removed prior to paper manufacture due to it lossy nature.

Load Factor: The ratio of the average to the peak load over a specified period of time.

Loss Factor: Average losses divided by peak losses.

Maintenance Test: A field test made during the operating life of cable system.

Magnetic Field: The force field surrounding any current carrying conductor.

mil: Unit of measure of a conductor equal to 0.001 inch.

Monomer: A single molecule of low molecular weight used as a starting material to produce molecules of higher molecular weight (called polymers) by the process referred to as polymerization.

Mouse: A device that is attached to one end of a line and blown into a duct or pipe for use in installation of a pulling line. Usually consists of a series of rubber gaskets sized to fit the duct or pipe.

Mutual Inductance: The common property of two electric circuits whereby an electromotive force is induced in one circuit by a change of current in the other circuit.

Nominal: A term used to describe functional behavior as being within expected norms or as designed.

Ohm: The SI unit of electrical resistance; 1 ohm equals 1 volt per ampere.

Organic: Matter originating from plant or animal life; composed of chemicals—such as carbon and hydrogen.

Oscillation: The variation, usually with time, of the magnitude of a quantity that is alternatively greater or smaller than a reference.

Oscillograph: An instrument for recording or making visible the oscillations of an electrical quantity.

Osmosis: The diffusion of fluids through a membrane.

Oxidize: (1) To combine with oxygen, (2) to remove one or more electrons, (3) to dehydrogenate.

Oxygen Index: A test to rate the flammability of materials in a mixture of oxygen and nitrogen.

Ozone: A form of oxygen, O^3, produced by a high electrical stress; active molecules of oxygen.

Parameter: The characteristic of a circuit from which other voltages or currents are referenced with respect to magnitude and time displacement—usually under steady-state conditions.

Partial Discharge: Decomposition of air in voids that may be present within the insulation. This leads to generation of ions and electrons that may eventually cause electrical stress and failure of the insulation.

Permeability: (1) The passage or diffusion of a vapor, liquid, or solid through a barrier without physically or chemically affecting either, (2) the rate of such passage. The passage of a liquid, gas, or vapor through a solid such as a polymer; generally reported as weight per unit time for a controlled sample size.

Permittivity: The capacitance of a dielectric in comparison with the capacitance of a vacuum where both capacitors have identical geometry. Also referred to as dielectric constant and inductive capacitance.

Phase Angle: The measure of the progression of a periodic wave in time or space from a chosen instant or position.

Phase Conductor: Any of the main conductors of a cable other than the neutral.

Phase Sequence: The order in which the successive members of a periodic wave reach their positive maximum values.

pH: An expression of the degree of acidity or alkalinity of a substance on a scale of 1 to 10. Acid is less than 7.0, neutral is 7.0, and alkaline is over 7.0.

Pig: (1) A device to isolate a portion of a pipeline to permit the local application of a test pressure, (2) an ingot of metal, such as lead.

Pilot Wire: An auxiliary insulated conductor in a circuit used for control or data transmission.

Plasticizers: Chemical agents added during compounding of certain polymeric materials (most commonly PVC [polyvinyl chloride]) to make them more flexible and pliable.

Polarization Index: The ratio of insulation resistance after two different time intervals—typically 10 minutes as compared with the measured value at 1 minute.

Polyethylene: A polymeric material employed as an insulation or jacket that comprises many ethylene molecules.

Polymer: A high molecular weight compound whose structure can usually be represented by a repetition of small units of that compound.

Polyolefin: A term that encompasses all insulation materials that are composed of carbon and hydrogen. These include polyethylene, cross-linked polyethylene, polypropylene, ethylene–propylene copolymers and terpolymers, and ethylene–alkene polymers.

Pothead: (1) A termination of a cable (potential head), (2) a device for sealing the end of a cable while providing insulated egress for the conductor or conductors. Most commonly associated with the porcelain housings for paper-insulated cables.

Power Factor (power): The cosine of the phase angle between the voltage and the current. Power factor is of interest because it is the measure of useful work. A unity power factor means that all of the current is used for useful work.

Power Factor (cable): A typical cable has a power factor of about 0.1 or less—meaning that it is almost a perfect capacitor and the majority of the current consumed by the charging current of the cable is not "useful" power. For cable purposes, the power factor is expressed as the tangent of the angle delta between the current and the voltage. For the small angles found in typical medium voltage power cables, the sin δ, tan δ, and cos Θ are essentially equal.

Power Loss: Losses due to internal cable impedances, such as the conductor I^2R and the dielectric losses in the insulation. These losses create heat.

Pulling Compound: The lubricating compound applied to the surface of a cable to reduce the coefficient of friction during installation in conduits and ducts.

Pulling Eye: A device attached to the end of a cable to facilitate field connection of the pulling ropes.

Quadruplexed: Four conductors twisted together.

Relative Capacitance: The ratio of the material's capacitance to that of a vacuum of the same configuration. Also known as specific inductive capacitance (SIC).

Reverse Lay: Reversing the direction of lay. For multiple-conductor aerial cables, a reversal in lay at a specified distance to facilitate field connections.

Rockwell Hardness: A measure of hardness of a material to indentation by a diamond or steel ball under pressure at two levels of stress.

Screen Pack: A series of metal screens used in an extruder for straining out impurities.

Semiconducting: A conducting medium where the conduction is through electrons. The resistance of these materials is generally in the range between that of conductors and insulators.

Shield: An electrically conducting layer that provides a smooth surface with the surface of the insulation. In Europe, this is called a "screen."

Sidewall Bearing Pressure (SWBP): The normal force on a cable under tension at a bend. This is a force that tends to flatten or crush the cable and is usually given as an allowable force for a given distance.

Silane: A silicone-based monomer that can be used to impregnate polyethylene or cross-linked polyethylene insulated cables and react with water, and undergoes in-situ cross-linking.

Skin Effect: The tendency of current to crowd toward the outer surface of a conductor that increases with conductor diameter and frequency of the applied current.

Sol Fraction: The uncross-linked portion of "cross-linked" polyethylene or EPR.

Specific Inductive Capacitance: The capacitance of a dielectric as compared with the capacitance of a vacuum where both capacitors have identical geometry. Also referred to as dielectric constant and permittivity.

Strand, Sector: A stranded conductor formed into sectors of a circle to reduce the overall diameter of the cable.

Strand, Segmental: A stranded conductor formed of sectors that are insulated from one another to reduce the AC resistance of the conductor.

Strand, Unilay: A stranded conductor having a unidirectional lay of the various wires. Frequently used in low voltage power cables.

Stress Relief Cone: A mechanical component of a termination to reduce electrical stress levels on a shielded cable, originally in the shape of a cone.

Stripping Strength: Refers to the ease (or lack of) when the insulation shield is "peeled" from the insulation surface.

Tandem Extrusion: Extruding two or more layers on a conductor where the extruders are in close proximity to one another.

Thermoplastic: A polymeric insulation material that possesses the property of softening when heated.

Thermosetting: Polymeric insulation materials that, upon curing (cross-linking), undergo an irreversible chemical and physical property change as a result of the cross-linking process. Thermoset materials cannot be refabricated.

Tinned: A strand having a thin coating of pure tin or an alloy of tin and lead. Used over copper to reduce the effect of sulfur from certain rubber compounds and to facilitate solder connections. Many aluminum connectors are also tin-plated to minimize the formation of aluminum oxide.

Treeing: An aging-induced defect in insulation materials that imparts a tree-like appearance to the affected region. Treeing is generally referred to as electrical treeing or water treeing.

Triplexed: Three conductors or cables that are twisted together.

Voltage Rating: The designated maximum permissible phase-to-phase AC (or direct current) voltage at which a cable is designed to operate.

Vulcanize: To cure (cross-link) by chemical reaction. Produces changes in the physical properties of the material by the reaction of an additive (originally sulfur) at an elevated temperature. A term initially applied to elastomers.

Water Tree: An aging-induced defect in insulation materials that results from the application of voltage stress in the presence of water. This may result at normal operating stress over periods of time that are relatively short as compared with electrical treeing.

24.2 ACRONYMS

A	Common abbreviation for ampere
AC	Alternating current, an electric circuit that periodically reverses direction
ACLT	Accelerated cable life test
AWG	American Wire Gauge
AWTT	Accelerated water treeing test
BIL	Basic impulse insulation level
BR	Butyl rubber (isobutylene–isoprene copolymer)
CPE	Chlorinated polyethylene rubber
CSPE	Chlorosulfonated polyethylene rubber (Hypalon)
CV	Continuous vulcanization
DAC	Damped AC voltage
AC	Direct current
DF	Dissipation factor, also tan δ (delta)
DiBenzo GMF	Quinone dioxime dibenzoate
DOE	U.S. Department of Energy
EC	Alloy designation for aluminum electrical conductors
EDG	Emergency diesel generator
EPDM	Ethylene propylene diene methane. An insulation material composed of three monomers, ethylene, propylene, and a third monomer generally referred to as a diene-monomer
EPR	An insulation material composed of a copolymer of ethylene and propylene
EPRI	Electric Power Research Institute
EPR	Ethylene propylene rubber
FLECTOL H	Polymerized trimethyl quinoline antioxidant
GMF	Quinone dioxime
GR-S	Government Rubber-Synthetic, now SBR, styrenebutadiene
HDPE	High-density polyethylene
hi-pot	High potential
HMWPE	High molecular weight polyethylene
HPFF	High-pressure fluid filled cable

HTK	A Kerite Company insulation polymer designation
Hz	Hertz, number of cycles per second
ICEA	Insulated Cable Engineers Association
IEC	International Electrotechnical Commission
IEEE	Institute of Electrical and Electronics Engineers
IET	Institution of Engineering and Technology (UK), formerly IEE
INPO	Institute for Nuclear Power Operations
IPCEA	Prior to 1979: Insulated Power Cable Engineers Association
IRC	Isothermal relaxation current
kcmil	Thousand circular mils, formerly MCM
kV	Kilovolts
LCO	Limiting Condition for Operation
LDPE	Low-density polyethylene
LER	Licensee event report
LOCA	Loss of coolant accident
LPFF	Low-pressure fluid filled cable
MCM	Thousands of circular mils (also KCM), an older term for conductor area
mil	1/1,000th of an inch
MBTS	Mercapto Benzo Thiazole
mm	Millimeter. Unit of measure equal to 0.001 meter
MV	Medium voltage, generally 5 to 46 kV
MW	Megawatt, equal to 1,000 kilowatts and 1,000,000 watts
NEC	National Electrical Code
NEI	Nuclear Energy Institute
NEMA	National Electrical Manufacturers Association
Nordel	DuPont trade name for EPR rubbers
NPRDS	Nuclear Plant Reliability Data System
NRC	U.S. Nuclear Regulatory Commission
NRR	Nuclear Reactor Regulation
pC	Pico coulomb
PD	Partial discharge
PDEV	Partial discharge extinction voltage
PDIV	Partial discharge inception voltage
PE	Polyethylene
PILC	Paper-insulated, lead covered cable
PSE	Plant support engineering
PVC	Polyvinyl chloride
SSC	System, structure, or component
SR 350	Sartomer 350 monomer
tan δ	tangent δ (a loss factor of insulation)
TDR	Time domain reflectometry

TR-XLPE	Tree-retardant XLPE
UD	Underground distribution
UL	Underwriters Laboratory
URD	Underground residential distribution
V_O, also U_O	Phase-to-ground voltage
Vistalon	EXXON trade name for EPR rubbers
XLPE	Cross-linked polyethylene
ZnO	Zinc Oxide
μs	Microsecond

Index

F

R

S

T

U

V

W

X

Printed in the United States
by Baker & Taylor Publisher Services